# All-Russian Research-To-Practice Conference “Ecology and Safety in the Technosphere” 2017



IOP Conference Series: Earth and Environmental Science
Volume 66

Yurga, Russia
6 - 7 March 2017

**Editors:**

**Dmitry A. Chinakhov**

ISBN: 978-1-5108-4072-0
ISSN: 1755-1307

**Printed from e-media with permission by:**

Curran Associates, Inc.
57 Morehouse Lane
Red Hook, NY 12571

**Some format issues inherent in the e-media version may also appear in this print version.**



Printed by Curran Associates, Inc. (2017)

For permission requests, please contact the Institute of Physics
at the address below.

Institute of Physics
Dirac House, Temple Back
Bristol BS1 6BE UK

Phone: 44 1 17 929 7481
Fax: 44 1 17 920 0979

techtracking@iop.org

**Additional copies of this publication are available from:**

Curran Associates, Inc.
57 Morehouse Lane
Red Hook, NY 12571 USA
Phone: 845-758-0400
Fax: 845-758-2633
Email: curran@proceedings.com
Web: www.proceedings.com

# Table of contents

## Volume 66

**All-Russian research-to-practice conference "Ecology and safety in the technosphere"**

**6–7 March 2017, Yurga, Russian Federation**

**Accepted papers received: 11 May 2017**
**Published online: 17 May 2017**

### Preface



### Papers

# PREFACE

All-Russian research-to-practice conference "Ecology and safety in the technosphere" will make a contribution to the rethinking of the relationship between man and nature in today's world, to finding the best ways to overcome the negative effects of the environmental crisis and to solve security issues.

The conference aims at discussing research development and introducing innovations in the field of environmental and technosphere safety; searching for collaborative research areas and strengthening cooperation between young researchers and professionals from educational and scientific institutions in Russia; creating a platform for creative communication and exchange of experiences between young researchers and professionals in the issues of environmental and technosphere safety. It will also focus on intellectual and creative development of young researchers and specialists; demonstrating the best achievements of young researchers and professionals in scientific and technological activities; acquisition scientific debate and public speaking skills by young researchers; increasing the prestige of education, research and innovative activity in the field of environmental and technosphere safety; systematization of scientific data on the topical issues of ecological and technosphere safety, developing evidence-based approaches to their solution.

The Organizing Committee of the Conference thank all the participants of the Conference for their fruitful work and personal contribution to the development of these Conference Proceedings.

*Dmitry A. Chinakhov, PhD,*
*Conference Convenor,*
*Vice-Chancellor, Yurga Institute of Technology*
*of National Research Tomsk Polytechnic University*

# Peer review statement

All papers published in this volume of *IOP Conference Series: Earth and Environmental Science* have been peer reviewed through processes administered by the proceedings Editors. Reviews were conducted by expert referees to the professional and scientific standards expected of a proceedings journal published by IOP Publishing.

# Direction of rational use of water at livestock facilities

**A A Potseluev[a], I V Nazarov**

Azov-Black Sea Engineering Institute FGBOU HPE «DGAU» in Zernograd
21 ul. Lenina, Zernograd, Russia, 347740

E-mail: [a] niv671@rambler.ru

**Abtstract.** The article notes the world water shortage problem. Against this background, Russia's agricultural production is considered, in particular the livestock sector as the main consumer of water resources. The structure of the main technological processes at livestock facilities is given and possible technological damage is indicated in case of the lack of technological processes for servicing animals and poultry with water. The direction of rational use of water based on the introduction of new technical and technological solutions of water supply systems and means is substantiated. Constructive solutions of systems and facilities that help to reduce water consumption are presented, and as well a possible positive effect.

## 1. Introduction

There is currently a shortage of water resources in many countries [1,2]. Therefore, there is a need to find technical and technological and organizational solutions for the rational use of water in all spheres of human life. At the same time, it should be noted that in relation to Russia, agriculture is the largest water consumer. In comparison with industry, agriculture consumes 4-5 times more water. [3, 4] Providing water is particularly important for livestock facilities. This is largely due to the fact that the lack of water directly affects the productivity of biological objects (animals, birds), and in some cases leads to their diseases and death [5]. The lack of water in the performance of separate technological processes for servicing animals (removal of excrements) has a negative impact on the ecology in the zone of location of the cattle-breeding facility. Therefore, under conditions of water resources shortage, rationalization of water consumption at livestock facilities within the framework of zoological and technical requirements is one of the key problems of modern livestock production.

Inside the livestock premises, that is a separate livestock facility, the water discharge is mainly distributed according to the technological processes of animal drinking, milking (large and small cattle), sanitary treatment of the skin of animals and the process of sanitary disposal of technological equipment. At the same time, the quality of the performance of technological processes for servicing animals and water consumption largely depends on the operated technical and technological solutions of the water supply system for technological processes, and some of its structural elements [5].

## 2. Results and discussion

The water supply system should take into account the physiological characteristics of the served animals, the actual water flow including its temperature in accordance with the zoological and technical requirements, and exclude or reduce the non-productive water flow. In order to fulfill these requirements, a multi-purpose water supply system for technological processes was developed Figure 1.

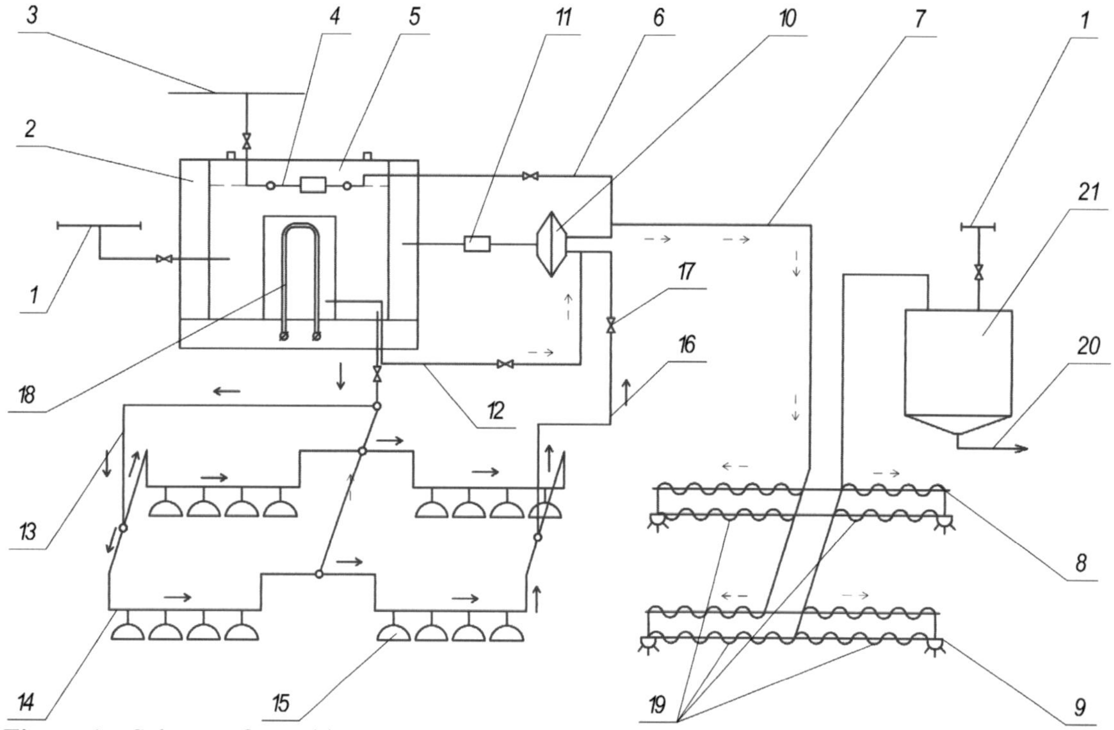


**Figure 1** – Scheme of a multi-purpose water supply system for technological processes of servicing animals: 1 – vacuum pipe of milk line; 2 – vacuum chamber; 3 – an inlet pipeline; 4 – valve-float device; 5 – storage capacity; 6 – return piping; 7 – the main piping for supply of wash water; 8 – working piping for supplying the wash water; 9 – washing device; 10 – pneumatic exciter; 11 – pulsator; 12 – piping for the wash water supply; 13 – the main piping for drinking water supply; 14 – distribution piping of drinking water; 15 – auto drinkers; 16 – return piping of drinking water; 17 – locking devices; 18 – electric heater; 19 – working piping for the wash water removal; 20 – main piping for contaminated water removal; 21 – vacuum tank for the collection of contaminated water

This system provides the preparation of water for use (heating up to the set temperature), its feeding into the zoning (means of automatic drinking, devices for washing the udder of the cow), the collection of nonproductive water (contaminated), followed by its filtration and treatment for reuse.

Considering the system in relation to the served species, it should be noted that it can be transformed and used by separate units (subsystems). When using the group means of automatic drinking of flowing type in the automatic drinking subsystem, the non-productive water flow increases. [5-14] It can be up to 3 $м^3$ per day from one group drinker.

Increased non-productive water flow is associated with manual periodic regulation of the water flow inside the drinker. With the adopted method of regulation one can not control the 24-hour monitoring of the water flow, which is directly dependent on the outside temperature of the surrounding air. Therefore, in order to rationalize the water discharge, we suggest their upgrading with the automatically regulated water passage in the winter period of their operation Figure 2, 3.

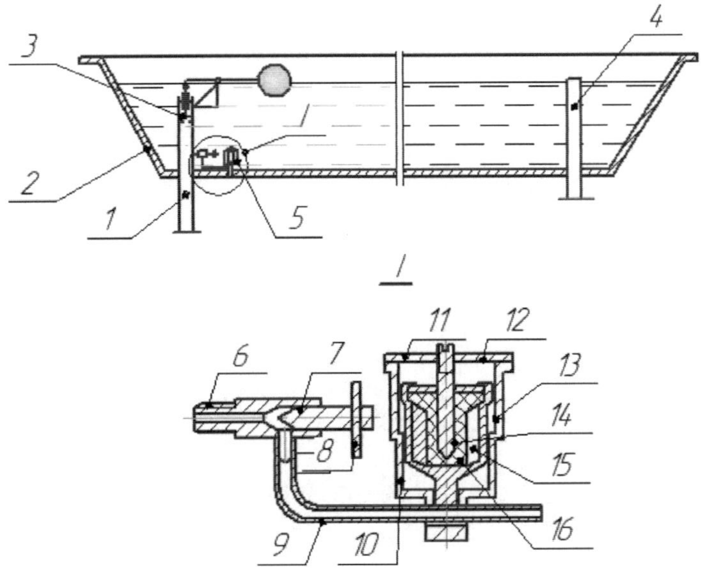


**Figure 2** – A device for regulating the water flow in the zone of its entry into the drinking: 1 – piping; 2 – water trough; 3 – valve device; 4 – drainage branch pipe; 5 – control device for draining the water; 6 – body; 7 – shut-off valve; 8 – handle; 9 – pressure hose; 10 – body; 11 – attachment point; 12 – cover; 13 – body of the thermosensitive element; 14 – rod; 15 – filler; 16 – rubber capsule

The main element of the device is a thermo-power sensor with a solid filler, one of its main statistical ones is the increment of the stroke of the rod, depending on the water temperature change in the water bowl. Studies have established that the operating temperatures of the sensor are in the range of 1-15 °C, and the developing effort reaches 16 kg.

To reduce water consumption and improve comfort of the technological process of servicing the cow, the subsystem for sanitary disposal (washing) the udders of animals should be equipped with containers for washing water and high pressure air, a system of pulsating feed of the wash water, lines for transporting water and air to the washing device [5,15,16]. To increase the effectiveness of the action of wash water on bacterial contamination, the skin of the udder, we consider it suitable to heat water and divide it by acidity into acidic and alkaline. Studies have established that the microflora of the skin of the udder is mainly represented by microorganisms, which refer to group 17 (gram-positive cocci) of the genus Micrococcus and Staphulococcus according to Berji's the classification. It was also found that the spore microflora is suppressed both with acidic and alkaline solution. The most effective ones are acidified solutions (pH 6.0 and below, as well as alkaline solutions with a pH of 8.0 or more. They reduce bacterial contamination by 11 or more times. However, the use of alkaline solutions contributes to the appearance of an adverse side effect (drying of the skin of the udder, followed by the appearance of cracks on its surface). It is advisable to perform the treatment of the udder with a water-air mixture twice (before putting on the teat cups and after removing them).

IOP Conf. Series: Earth and Environmental Science **66** (2017) 012001 doi:10.1088/1755-1315/66/1/012001

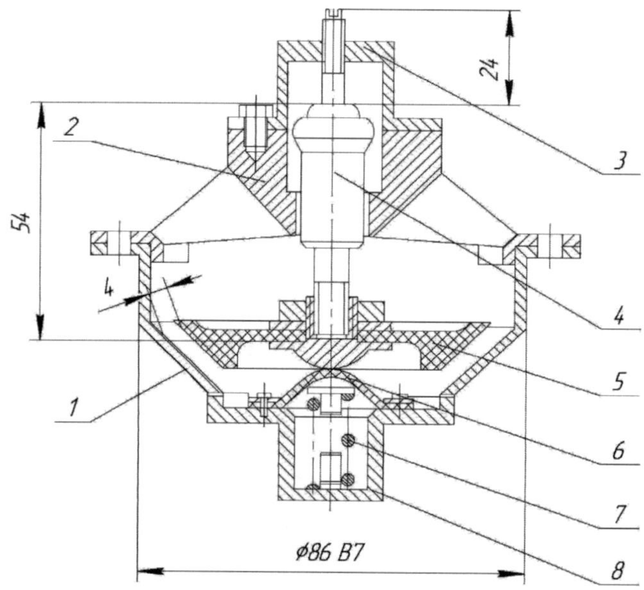


a)

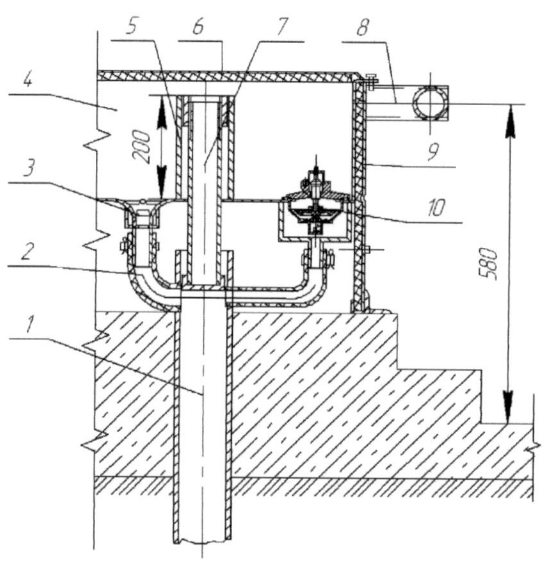


b)

**Figure 3** – Construction (a) and installation of a control device for the water flow in the drainage zone (b): a – the construction of the device; 1 – body; 2 – cover; 3 – bracket, 4 – thermo sensor; 5 – valve; 6 – membrane; 7 – spring; 8 – damping device; b – scheme of the device installation; 1 – drain piping; 2 – drain hose for water; 3 – drain plug; 4 – water trough; 5 – bypass pipe; 6 – the cover of the bowl; 7 – drainage branch pipe; 8 – the protective cover of drinking bowl; 9 – the body of drinking bowls; 10 – thermo-power device

Analysis of the proposed structural improvements of the subsystem (Figure 4) and technological requirements for the udder washing process shows that the subsystem under consideration can be unified with the automatic drinking subsystem (a single transporting material is water; a compatible temperature regime for water preparation is up to 30...45°C) and a ventilation system for livestock facilities (the required component of the mixture is air). In this case, the design of the storage tank should be three-section (the section for placing water for needs of automatic drinking, the section for placing the activated water with pH 6.0, the section for placing the activated water with pH 8.0), while the body is double-walled with a cavity for accumulation of air of increased pressure.

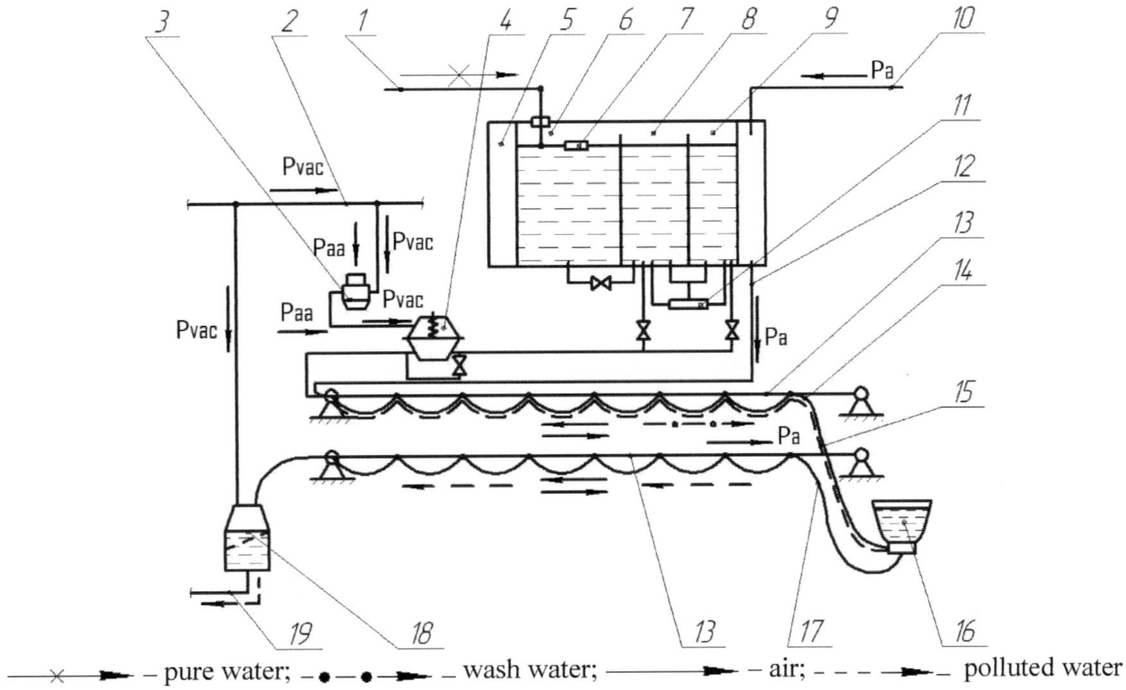


**Figure 4** – The scheme of the subsystem of washing the udder of cows: 1 – fresh water delivery line; 2 – vacuum line of the milking plant; 3 – pulsator; 4 – pneumatic exciter; 5 – air cavity of the water tank; 6 – cavity for drinking water; 7 – valve-float device; 8, 9 – cavities for activated water in pH; 10 – airline ventilation system; 11 – heat power unit; 12 – compressed air delivery line; 13 – the suspension; 14 – piping for wash water supply; 15 – piping for compressed air supply; 16 – washing device; 17 – piping for the suction of polluted water; 18 – capacity for the collection of contaminated water; 19 – piping for polluted water discharge; $P_a$ – air of increased pressure; $P_{aa}$ – atmospheric air; $P_{vac}$ – a vacuum

## 3. Conclusions

As a result of the analysis of the importance of water resources in various spheres of human activity and the dynamics of changes in the natural water balance, the following conclusions should be drawn:

1. the rational use of water is one of the ways to maintain the water balance within a particular economic entity;
2. Animal husbandry is one of the main consumers of water when performing technological processes for servicing animals;
3. The main water flow in the maintenance of animals goes to the processes of automatic drinking and milking;
4. Development and introduction into production of a multifunctional system of water supply for technological processes of servicing animals and resource-saving constructive solutions of the basic elements of the system makes it possible to ensure rational use of water;
5. The offered technical and technological solutions of systems and means of water supply for technological processes of servicing animals allow to reduce non-productive water discharge for a

drain of 3 $m^3$/day up to 0.8...1.0 $m^3$/day, the production water consumption for a single treatment of the udder of one cow from 2 litters to 1.5 water. At the same time, a positive concomitant effect is possible: improving the quality of milked milk; decrease in the incidence of the udder of the cow; decrease in the risk of morbidity of attendants.

**References**

[1] Medvedev D.A.: National priority «Development of the agro-industrial complex» is the project of enterprising people / Agromarket-2007. No. 1. P. 9. Transcript of the speech by D. Medvedev.
[2] Ulanova O. The role of the national project «Development of the agroindustrial complex» in ensuring food security of the Russian Federation / O. Ulanova / International agricultural journal. 2008. No. 5. P. 54-56.
[3] Danilov-Danilyan V.I. Water Consumption: Environmental, Economic and Political Aspects / V.I. Danilov-Danilyan, K.S. Losev. Moscow: Nauka, 2006. 218 p.
[4] Nature and resource complex of the Russian Federation: analytical report / Ed. O.V. Komarova. - Moscow: NIA-Nature.
[5] Potseluev A.A. Resource-saving water supply systems for technological processes of servicing cattle: Diss. .... doctor. tech. sciences. Zernograd, 2011.
[6] Vandeho Ven J.Z.J.Drinking Groughs for Animals//Patent specification. 873, 977. 1961.
[7] Doker M.S. Jmprovements in or relating to Cattle Deinking Bowls//Patent Specification. 928, 287. 1963.
[8] Zundin Z.A. Jmprovements in or relating to a Drinking Groughs for Animals//Patent Specification. 882, 414. 1961.
[9] Zimited F.F. Jmprovements in or relating to Animal Drinking Appliances//Patent Specification. 1, 070, 888. 1967.
[10] Bertazzoni M.A. Dispositif pour Jbuverture et la fermeture de la canalization d'eau desauges d'abreuvair de bestiaux – Brever D'invention. 1, 160, 807. 1958.
[11] Bordet M.A. Abreuvoir automatigue – Brevet D'invention. 1, 240, 462, 1960.
[12] Spenser T.M.N/ Abreuvoir – Brevet D'invention. 1, 949, 205, 1964.
[13] Fonderies et Ateliers de Construcctions Fils. Perfectionnements apportes aux abreuvoris automatigues anti – gel/ - Brevet Dinvention. 1, 489, 098, 1967.
[14] Clay Eguipment Corporation. Circulating Water Bowl//Patent Specification. 936, 576. – 1962.
[15] Ardzan G.M., Kemp G.H., Clough P.A., Dodd F.N. Cineradoi gzaphic observations on machine milking. Y.Dairi, 1958. 25. P. 154–158.
[16] Bartmann R Mechenisierte milchgewinnung.2 Uberarb. Und erg Auft. Berlin. Deutgch. Landwirtscheftsverl, 1966. 15. 9 p.

IOP Conf. Series: Earth and Environmental Science **66** (2017) 012002 doi:10.1088/1755-1315/66/1/012002

# Distribution of Heavy Metals in Soil of Urban Ecosystems by the Example of the Oryol City

**S Vorobyov [a], D Kozlov [b], T Poturaeva [c], I Chernyaeva [d]**

Oryol State University,
Komsomolskaya Street 95, Oryol, Russian Federation, 302026

E-mail: [a] vorser323@rambler.ru, [b] dimoktank@mail.ru, [c] tanpo77@mail.ru,
[d] schunya87@yandex.ru

**Abstract.** The article deals with the impact of vehicle emissions on the distribution of heavy metals in the soils of urban ecotopes - land urban areas with similar parameters of the biosphere, the technosphere, demographic and urban development factors by the example of the Oryol city.

## 1. Introduction

Today, the problem of environmental protection is based on the concept of sustainable development, which is used throughout the world and serves as the basis for the development of environmental protection strategies for most countries in the world, including Russia [1, 2, 3]. At the same time, environmental problems are becoming increasingly important especially in urban areas.

City - a specific human creation, adaptation for which is linked to human health problems [4]. The human as a biological species is adapted to life in other conditions. But since man is also a social being, its development by is associated with social needs [5]. Satisfaction of social needs and transformation of a environment by human is leads to a violation of the balance. Than greater the transformation of the environment that stronger the negative impact on the biosphere. The human is also a part of the biosphere that is way unbalance of the environment is effect on his health [6]. There is a paradox - a human created urbanized areas for maximum comfort, but pays for it with his health. In the Orel region of the Russian Federation the incidence of cardiovascular diseases of the urban population is 15% higher than of the rural population [7]. This fact is due by the high concentration of industry and vehicles in a limited area. Besides, in large cities 60-80% of atmospheric air pollution falls on motor vehicles [7]. One of the most dangerous types of pollutants in emissions of motor vehicles is a heavy metals. The number of motor vehicles is growing so rapidly that the reduction of emissions that achieved by the cleaners and high-quality fuel is offset by an increase in the number of cars [8]. Vehicles is very mobile, so it easily penetrates into the residential areas, while large industrial enterprises are located outside the city and are separated by sanitary protection zones [9].

Cities are now becoming the main systems for life, so it is important to study and forecast their effects on humans, their environment and the biosphere processes.

## 2. Results and discussions

As an example, we have analyzed the land territory Orel city as a typical regional center of Central Federal District of Russia. To detect anthropogenic impacts on urban ecosystems we have identified ecotopes - land urban areas with similar parameters of the biosphere, the technosphere (polluters), demographic and urban development factors. As these areas we have chosen the following areas: the area adjacent to the housing number 16 (Moskovskaya street 77) Oryol State University (area 1), the park near main building Oryol State University (Komsomolskaya street 95) (area 2), area adjacent to the Museum of Oryol writers (area 3). All of these areas are characterized by the similar buildings (five-storey buildings, the lack of high-rise buildings, flat terrain), the characteristics of human impact (prevalence of vehicle emissions from major highways), similar areas (about 4 ha), the same composition of green spaces, the same number of inhabitants. All of these areas are characterized by the same set of factors of anthropogenic impact and varying degrees of quantitative manifestation of these factors. As a control area that not feel the impact of these factors, we have chosen the area of Medvedevsky forest (area 4).

**Figure 1** – Location of Area 1

**Figure 2** – Location of Area 2

**Figure 3** – Location of Area 3

**Figure 4** – Location of Area 4

All ecotopes are differ by only in one indicator - traffic.To determine the number of vehicles for the subsequent calculation of emissions, at study sites were counted in the rush hour, within a year. Noted the number of cars in 30 minutes, followed by recalculation of 1 hour. Car traffic was divided into categories: passenger road transport; cargo, with a diesel engine; truck, with a carbureted engine; bus, diesel engine; with carburetor engine bus; trolley bus; minibuses. Separation was made to more accurately assess the environmental burden, as each of these categories is characterized by a specific set of emissions. Traffic is amounted 2067, 1528, 1231 cars in hour, for areas 1, 2 and 3 respectively.

Table 1 - Number of vehicles in the investigated areas, pcs / hour.

| Vehicle category | Number of vehicles, pcs / h. | | |
|---|---|---|---|
| | Area 1 | Area 2 | Area 3 |
| Car | 1560 | 1100 | 903 |
| Diesel truck | 6 | 12 | 5 |
| Gasoline truck | 48 | 24 | 7 |
| Diesel bus | 3 | 2 | 12 |
| Gasoline bus | 120 | 95 | 86 |
| Tolley bus | 55 | 45 | 68 |
| Minibuses | 275 | 250 | 205 |
| Total | 2067 | 1528 | 1231 |

The main source of contamination at all study sites were selected vehicles, as it accounts for a significant portion of the emissions in the city of Oryol. To determine the amount of pollutants from vehicle emissions has been used a technique developed by A. Voeikov [10].
We obtained the following results on the emissions of vehicles on the study areas (Table 2):

Table 2 - Distribution of total and maximum single emission investigated areas

| Investigated area | Indicator | |
|---|---|---|
| | Total emissio, t/year | Maximum single ejection, g/s |
| Area 1 | 13454 | 15.2789 |
| Area 2 | 10274 | 1.0471 |
| Area 3 | 9338 | 0.8971 |

When studying the gross forms of heavy in urban and forest soils, we obtained the following results (Table 3).

Table 3 - Content of gross forms of heavy metals in soil (mg / kg)

| Investigated area | Lead, Pb | Zinc, Zn | Copper, Cu | Nickel, Ni |
|---|---|---|---|---|
| Area 1 | 65.66 | 101.4 | 107.05 | 22.9 |
| Area 2 | 52.35 | 138.35 | 107.25 | 22.2 |
| Area 3 | 39 | 226.15 | 138.6 | 22.75 |
| Area 4 (control) | 30.34 | 88.05 | 64.5 | 16.65 |
| MPC | 130 | 220 | 132 | 80 |

From the data given in the table, it can be seen that the concentration of gross forms of heavy mettals in the soil is no more than the MPC. The highest values of lead are noted in the soil of Area 1, the smallest - in the control section (Area 4). The difference between these data is 53.8%. The data obtained indicate a close direct relationship between the content of lead in the soil and the traffic ($r = 0.97$).
Gross zinc forms do not exceed the maximum permissible concentration values at most sites, except for the concentration at Area 3, where concentration more than MPC. The concentration of zinc in Area 3 is higher than in the remaining areas in 1.6–2.2 times and higher than in the control area in 2.6 times. The correlation coefficient $r = -0.74$, which indicates a rather high inverse relationship between the intensity of the traffic and the concentration of the gross forms of this heavy metal.
The highest values of copper concentration are recorded at Area 3 in 2.1 times higher than in the control area. The value of the correlation coefficient between the concentration of the total copper forms and the amount of electric transport (trolley buss) is high and is $r = 0.99$.

The distribution of gross forms of nickel in soils of different areas differs uniformity. Its concentration in urban soils is higher than in forest soils, but does not exceed the MPC values. The biggest values in the Area 1.

Analysis of mobile forms of heavy metals in soil is also showed a different dependence of their concentration on the traffic (Table 4).

Table. 4 - Content of mobile forms of heavy metals in soil (mg / kg)

| Investigated area | Lead, Pb | Zinc, Zn | Copper, Cu | Nickel, Ni |
|---|---|---|---|---|
| Area 1 | 48 | 22.39 | 0.34 | 3.35 |
| Area 2 | 2.83 | 28.07 | 0.24 | 3.55 |
| Area 3 | 10.16 | 19.16 | 1.72 | 3.45 |
| Area 4 (control) | 2.8 | 8.13 | 0.14 | 2.64 |
| MPC | 6 | 23 | 3 | 4 |

The biggest correlation dependence (r = 0.99) is characteristic for a lead. For other elements, the correlation coefficient, respectively, is: Zn-r = 0.17; Cu-r = –0.73; Ni-r = –0.62.

A comparative analysis of the data in Tables 3 and 4 showed that the pattern of the distribution of gross and mobile forms of heavy metals over the areas is basically identical. The difference is that the highest concentration of mobile forms is observed for a lead. Gross forms do not exceed MPC, but the content of mobile lead is above the MPC at all areas, respectively, at 8; 3.6 and 1.7 times. Mobile forms of Cu do not exceed the values of MPC, but in area 3 the concentration of this heavy metal is higher than on a other areas in at 5.1 to 7.2 times. For mobile forms of Cu, a rather high inverse dependence on the total number of vehicles, but if we consider separately the trolleybus transport in the investigated areas and its influence on the concentration of the this metal, the correlation coefficient is positive and high (r = 0.98). This close connection, as well as the fact that most qantity trolley buses on area 3, makes it possible to assume that in this area the soil complexes, while still coping with the Cu flow, and are able to accumulate this heavy metal (the values of the mobile form concentration are not exceed the MPC), but gradually, with increasing traffic, lose their effectiveness. Probably, the presence in the soil of an organic substance capable to fix a this metal to a stationary state is decreased. Over time and with increasing levels of pollution, this process is accelerated more and more. Therefore, in areas 1 and 2, the values of the concentration of mobile forms do not differ much from each other. It can be assumed that organic soil complexes actively absorb this element, and at Area 3 the concentrations differ sharply from the other.

The mobile forms of Zn is exceed the MPC values only at Area 2. The concentration values on the other investigated areas do not differ much between themselves and, although they do not exceed the MPC, but approach this mark.

The mobile forms of Ni do not exceed the MPC and differ in the relative uniformity of the distribution. The highest concentrations are in the area 2. Nickel is characterized by a low correlation coefficient of concentration and traffic (r = –0.62), and a negative sign indicates an inverse relationship between Ni concentrations and traffic.

The revealed dependencies between the content of gross and mobile forms of heavy metals are determined by the mobility of these elements in the soil. Data on the mobility of heavy metals, the ratio of the concentration of the mobile element to its gross content are given in Table 5.

Table 5 - Values of the mobility of heavy metals

| Investigated area | Lead, Pb | Zinc, Zn | Copper, Cu | Nickel, Ni |
|---|---|---|---|---|
| Area 1 | 0.73 | 0.22 | 0.003 | 0.150 |
| Area 2 | 0.42 | 0.20 | 0.002 | 0.16 |
| Area 3 | 0.26 | 0.08 | 0.012 | 0.16 |
| Area 4 (control) | 0.09 | 0.09 | 0.002 | 0.15 |

Analysis of the data in the table showed that the coefficient of mobility of all elements, more in urban soil than in the control. We can see the highest values the coefficient of mobility and the ratio to the control soil are characteristic for a Pb. The maximum value of the coefficient of mobility is in Area 1, in the soil of which more than 70% of lead is in mobile form, while in control soil it is only 9%. The lowest values of the coefficient of mobility are characteristic for the Area 3, where the share of trolleybus transport is high. The values of the coefficient of mobility on the investigated areas are very different from each other - the difference between the maximum and minimum values for urban areas is almost 3 times (Figure 5).

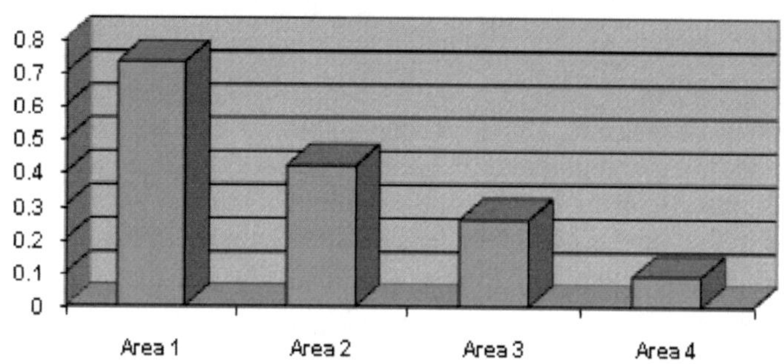


**Figure 5** - The values of the coefficient of mobility of Pb

The increase in the mobility of lead under the influence of traffic is particularly noticeable if we compare these indicators with forest soil in which the coefficient of mobility decreases, compared to the soil of Area 1 in **8** times.

The dynamics of distribution of the coefficient of mobility of zinc is similar to the previous one, but the values of the coefficient of mobility are smaller. The coefficient of correlation of the values of the coefficient of mobility of Zn and the total amount of vehicle in the investigated areas is $r = 0.85$, which allows to tolk about a rather high degree of influence of vehicles.

Higher values of the coefficient of mobility in urban soils in comparison with forest so testify to the effect of transport on the increase in the soil of mobile forms of zinc (Figure 6).

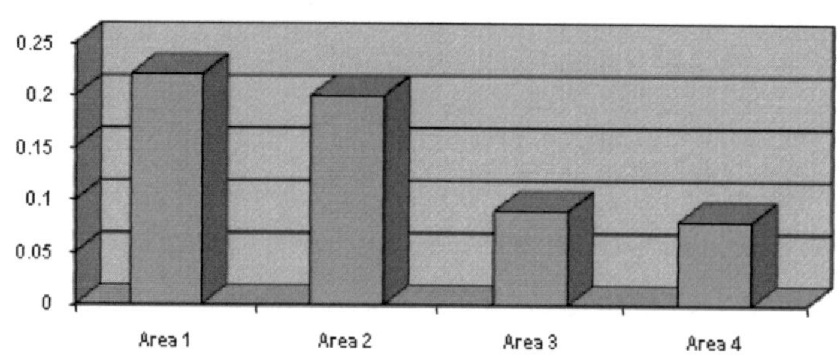


**Figure 6** - The values of the mobility coefficient Zn.

IOP Conf. Series: Earth and Environmental Science **66** (2017) 012002 doi:10.1088/1755-1315/66/1/012002

The values of the coefficient of mobility of copper are as follows. The greatest value is accounted for by Area 3 (where concentration of copper more than MPC) and differs from the values in the remaining areas by almost 4 times. The values of the mobility coefficient for the soils of the remaining areas do not differ much from each other, which indicates that this metal is in a relatively fixed state. The correlation coefficient of the total amount of vehicles and the coefficient of mobility of Cu is $r = -0.71$, the inverse relationship is high. But if we consider separately the category of trolleybus and its influence on the coefficient of mobility of Cu, then the correlation coefficient is $r = 0.98$, which indicates a very high degree of influence. The dynamics of the distribution of mobility coefficients completely coincides with the dynamics of the distribution of mobile forms of Cu (Figure 7).

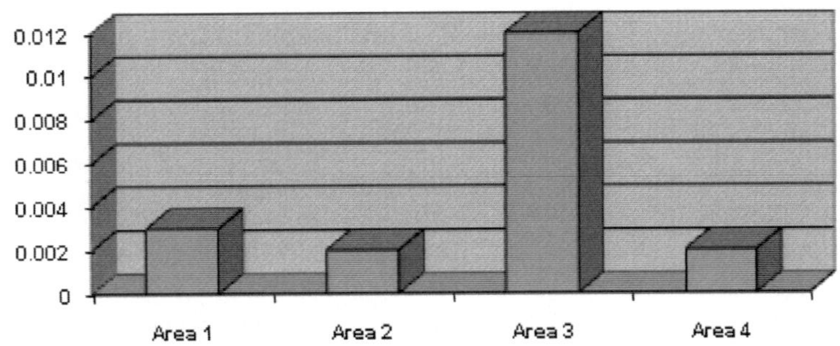


**Figure 7** - The values of the coefficient of mobility of Cu

In comparison with forest soil, the mobility of copper in area 3 is increased by a control in 6 times. In other areas the coefficient of mobility does not differ from the value in control area.
The values of the coefficient of mobility for nickel are characterized by a slight difference not only in the soils of experimental areas, but also on the value of the coefficient of mobility in the control area (Figure 8).

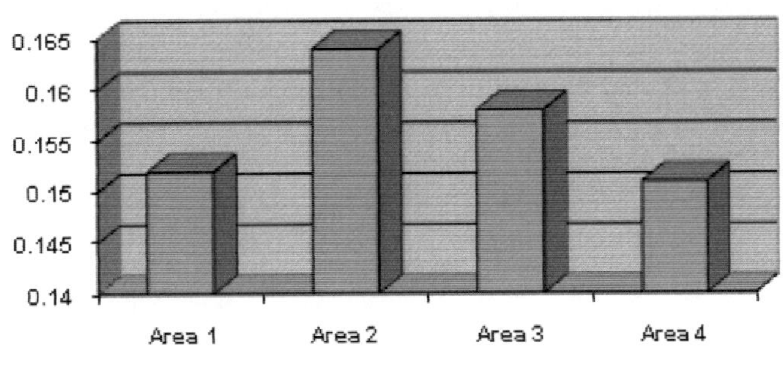


**Figure 8** - The values of the coefficient of mobility of Ni.

These data indicate that the influence of vehicles on the mobility of this metal is minimal, which is confirmed by a negative value of the correlation coefficient ($r = -0.64$).

## 3. Conclusions

The distribution of heavy metals in the soil is affected by traffic. In all three areas the values of mobile Pb more than MPC (from 2 to 10 times) do not exceed the gross forms, but more clearly show the dependence of concentration on traffic. Movable forms of Cu do not exceed MPC, gross forms is exceed MPC in Area 3. This is due to the fact that there are more trolleybuses in this area compared to other areas. Rubbing of the contact wires is a source of Cu. The mobile forms of Zn is exceed the MPC values at all areas, although the values of the gross forms is exceed the MAC values only in Area

3. The mobile and gross forms of Ni do not exceed the MPC and differ in the uniformity of the distribution. Thus, the distribution of heavy metal in urban ecotopes is affected by motor vehicles, except of Ni.

**References**

[1] Ilyichev V. About the dynamic formation of the urban livelihood system compatible with the biosphere. V. Ilyichev, S. Emelyanov, V. Kolchunov., N. Bakaeva. Applied mechanics and materials. V. 725-726, 2015, p. 1224-1230.
[2] Ciegis R. The Concept of Sustainable Development and its Use for Sustainability Scenarios. Inzinerine Ekonomika-Engineering Economics. The Economic Conditions Of Eenterprise Functioning. 2009. p. 28-37.
[3] Kates R. Environment: Science and Policy for Sustainable Development, Robert W. Kates, Thomas M. Parris, and Anthony A. Leiserowitz, 2005. 47, 3, p. 8–21.
[4] Stephen Appiah Takyi; Andrew D. Seidel Adaptive management in sustainable park planning and management: case study of the city of Vancouver Parks. Jurnal of Urban Ecology. Volume 3, Issue 1 January 2017.
[5] Gabriella Hancz, János Bíró The potential role of urban green infrastructure in sustainable municipal water supply management in a Hungarian town. Jornal of International Scientific Publications: Ecology&Safety. V. 10. 2016. p. 409-417.
[6] Peter M. Groffman. Moving Towards a New Urban Systems Science. Ecosystems January 2017, Volume 20, Issue 1, pp 38–43.
[7] Vorobev S. Using bioindication methods for assessing the environmental quality of urban settlements from the perspective of the concept of biosphere compatibility on an example of the Oryol city. Vorobev S, Kozlov D. Lambert academic publishihg: Saarbruken, Deutschland, 2014. p. 54.
[8] John Pastor. Ecosystem Ecology and Evolutionary Biology, a New Frontier for Experiments and Models. Ecosystems. March 2017, Volume 20, Issue 2, pp 245–252.
[9] Vorobev S. Application Biosphere Compatibility Concept To Evaluate The Quality Of Urban Environment By Bioindication Methods. Volume 50, 2017, Pages 1-8.
[10] Vorobev S. The influnce of meteorological factors on the pollution emission by vehicles in the city Orel // S. Vorobev, D. Kozlov - VI International scientific conference «21 century: fundamental science and technology» (North Charleston, SC, USA, 20-21 Aprylr 2015) Charleston, SC, USA: Science Book Publishing House, 2015. 11-14 p.

# Assessment of the Phytotoxicity of Soils on Main Streets of Voronezh City

**A A Voronin[a], M A Klevtsova[b], L A Lepeshkina[c], A A Mikheev[d], S A Yeprintsev[e]**

Voronezh State University,
Universitetskaya pl. 1, Voronezh, Russia, 394018

E-mail:[a]voronin@bio.vsu.ru, [b]klevtcova@geogr.vsu.ru, [c]lilez1980@mail.ru, [d]alexey.mikheev.88@gmail.com, [e]esa81@mail.ru

**Abstract**. This research looks at the results of a study assessment of phytotoxicity soils of the main streets of the city Voronezh. The experiments were performed under ecologo-analitic laboratory Voronezh State University conditions over a 21-day evaluation period. The phytotoxic effect germination was analyzed on *Avena sativa* L. seeds, germination energy, morphological parameters of seedlings, and plant biomass. The research soil samples had both an inhibitory and stimulating effect on the growth of the above-ground part of the seedlings and roots *Avena sativa*. The sections of main streets with increased phytotoxicity soils were revealed. It is recommended to carry out measures to reduce the phytotoxicity of urban soils.

## 1. Introduction

Soils hold a specific place in ecosystems and carry out a huge number of functions, the most important of which – environmental. Growth of the cities leads to reduction of a share of "live" soils and increase in a share of the broken lands that creates threat to health and human life in the urbanized environment. City soils concern to group of the anthropogenic transformed soils – urbanozem. Existence in the soil of pollutants and toxins suppresses growth and development of plants. City soils are characterized by property of phytotoxicity [1].

In Russia and other countries the study of environmental condition soils on the basis bioassay methods are very relevant and widely used to assess toxicity in agricultural [2-5] and urban [6] soils. Phytotesting has been relatively recently used in the environmental field to assess the quality of the environment in natural environments (water, soil). Methodical recommendations on the use of higher plant seeds in phytotesting have been developed [1,7]. The thresholds of phytotoxicity pollutants for various test plants are established [8,9].

The main contribution to the atmosphere pollution and soils of Voronezh is made by vehicles, the emissions make up more than 85% of the total amount pollutants. An inevitable factor is contamination with heavy metals, which easily accumulate and reduce the stability of the vegetation cover of urbanized areas. In the most depressed state is the vegetation cover of highways, streets with intensive traffic and boulevards the central part of the city.

The aim of this work is to study the assessment phytotoxicity of soils in the main streets of Voronezh. Objectives of the study: to conduct a laboratory study of the phytotoxicity urban soils; to assess the phytotoxic effect in terms of the germination Avena sativa L. seeds, the germination energy and the morphological parameters of *Avena sativa* seedlings, the biomass (green mass) of the plants; identify areas of main streets with increased phytotoxicity of soils.

## 2. Equipment and devices used in studies

Zonal soil is Voronezh black soil, which occupies lofty upland areas of the Voronezh. Gray forest-steppe soils are developed where previously there was an array of oak. Sod-forest sandy soils and meadow-black soil remained within the limits of the Voronezh on a very limited area and everywhere have undergone anthropogenic changes. The natural soil is in the central part of the city almost did not survive intact. Only in gardens, parks, and large yards presents less transformed anthropogenic transformed soils.

Long-term studies based on the analytical laboratory of Voronezh State University show that, the central part of Voronezh, Pb content exceeds the background level by 1.3-26 times and MAC (normative approved in law in Russia) (32.0 mg/kg) by 1.3-6.4 times. The Cd content exceeds the background values of 1.3-3.3, but does not exceed the MAC (0.5 mg / kg). Approximately 90% of the cases, the Zn content exceeds the background and 1.3-2.3 times the MAC (23.0% mg/kg). In the upper horizons of the urban soils Voronezh, the Cu content varies widely - from 6.0 to 284.0 mg / kg and exceeds the MAC values (3.0 mg/kg) by a factor of 3-95. In general, pollution of oil products, gross and mobile forms of heavy metals, Voronezh's urban areas are classified as low, medium and high pollution [10].

Phytotesting as a method of environmental monitoring is an integral method of analysis, which provides information on the toxicity of various media and substances. The method is based on the sensitivity of plants to exogenous chemical effects, which affects the growth and morphological characteristics. This method is accessible and simple for experiments, reproducible, reliable, economical and objective [1, 11].

To study the phytotoxicity of Voronezh soils, samples were taken in the middle of summer 2015 and 2016. In 10 points of main streets on both sides. At each point, two samples were taken at a distance of 1 and 5 m from the roadway. Sampling points of soils on the main street Moskovskijprospekt, public transport stops: «TorgovyjCentr» No. 1 (samples 1-4), «PamyatnikSlavy» No. 2 (samples 5-8), «Avtovokzal» No. 3 (samples 9-12), square «Politekhnicheskij» No. 4 (samples 13-16), «ProspektTruda» No. 5 (samples 17-20). The sampling points of soils on the main street of 9 Yanvarya are located at the intersection with the following streets: F. EHngel'sa No. 6 (samples 21-24), Kol'covskaya No. 7 (samples 25-28), ProspektTruda No. 8 (samples 29-32), Mashinostroiteley No. 9 (samples 33-36), Krasnodonskaya No. 10 (samples of 37- 40). As a background (control), the soil (samples 41, 42) specially protected natural area of the suburban area was selected. A total of 42 soil samples were taken in the study in duplicate.

*Avena sativa* L. was used as a test culture. Intact seeds of *Avena sativa* were selected for the experiments, whose germination capacity was not less than 95%. Phytotesting was carried out on the basis of the ecological-analytical laboratory of the Voronezh State University. Seeds of *Avena sativa* were sown in special containers with soil samples. Observations lasted 21 days. The percentage of seed germination was noted. Germination was considered as a percentage of germinating seeds in experimental containers, with respect to control. On the 10th day, the morphometric parameters of the seedlings were measured: the height and green mass of the aerial part, the length and the living mass of the root.

Phytotoxic effect (PE) can be calculated using different approaches. More commonly used indicators of germination of seeds and plant biomass. Phytotoxic effect of PE (%) is calculated by the formula:

$$PE = \frac{M_o - M_x}{M_o} 100\% \quad (1)$$

whereMo – is the mass of the control plant (or all plants per container with a control soil sample); Mx – is the mass of the plant (s) grown on a phytotoxic soil.

## 3. Results and discussion

It was revealed that on 8 soil samples germination of seeds *Avena sativa* was equal to or greater (90-100%) than the control value. On 11 soil samples, seed germination was 70-89%. The lowest percentage of seed germination (less than 50%) was recorded on 13 of the studied soil samples, which corresponded to the following items: public transport stops «TorgovyjCentr» No. 1, square «Politekhnicheskij» No. 4, «ProspektTruda» No. 5; the intersection of the streets of the 9 Yanvarya with F. EHngel'sa No. 6, ProspektTruda No. 8 and Mashinostroiteley No. 9.

In some samples, the phenomenon of hormesis (Figure 1) was noted - the improvement of seed germination, growth and survival of test plants at low concentrations of pollutants. The maximum germination of seeds was noted on soil samples from points: «Avtovokzal» No. 3 (samples 9-12), square «Politekhnicheskij» No. 4.

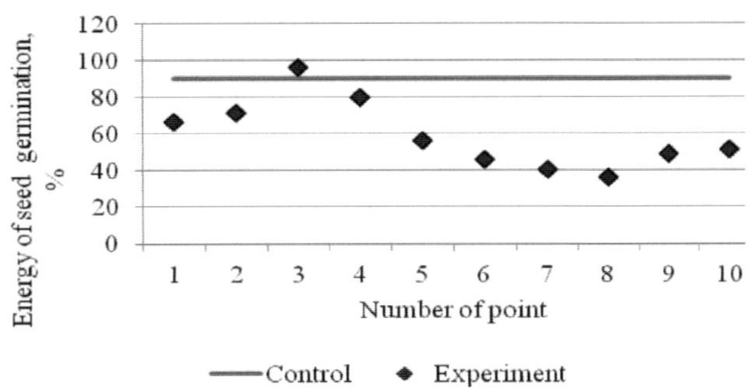


**Figure 1** – Energy of seed germination *Avena sativa* on the 4th day (2015)

For 13 soil samples, the percentage of seed germination *Avena sativa* is equal to or greater than the control (90-100%). In 16 samples, the proportion of germinating seeds was 70-89%. The lowest percentage of seed germination (less than 50%) is noted on soil samples of the following point: «TorgovyjCentr» No. 1, «ProspektTruda» No. 5, the intersection of the streets of the 9 Yanvarya with F. EHngel'sa No. 6 and Mashinostroiteley No. 9 (figure 2).

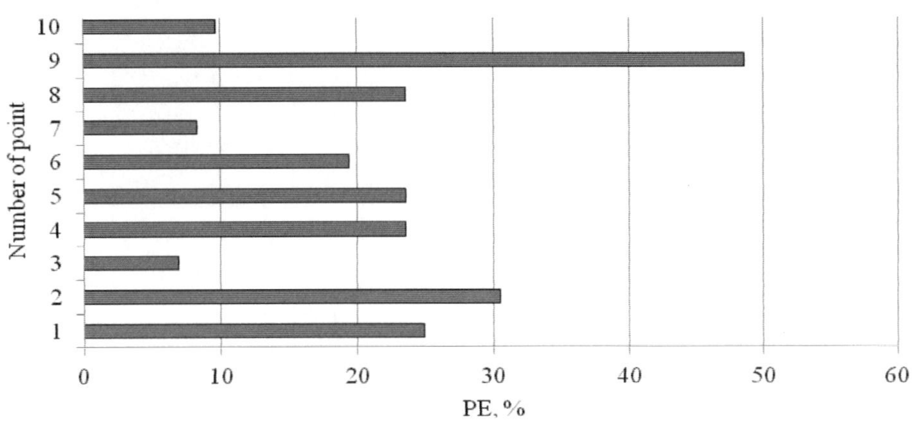


**Figure 2** – Phytotoxic effect on the germination of *Avena sativa* seeds(2016)

Considering the values obtained in the control soil samples (table 1), a comparative analysis of the results of the study.

**Table 1**–Morphometric parameters of *Avena sativa* seedlings (on control soil samples 41, 42) 2015, 2016

| Germination, % | Above-groundpart | | Roots | |
|---|---|---|---|---|
| | 2015 | | | |
| | height, см | weight, g | length, см | weight, g |
| | $X_{average} \pm m_x$ | $X_{average} \pm m_x$ | $X_{average} \pm m_x$ | $X_{average} \pm m_x$ |
| 90 | 17±0.56 | 0.84±0.03 | 8.65±0.45 | 0.55±0.01 |
| | 2016 | | | |
| | 16.5±0.53 | 1.01±0.08 | 13.0±0.60 | 0.58±0.03 |

The investigated soil samples had both an inhibitory and stimulating effect on the growth of the above-ground part and the root system of *Avena sativa* seedlings (Fig. 3).

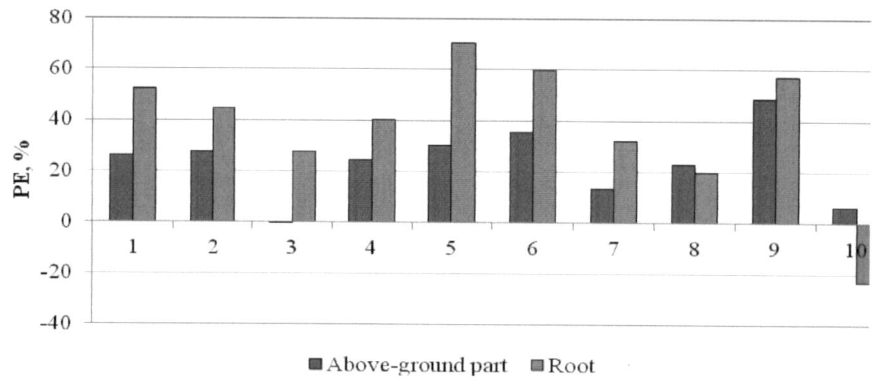


**Figure 3** – Phytotoxic effect on biomass of Avena sativa seedlings (2016)

The most sensitive organ of plants to the environmental conditions of growth are the roots. Therefore, along the length of the roots, one can also judge the phytotoxicity of the soil samples under study.

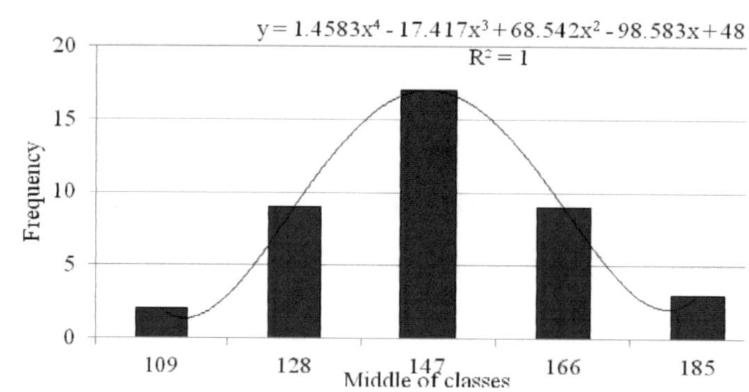


**Figure 4** - The length distribution curve (mm) of the root system of *Avena sativa* seedlings (2016)

The distribution of values of the root length index can be described by the equation of a polynomial curve of the fourth degree. The distribution of the length index of the aboveground part of plants is characterized by a polynomial curve of the third degree. These models are considered adequate, because the criterion of adequacy (approximation) $R^2>0.95$ (Fig. 4,5).

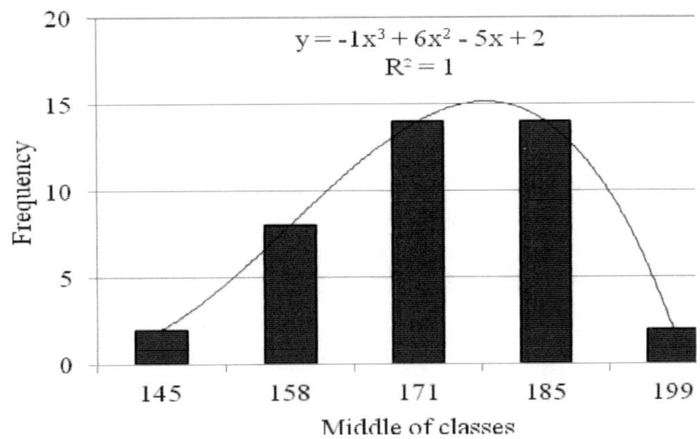


**Figure 5** – The height distribution curve (mm) of the above-ground part of the *Avena sativa* seedlings (2016)

The longest root length of Avena sativa seedlings was formed on soil samples of the following points: «TorgovyjCentr» No. 1, «Avtovokzal» No. 3, square «Politekhnicheskij» No. 4, «ProspektTruda» No. 5. The maximum height of seedlings was noted for soil samples selected in the vicinity of public transport stops «TorgovyjCentr» No. 1, «PamyatnikSlavy» No. 2, «Avtovokzal» No. 3, the intersection of the street 9 Yanvarya with F. EHngel'sa No. 6 and ProspektTruda No. 8.

The results of phytotesting of soil samples confirmed technogenic contamination of the soils the main streets of Voronezh.

## 4. Conclusions

In general, most *Avena sativa* seedlings stimulated the growth of the root system. On average, 70% of the seedlings grown on soils from different points showed a significant increase in the length of the root system by 20-25% compared to the control. The same trend is observed for the mass of the roots. In 55% of the seedlings, the mass of the root system exceeds the values of the control sample. In 62.5% of the seedlings, the growth of the above-ground part was inhibited. The data obtained are consistent with the results of studies phytotoxicity soils in Voronezh in 2002-2009. In experiments with *Triticumaestivum* L. and *Raphanussativus* L. The study shows that the more active formation of the root system in conditions of pollution occurs due to the reduction of the biomass of the aerial organs and is considered as the forced need of the seedling to strengthen the root capacity and protect the vulnerable assimilation system of the aerial part from toxicity polluting substances.

The sections of the main streets of Voronezh with increased phytotoxicity of soils are determined: the stop public transport «ProspektTruda», the intersection of the street 9 Yanvarya with F. E ngel'sa street, Krasnodonskayastreet, Kol'covskaya street and ProspektTruda street. Avena sativa seedlings grown on the soils of the indicated points showed maximum inhibition of the above-ground part of the plant and an increase in the length and mass of the root system.

The results should be taken into account when planning the works on the greening of the main streets of Voronezh. The creation of sustainable lawns and flower beds is possible only after carrying out measures to reduce the phytotoxicity of urban soils.

**References**

[1] Kabirov, R.R., Sagitova, A.R., Sukhanova N.V. Development and of a multicomponent test system for evaluating the toxicity of soil in an urban territory // Russian Journal of Ecology. 1997. T. 28. No.6.P. 360-363.

[2] Aktaruzzaman, M., Fakhruddin, A.N.M., Chowdhury, M.A.Z., Fardous, Z., Alam, M.K. Accumulation of heavy metals in soil and their transfer to leafy vegetables in the region of Dhaka Aricha Highway, Savar, Bangladesh // Pakistan Journal of Biological Sciences, Volume 16, Issue 7, 2013, P. 332-338.

[3] Du, Y.-L., He, M.-M., Xu, M., Yan, Z.-G., Zhou, Y.-Y., Guo, G.-L., Nie, J., Wang, L.-Q., Hou, H., Li, F.-S. Interactive effects between earthworms and maize plants on the accumulation and toxicity of soil cadmium // Soil Biology and Biochemistry , Volume 72, May 2014, P. 193-202.

[4] Pospelova, O.A., Mandra, Y.A., Stepanenko, E.E., Okrut, S.V., Zelenskaya, T.G. Identification of technogenic disturbances of urban ecosystems using the methods of bioindication and biotesting // Biosciences Biotechnology Research Asia,Volume 12, Issue 3, December 2015, P. 2241-2251.

[5] Silva, E.B., Fonseca, F.G., Alleoni, L.R.F., Nasimento, S.S., Grazziotti, P.H., Nardis, B.O. Availability and toxicity of cadmium to forage grasses grown in contaminated soil // International Journal of Phytoremediation, Volume 18, Issue 9, 2016, P. 847-852.

[6] Aleksandrovich, G.E. Zinc as the environmental factor of urban ecosystems Ecology, Environment and Conservation, Volume 22, Issue 1, 2016, P. 433-436.

[7] Gheorghe, S., Vasile, G.G., Stoica, C., Lazar, M.N., Lucaciu, I., Banciu, A. Phytotoxicity tests applied on sewage sludge resulted from urban wastewater treatment plants // Revista de Chimie, Volume 67, Issue 8, August 2016, P. 1469-1473.

[8] Sharma, P., Dubey, R.S., Lead toxicity in plants // Brazilian Journal of Plant Physiology, Volume 17, Issue 1, January 2005, P. 35-52.

[9] Dan, T., Hale, B., Johnson, D., Conard, B., Stiebel, B., Veska, E. Toxicity thresholds for oat (Avena sativa L.) grown in Ni-impacted agricultural soils near Port Colborne, Ontario, Canada // Canadian Journal of Soil Science, Volume 88, Issue 3, May 2008, P. 389-398.

[10] Yeprintsev, S.A., Kurolap, S.A., Komov, I.V., Minnikov, I.V. Monitoring of factors of ecological safety of urbanized territories' population (by example of settlements of Voronezh region) // Life Science Journal, Volume 10, Issue 12 SPL.ISS., 2013, Article number 136, P. 846-848.

[11] Füleky, G., Barna, S. Biotesting of heavy metal pollution in the soil // Carpathian Journal of Earth and Environmental Sciences, Volume 3, Issue 2, 2008, P. 93-102.

# Study of Biometric Datum of Green Manure Crops in the Process of Biological Soil Reclamation on the Territory of Coal Producer Ojsc "Mine No 12"In The Kemerovo Region

**M A Yakovchenko[a], L G Pinchuk[b], A A Kolosolapova[c], D N Alankina[d]**

FSBEI HPO Kemerovo State Agricultural Institute,
Markovtseva Street, 5, Kemerovo, Russia, 650056

*E-mail*: [a]mara.2002@mail.ru, [b]Ludmilapinchuk@mail.ru, [c]annadbim@mail.ru, [d]alankina92@mail.ru

**Abstract.** The article presents the results of a study of green manure crops of different kinds, it is determined that the best growth results were obtained with incorporation of hydrogel in the substratum, and particularly in the clay, so as due to the amount of moisture in the substratum, clay is more hygroscopic and physico-chemical properties of hydrogel significantly increase moisture capacity of the substratum. Sowing field germination of all crops is much higher in the clay then in soil. Territories with the hydrogel usage showed a greater plant density per $1m^2$. Almost all crops with the major growth of herb were sowed in the clay with hydrogel addition, the crops height increased by 2.5 times. The only exception was Rye, the difference in height between its plants in "soil + hydrogel" and "clay+hydrogel" was less than 1%. It was registered that the root growth of Phacelia in "clay+hydrogel" increased by 2.5 times. While the root growth of Rump, on the contrary, increased in mellow soil with hydrogel by 43%. Other kinds of crops did not perform any difference in their root length. The increase of herbage in mixtures of green manure crops was negligible, whereas mono-sowing of such crops as Esparcet, Rump and Buckwheat showed the greatest increase of herbage in comparison to the soil lots and other sowing variants. The greatest increase of herbage among lots without hydrogel addition was performed on the clay ones: Esparcet - 250%, Buckwheat - 172% Rump - 123% Phacelia - 77.5%. The best results of herbage accumulation were showed by Esparcet, Buckwheat, Rump, Phacelia.

## 1. Introduction

The main objectives of biological reclamation are resumption of the process of soil formation, increase of self-purification soil capacity and reproduction of biocenoses. In order to develop effective methods of biological reclamation it is very important to study the processes of vegetational cover evolution in various natural areas and technological conditions. Formation of vegetational cover on the dumps of overburden operations is very slow. It takes from 5 to 10-15 years because of the complex, time-varying dump relief, lack of nutrients for rock development, water and heat unstable mode [1].

The rate of soil and soil layer formation depends on the properties of parent rock material, water and heat conditions, topography and natural climatic conditions of the area, vegetation composition of species and duration of the natural land reclamation [2,3].

The most effective method of biological reclamation of disturbed areas is the creation of a multi-species vegetation cover including perennial grasses and resistant species of bushes and trees. Multistage structure of disturbed land protects it from erosion and deflation, while leaf litter and root systems give a large increase of organic substance [4,5].

Experimental studies, carried out by soil scientists in the Kemerovo region for 40 years, show that the full restore of soil functions is impossible to get in the historically foreseeable future. The maximum score of the experimental stages was 90%, and the average fertility of reclaimed land was about 30% [6,7].

Particularly relevant issue for the reclamation of the Kemerovo region, where natural cover is permanently destroyed by mining, road construction, building of industrial and other objects. On average the annual area of disturbed lands in the region reaches 1118 hectares. According to estimates of the Institute of Coal and Coal Chemistry SB RAS the coal mining companies have violated 9.17 ths. km, which is 9.6% of the total land fund of the Kemerovo region. Moreover, new licenses to industrial land usage are issued every year. If this situation does not change, in a few decades Kuzbass won't have any coal mining unaffected land area. The problem is complicated by the fact that most of the disturbed areas occupy the highly productive soils of great economic importance [8,9,10].

The aim of this research was to study the biometric characteristics of the green manure crops (traditional and non-traditional) in different conditions.

The research objectives were:

1. To determine the crop density.
2. To determine the crop height.
3. To measure crop root length in different experimental conditions.
4. To study changes of herbage of green manure crops.

The object of the study was traditional and non-traditional green manure plants.

The subject of the research was vegetative reclamation technology.

## 2. Conditions, equipment and methods of the study

Land for reclamation is located in the South part of the forest-steppe zone of the region that is characterized with insufficient precipitations especially in the first summer half. Vegetation cover of the territory is the steppe vegetation with predominance of grain and herbaceous species.

Land parcels for reclamation are located on the claim of the strip mine, and surrounding territories belong to urban locality of the Prokopyevsky District.

OJSC «Mine No 12» is an operating coal mining enterprise in Kiselevsk, Kemerovo region.

Field of the Mine No 12 is on the southeast from the Kiselevsk's coal deposit in Kuzbass, that is the center of Kiselevsk, the Kemerovo region, Russia.

OJSC «Mine No 12» is one of the core enterprises of Kiselevsk. Number of employees at the mine is 1755 people.

Green manure crops are a wonderful fertilizer. The organic matter, which is formed from the junction of sunlight, air and water, is much more efficient than manure. In addition, green manure crops accumulate nitrogen, phosphorus, potassium, and many macro- and microelements, while the cost of the crops is relevantly low.

The root system of green manure crops deeply hoes the soil providing an excellent water and air exchange. So these crops are also called "bio- plough".

Many green manure crops are honey plants. In addition, their root excretions are useful, e.g., excretions of mustard repel larvae of the May beetle, wireworms, and mole crickets.

## 3. Results and discussion

On June 7th, 2016 scientists and research workers of KSAI started an experiment on the territory of coal strip mining of OJSC «Mine No12» in Kiselevsk (Figure 1).

The area of each experimental plot was 1 $m^2$. Experience was laid out in a 3-fold replication. The size of the guard band was 0.5 m. The total area of the experimental plot was 200 $m^2$.

Seeding rate of green manure: Buckwheat – 90 kg / ha; Meadow Clover – 20 kg / ha; Rye – 20 kg / ha; Rape – 20 kg / ha; Phacelia – 20 kg / ha.

Seeding rate of traditional crops: Rump – 20 kg / ha; Espartset- 20 kg / ha.

The consumption rate of hydrogel – 50 g / $m^2$.

**Figure 1** - Experimental plot OJSC «Mine No12»

On June 16th, 2016 scientists and research workers of KSAI visited the Experimental plot OJSC «Mine No12» for the third controlled time.

They studied plants density in the seedling phase on the experience lots of 1 $m^2$. Results of calculations are shown in Table 1 and Figure 2.

**Table 1** - Plants density, pcs. /$m^2$

| crop | soil | soil + hydrogel | clay | clay + hydrogel |
|---|---|---|---|---|
| Rye | 203 | 233 | 318 | 343 |
| Rape | 200 | 210 | 233 | 256 |
| Rump | 53 | 63 | 101 | 151 |
| Phacelia | 50 | 58 | 51 | 125 |
| Espartset | 41 | 68 | 101 | 153 |
| Buckwheat | 50 | 60 | 103 | 155 |
| Meadow Clover | 50 | 55 | 151 | 168 |

Field germination of all the crops is relevantly higher on the clay lots than on the soil ones. Lots with hydrogel showed greater plant density per 1$m^2$. This is due to the amount of moisture in the substratum at the sowing stage. Clay soil is more hygroscopic and physico-chemical properties of hydrogel significantly increase moisture capacity of the substratum.

IOP Conf. Series: Earth and Environmental Science **66** (2017) 012004 doi:10.1088/1755-1315/66/1/012004

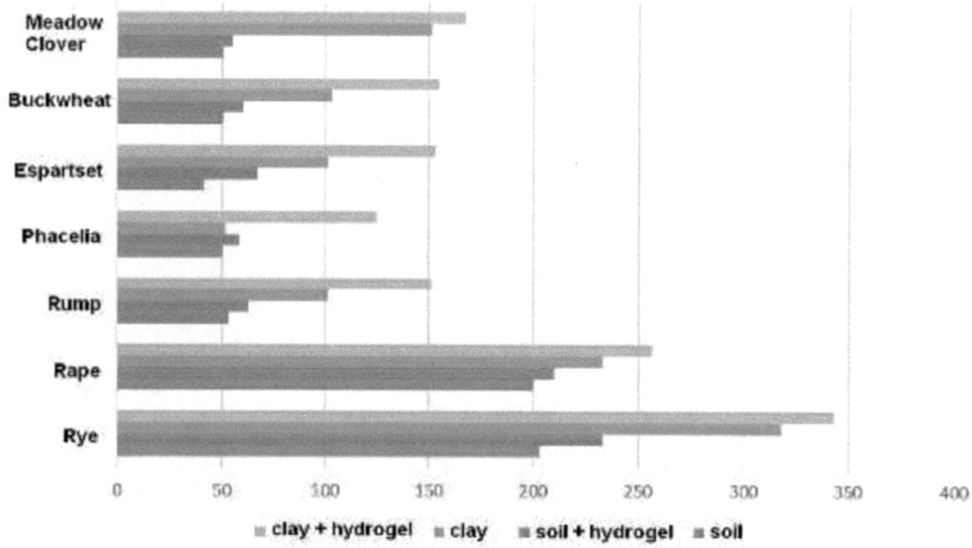


**Figure 2** – Plants density, pcs. /$m^2$

All green manure crops had low germination on the soil lots, although the introduction of hydrogel increased this index at the average by 15%. The greatest plant density on all types of the lots showed Rye and Rape. The results obtained in the course of the study, physical and chemical conditions of anthropogenic disturbed lands are optimal for these kinds of plants.

There were evaluated biometric characteristics of the plants, such as plant height and root length. Measurement of the plant parts was carried out before mowing. The results are given below in the Figure 3-6.

For all studied parameters the best values were showed by plants on lots with hydrogel. The greatest plant height was obtained on the lot "clay+ hydrogel" by Buckwheat (75.3 cm) and Espartset (57 cm). The use of hydrogel on clay substratum had a positive impact on the plants herbage.

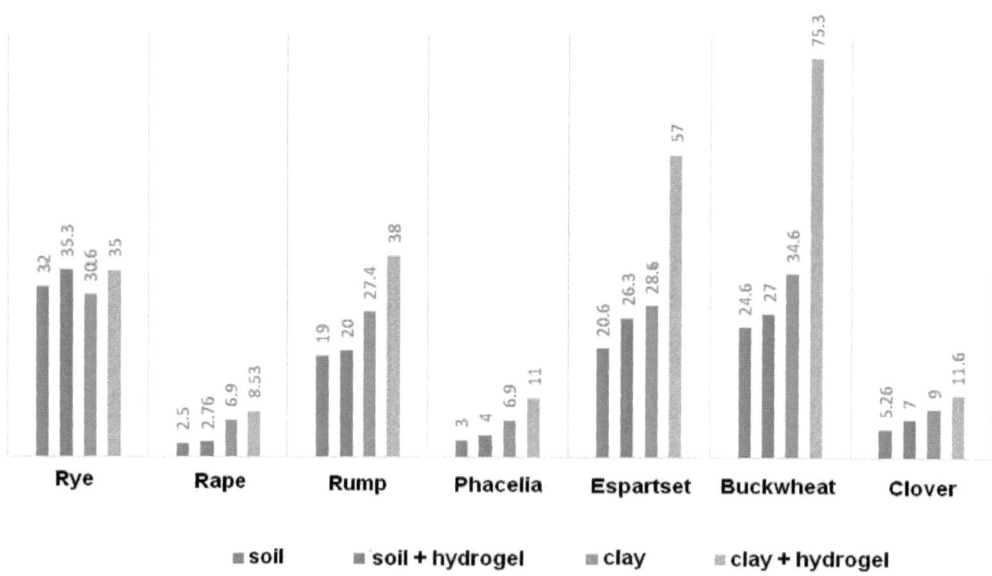


**Figure 3** – Plant height, sm

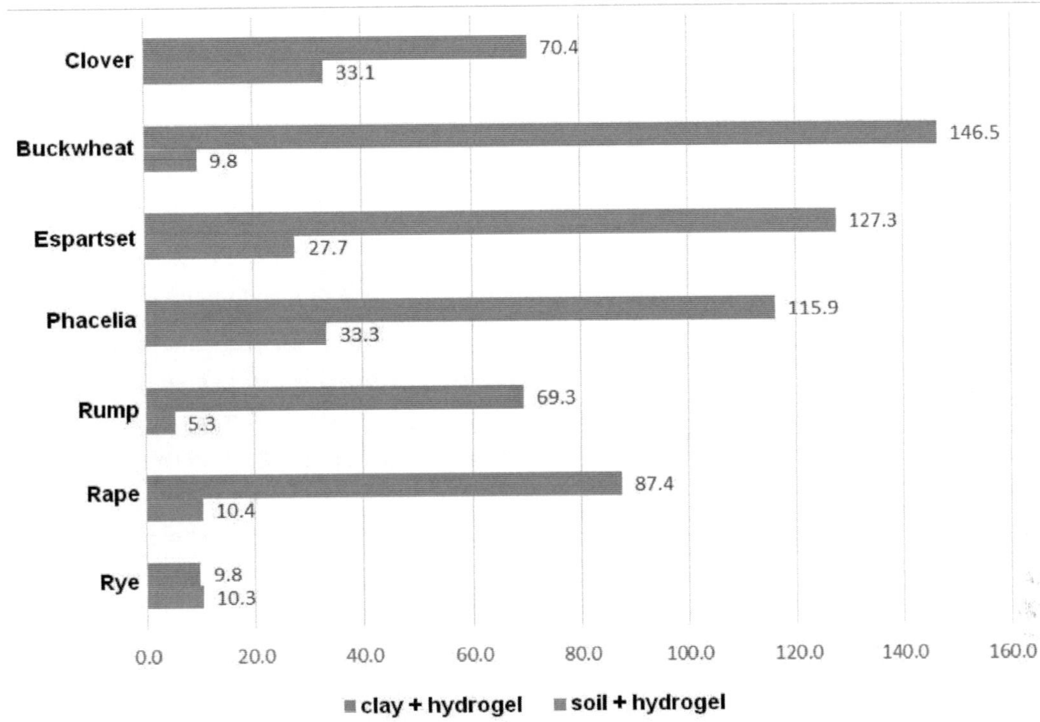


**Figure 4** – Increase of plant height on lots with hydrogel in comparison to similar lots, %

To determine the increment of plant height in percentage lots with hydrogel were compared with similar lots without hydrogel addition. Almost all crops with the major growth of herb were sowed in the clay soil with hydrogel addition, the crops height increased by 2.5 times. The only exception was Rye, the plants height in "soil + hydrogel" and "clay+hydrogel" increased by 10.3% and 9.8% respectively. While Buckwheat increased by 3 times, giving an increase in 146%, Espartset - by 127%, Phacelia – by 116%.

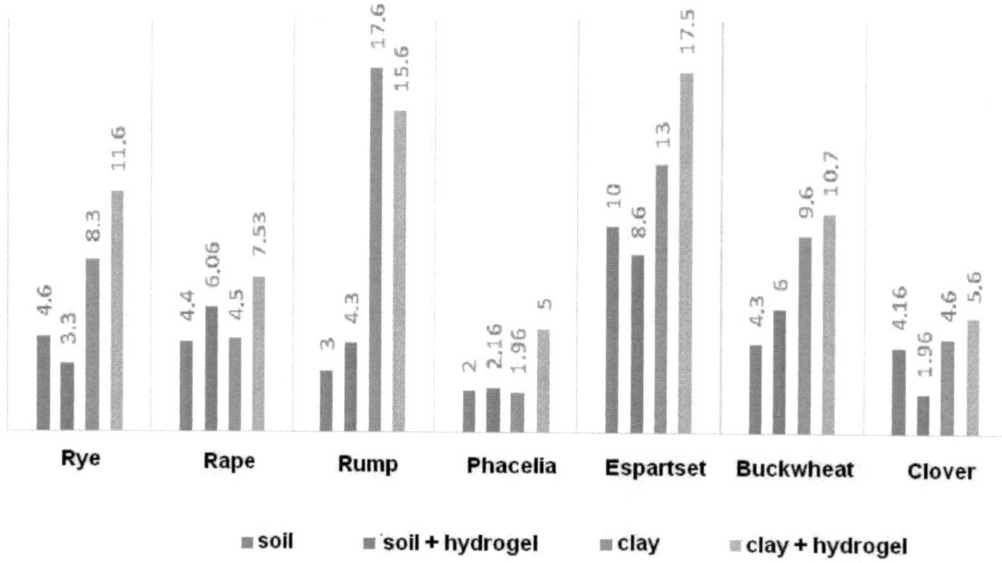


**Figure 5** - Root length, sm

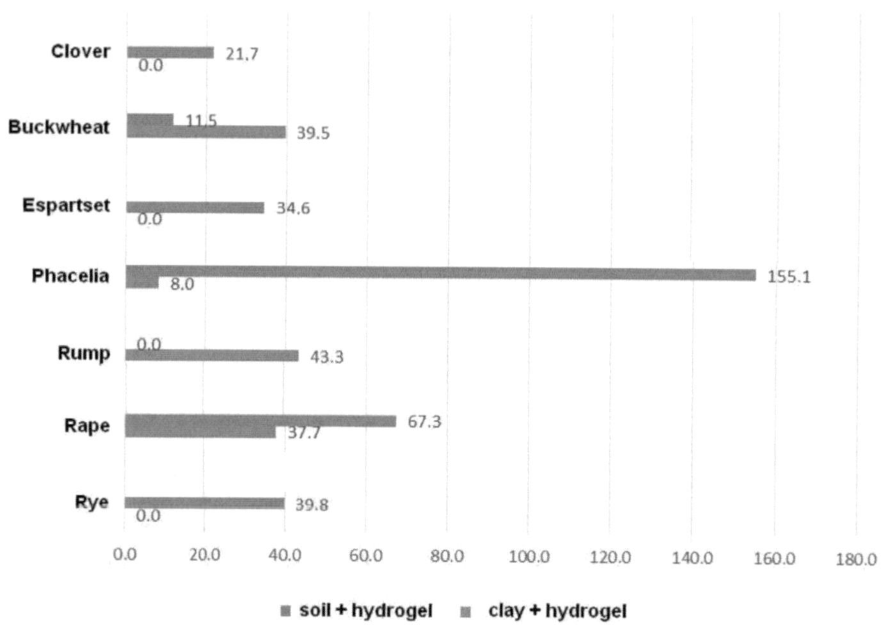


**Figure 6** – Increase of root length in lots with hydrogel in comparison to similar lots %

The roots of green manure crops are of great importance in the restoration of physical and chemical properties of technogenically disturbed and mining soils. Plants have an impact on the soil's quality, loosen soil layers with roots, reduce leaching of mineral elements, release poorly soluble phosphates, enrich soil with potassium, and low soil pH to neutral.

Roots of such crops as Clover, Espartset, and Rye did not showed significant increase in growth on the "clay+ hydrogel" lot. This combination with hydrogel was the most favorable for Phacelia's root growth – it increased in 2.5 times. Rump’s root, on the contrary, increased its growth on the “soil+ hydrogel” lot - by 43%.

Active growth of herbage of the crops inhibits the germination of weeds. Plowing of plants herbage improves soil structure, reduces volume weight of the plowing layer and the bulk-density of the soil. As a result plowing activity significantly increases water infiltration and water-holding capacity of the soil, thereby the land runoff of precipitation reduces and the soil moisture content increases dramatically. Therefore, life’s activity of soil microorganisms improves greatly. Moreover, plants’ stems and leaves enrich the soil with nutrients, accumulated during the growing season.

In the course of the research the crops herbage growth on different soil lots has been studied, the results are shown in Table 2.

The increase in herbage on lot with crops mixtures was 4.5% - for Rump with Espartset and 8.10% - for Buckwheat with Clover, whereas mono-sowing of such crops as Esparcet, Rump and Buckwheat showed the greatest increase of herbage in comparison to the soil lots and other sowing variants (Table 2 ). The greatest increase of herbage among lots without hydrogel addition was performed on the clay ones: Esparcet - 250%, Buckwheat - 172% Rump - 123% Phacelia - 77.5%.

**Table 2** – Increase of plants herbage

| Experimental lots | Increase | | | |
|---|---|---|---|---|
| | $g/m^2$ | | % | |
| | soil | clay | soil | clay |
| Espartset + Rump | 0.37 | 0.49 | 4.54 | 5.05 |
| Rye | 0.40 | 0.84 | 12.35 | 17.80 |
| Rape | 0.12 | 0.87 | 1.56 | 9.75 |
| Rump | 0.82 | 85.00 | 45.05 | 123.19 |
| Phacelia | 0.20 | 9.60 | 2.50 | 77.42 |
| Espartset | 5.00 | 155.20 | 15.15 | 249.52 |
| Buckwheat | 14.40 | 664.90 | 9.63 | 172.21 |
| Meadow Clover | 0.67 | 7.30 | 7.83 | 23.47 |
| Buckwheat + Clover | 14.70 | 21.30 | 10.76 | 8.75 |

## 4. Conclusions

1. The study shows that the best growth results of green manure crops of different kinds have been obtained with incorporation of hydrogel in the soil substratum, and particularly in clay, so as due to the amount of moisture in the substratum, clay is more hygroscopic and physico-chemical properties of hydrogel significantly increase moisture capacity of the substratum.

2. Sowing field germination of all crops is much higher in the clay then in soil. Territories with the hydrogel usage showed a greater plant density per 1m2. Almost all crops with the major growth of herb were sowed in the clay with hydrogel addition, the crops height increased by 2.5 times. The only exception was Rye, the difference in height between its plants in “soil + hydrogel” and “clay+hydrogel” was less than 1%.

3. The root growth of Phacelia in “clay+hydrogel” increased by 2.5 times. While the root growth of Rump, on the contrary, increased in mellow soil with hydrogel by 43%. Other kinds of crops did not perform any difference in their root length.

4. The increase of herbage in mixtures of green manure crops was negligible, whereas mono-sowing of such crops as Esparcet, Rump and Buckwheat showed the greatest increase of herbage in comparison to the soil lots and other sowing variants. The greatest increase of herbage among lots

without hydrogel addition was performed on the clay ones: Esparcet - 250%, Buckwheat - 172% Rump - 123% Phacelia - 77.5%.

Thus, on the ground of the undertaken study it can be argued that the hydrogel introduction in anthropogenic disturbed soil favorably affects the growth and development of the green manure crops.The best results of herbage accumulation were showed by Esparcet, Buckwheat, Rump, Phacelia.

**References**

[1] Yakovchenko M. The Study of Soil Protection in the Sistem of the Cultivated Lands of Kemerovo Region / IOP Conf. Series: Materials Science and Engineering 91(2015)012078 doi:10.1088/1757-899x/91/1/012078.

[2] Růžek L. Chemical and biological characteristics of reclaimed soils in the Most region (Czech Republic) / Plant Soil Environ., 49, 2003 (8):346-351.

[3] Micanova O. Utilization of Microbial Inoculation and Compost for Revitalization of Soils / Soil and Water Res., 4,2009 (3): 126-130.

[4] Zhonqiu Zhao. Soils development in opencast coal mine spoils reclaimed fo 1-13 years the West-Northern Loess Plateau of China / European Journal of Soil Biology 55 (2013) 40-46.

[5] Blaylock M.J. Enhanced accumulation of Pb in Indian mustard by soil-applied chelating agents / M.J. Blaylock, D.E. Salt, S. Dushenkov et al. // Environ. Sci. Technol. – 1997. – V.31. - No3. – P.860-865.

[6] Remes J. Biological activity of anthropogenic soils after spoil-bank forest reclamation / Journal of Forest Science, 53, 2007 (7): 299-307.

[7] PietrzykowskiM. Soil and plant communities development and ecological effectiveness of reclamation on a sand mine cast / Journal of Forest Science, 54, 2008 (12): 554-565.

[8] Růžek L. Slit seeded grass-legume mixture improves coal mine reclamation / Plant Soil Environ., 58,2012 (2): 68-75.

[9] Ibanez J.J. Future of soil science / J.J. Ibanez // The future of soil science / Ed. A.E. Hartemink. – Wageningen: IUSS,2006. – P.60-62.

[10] Pietrzykowski M. Linking heavi metal bioavailability (Cd, Cu, Zn and Pb) in Scots pine needles to soil properties in reclaimed mine areas / Science of the Total Environment 470-471 (2014) 501-510.

# Vegetation Use for Resolving Electromagnetic Compatibility and Ecology Issues

**M Yu Zvezdina[a], Yu A Shokova, L V Cherckesova, T M Golovko and A A Cherskaya**

Don State Technical University, Radioelectronic Department
1 Gagarin Square, Rostov-on-Don, 344000, Russian Federation

*E-mail:* [a]zvezdina_m@mail.ru

**Abstract.** The wide spread of Information and Communication Technologies and the development of Internet-enabled mobile applications have aggravated electromagnetic compatibility and ecology problems. Inability to excite electromagnetic field of a desired structure and strength with traditional approaches actualizes additional actions, including providing diffraction on propagation path, to resolve these issues. Diffraction on a stand-alone obstacle along the propagation path and the one on set of obstacles near receive antenna location can be considered as the additional actions in ultrashort band. The accomplished studies have shown that one the most effective means to lower electromagnetic field strength is to shield the receive antenna with vegetation from jamming radio equipment. Moreover, vegetation resolves electromagnetic ecology issues, for the energy flux density can be lowered by about two orders of magnitude.

## 1. Introduction

Up-to-date society development is accompanied by an extensive use of Information and Communication Technologies [1]. The aggravation of electromagnetic compatibility and ecology issues are the disamenities of this process. In addition, the hardest problems of the electromagnetic compatibility lie within radio spectrum overflow [2]. These problems arise when several communication services and similar services operators use adjacent service bands. For instance in [3] it is shown, that rapid evolution of land mobile telecommunications of GSM-900 standard aggravated electromagnetic compatibility with short-range navigation and landing devices working in P-GSM band (890-915/950-960 MHz) essentially. Moreover, solving this problem by switching into E-GSM band (880-890/925-935 MHz) has severe constraints.

Electromagnetic ecology issues aggravation is caused by expansion of Internet-enabled mobile applications [4]. As a result, in high-populated areas a complete cover of non-ionizing radiation from base station aerials is formed [5]. Thus, electromagnetic field of base station antenna transmitters has a constant round-the-clock health impact on residents, including children, elderly and disabled. The consequences of this impact can be harmful under specific conditions [6].

Solving the issues of electromagnetic compatibility and ecology usually starts with known structure management techniques of electromagnetic field directly excited by antenna. As an example, construction design can presume composite materials application upon antenna carcass, as it I shown in [7, 8]. For this technique the results and extended recommendations on electromagnetic field structure management of an antenna are given in [9]. It should, however, be noted that desired field structure cannot always be managed with conventional techniques.

In the paper the applicability of additional actions based on consideration of ground-wave propagation characteristics in ultra-short band is estimated. As it is shown in ITU recommendations [2, 10, 11], signal strength can be lowered on propagation path by artificial diffraction effects, produced when first Fresnel zone of a radio wave is tangent to a stand-alone obstacle or when energy is scattered near receiver on the set of obstacles. In the paper the first diffraction source is simulated by a stand-alone high-rise building, while another one is simulated by vegetation. The applicability of these actions for electromagnetic ecology issues resolution is estimated simultaneously. At that, the obtained field strength is estimated according to regulatory documents [12-15].

Thus, the paper objective is the estimation of objects (buildings, vegetation) effect on electromagnetic strength attenuation near the receiver.

## 2. Calculation methodologies

Diffraction loss estimation on objects located on propagation path near the receiver is analyzed with ITU recommendations [2, 10, 11]. The vegetation effect on energy flux density is estimated through electromagnetic environment studies near residential building with regulations [14, 15] and the methodology given in monograph [16]. All the results regarding electromagnetic ecology were obtained with «preventive principle» recommended by the World Health Organization [17]. The principle calls for preventive actions to reduce possible harmful influence of radiation on human, even if the influence is not proven and is presumed out of the current scientific knowledge level.

Efficiency of additional actions will be estimated in two steps. The first step research is aimed to estimate the diffraction effect on electromagnetic compatibility issues, while the second stage is linked to electromagnetic ecology issues.

### *2.1. Stand-alone obstacle diffraction effect*

To estimate the effect of obstacle in the form of a building a model of double isolated knife-edge edges will be used, as it is recommended in [10]. The problem geometry is shown on Figure 1 taken from [10].

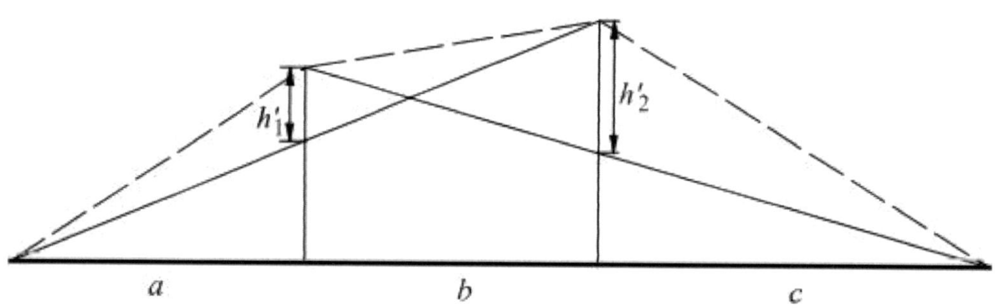


**Figure 1.** Method for double isolates edges

This method consists of applying single knife-edge diffraction theory successively to the two obstacles, with the top of the first obstacle acting as a source for diffraction over the second obstacle. The first diffraction path, defined by the distances *a* and *b* and the height $h'_1$, gives a loss $L_1$ (dB). The second diffraction path, defined by the distances b and c and the height $h'_2$, gives a loss $L_2$ (dB). The total diffraction loss is then given by:

$$L_{ds} = L_1 + L_2 - L_c. \quad (1)$$

$L_1$ and $L_2$ are calculated using formulae [10]:

$$L_{1,2} = -20\lg\left(0.5\sqrt{(1-C(\nu)-S(\nu))^2+(C(\nu)-S(\nu))^2}\right), \tag{2}$$

where

$$\nu = -h\sqrt{\frac{2}{\lambda}\left(\frac{1}{d_1}+\frac{1}{d_2}\right)}, \tag{3}$$

$d_1$ and $d_2$ - distances of the two ends of the path from the top of the obstacle;
$\lambda$- wavelength (m);
$h$ - height of the top of the obstacle above the straight line joining the two ends of the path. If the height is below this line, $h$ is negative;

$$h'_1 = h_1 - h_2\frac{a}{a+b}; \qquad h'_2 = h_2 - h_1\frac{c}{c+b}. \tag{4}$$

A correction term $L_c$ (dB) must be added to take into account the separation $b$ between the edges. $L_c$ may be estimated by the following formula:

$$L_c = \left(12 - 20\lg\left(\frac{2}{1-\frac{\alpha}{\pi}}\right)\right)\left(\frac{q}{p}\right)^{2p}, \tag{5}$$

where

$$p = h_1\sqrt{\frac{2}{\lambda}\frac{a+b+c}{(b+c)a}}; \; q = h_2\sqrt{\frac{2}{\lambda}\frac{a+b+c}{(a+b)c}}; \; \tan\alpha = \sqrt{\frac{b(a+b+c)}{ac}}.$$

*2.2. Diffraction effect on set of objects*

To analyze the diffraction loss on objects near the receiver the ITU recommendations [2] will be used. The problem geometry and legend are shown on Figure 2 taken from [2].

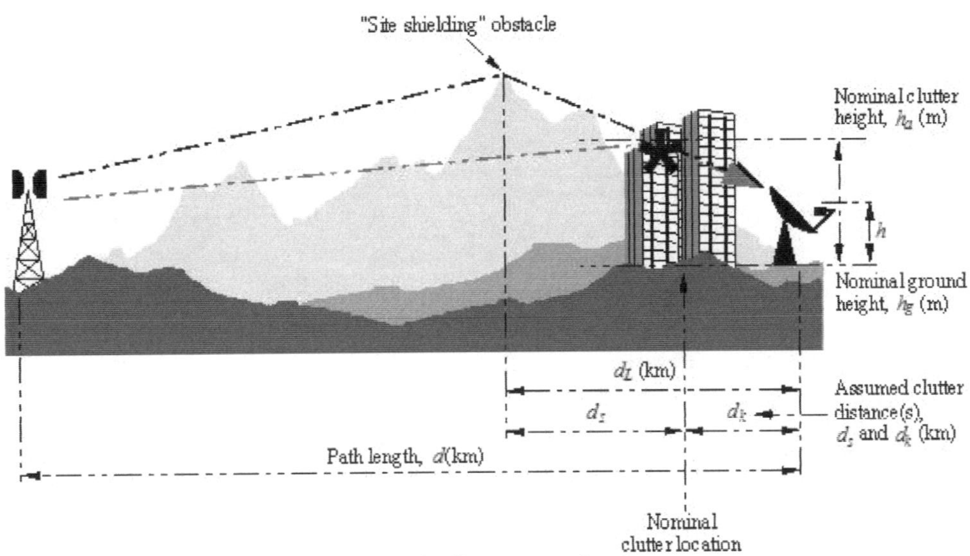


**Figure 2.** Geometry of tasks

According to ITU [2] recommendations, benefit, in terms of protection from interference, can be derived from the additional diffraction losses available to antennas which are imbedded in local ground clutter (buildings, vegetation, etc.). The maximum additional loss is 20 dB above 0.9 GHz, and progressively less at lower frequencies, down to 5 dB at 0.1 GHz.

The additional loss due to protection from local clutter is given by the expression:

$$A_h = 10.25\, F_{fc} \cdot \exp(-d_k)\left(1 - \tanh\left[6\left(\frac{h}{h_a} - 0.625\right)\right]\right) - 0.33 \text{ dB} \qquad (6)$$

where

$$F_{fc} = 0.25 + 0.375\{1 + \tanh[7.5(f - 0.5)]\},$$

$d_k$ - distance (km) from nominal clutter point to the antenna;
$h$ - antenna height (m) above local ground level;
$h_a$ - nominal clutter height (m) above local ground level;
$f$ – frequency (GHz).

The additional protection available is height dependent, and is therefore modelled by a height-gain function normalized to the nominal height of the clutter. Where the clutter parameters are more accurately known they can be directly substituted for the values taken from Table 1 [2].

**Table 1.** Nominal clutter heights and distances

| Clutter (ground-cover) category | Nominal height, $h_a$ (m) | Nominal distance, $d_k$ (km) |
|---|---|---|
| High crop fields<br>Park land<br>Irregularly spaced sparse trees<br>Orchard (regularly spaced)<br>Sparse houses | 4 | 0.1 |
| Village centre | 5 | 0.07 |
| Deciduous trees (irregularly spaced)<br>Deciduous trees (regularly spaced)<br>Mixed tree forest | 15 | 0.05 |
| Coniferous trees (irregularly spaced)<br>Coniferous trees (regularly spaced) | 20 | 0.05 |
| Tropical rain forest | 20 | 0.03 |
| Suburban | 9 | 0.025 |
| Dense suburban | 12 | 0.02 |
| Urban | 20 | 0.02 |
| Dense urban | 25 | 0.02 |
| High-rise urban | 35 | 0.02 |
| Industrial zone | 20 | 0.05 |

*2.3. Electromagnetic environment estimation*

To estimate electromagnetic environment according to offered methods of signal strength depression the correlations from [15] were used for frequencies over 300 MHz. To calculate energy flux density (EFD) Π let's use the following formula

$$\Pi = E^2 / 3.77, \quad (7)$$

where $E$ - electric field intensity amplitude (V/m).

For mobile communication systems maximum permissible level of radiation power, which does not cause permanent changes in human body, is equal to 10 μW/cm² [14, 15].

## 3. Results and discussion

### *3.1. Solving the electromagnetic compatibility issue*

In the research the transmitter power was assumed to be equal to 100 W. Antenna centers were placed at the following heights: 100 m for mobile communications base station antenna and 10 m for receiver. Obstacle distance from the transmitter was assumed to be equal 10 km; it consisted of two peaks of equal height spaced 100 m apart. As it is shown in [10], the assumed model describes high-rise building. In the research the building height was variated in the range from 30 to 90m. Moreover, clutter distance was variated in the range from 0.1 to 0.02 km; clutter parameters were taken from Table 2. Path length changed within line-of-sight range for used antenna height.

For spatial separation is required for radiation sources by electromagnetic compatibility rules, let us also assume that the path length is also variated in the range from 15 to 25 km. In Table 3 the results of loss forecast for the considered transmission path are given for a dense urban zone. It follows from Table 2 that the distance from transmitter will be equal to 200 m and clutter height will be equal to 20 m in this case.

**Table 2.** Loss structure forecast at reception point

| Path length, km | Loss, dB | | | | Signal strength at reception point, dB |
|---|---|---|---|---|---|
| | Total | Basic | Obstacle diffraction | Clutter | |
| Stand-alone obstacle (building) height 30 m | | | | | |
| 15 | 136.73 | 114.32 | 6.26 | 16.10 | -116.42 |
| 20 | 139.96 | 116.86 | 6.23 | 16.10 | -119.96 |
| 25 | 141.18 | 118.84 | 6.22 | 16.10 | -121.18 |
| Stand-alone obstacle (building) height 60 m | | | | | |
| 15 | 130.41 | 114.32 | 6.29 | 16.10 | -110.41 |
| 20 | 139.96 | 116.86 | 6.23 | 16.10 | -119.96 |
| 25 | 141.18 | 118.84 | 6.22 | 16.10 | -121.18 |
| Stand-alone obstacle (building) height 90 m | | | | | |
| 15 | 136.73 | 114.32 | 6.26 | 16.10 | -116.42 |
| 20 | 139.96 | 116.86 | 6.23 | 16.10 | -119.96 |
| 25 | 141.18 | 118.84 | 6.22 | 16.10 | -121.18 |

The analysis of table data shows, that basic loss contribution to total losses is about 83.9%, diffraction loss contribution is 4.5% and clutter loss near the receiver is 11.6%.

In the analysis process of clutter influence, the path length was considered to be equal to 25 km and the stand-alone obstacle height was equal to 30 m. Stand-alone objects were used as clutter sets of objects from the Table 2; the object properties were set equal to the properties of the set. The simulation results are shown in Figure 3.

The simulation results analysis shows, that the highest signal loss (in the range from 18.5 to 15.6 dB) caused by multipath effect can be observed in cities and close to deciduous trees. Mixed tree forest also causes signal loss, but the contribution of this clutter is only 7.03 dB. Stocky vegetation, low-rise buildings and sparsely built suburban areas have no significant influence on signal loss.

Thus, we can conclude that shielding the first Fresnel zone does not solve the receiver broadside radiation problem, which is one of electromagnetic compatibility issues. The most effective approach is linked to planting vegetation near the receiver in the direction of jamming radio equipment.

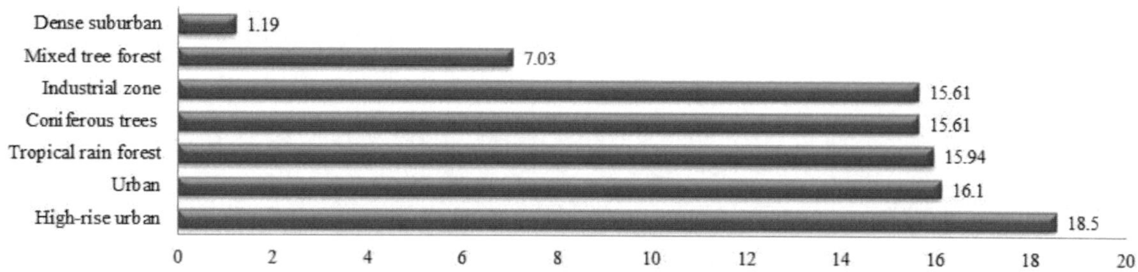


**Figure 3.** Forecasted clutter level for different objects

*3.2. Solving the electromagnetic ecology issue*

In [18-20] it was shown that sanitary protection zone size, in which the energy flux density is lower than the maximum permissible level of 10 $\mu W/cm^2$, is about 30 m for GSM antenna with transmission power 100 W. Let's consider the gain in energy flux density loss for different vegetation types. In Table 3 calculated gain values are given for reference point placed 10 m high (second floor of residential building).

**Table 3.** Achievable field loss at point of receive with vegetation clutter

| Clutter (ground-cover) category | Nominal height, $h_a$ (m) | Nominal distance, $d_k$ (km) | Signal strength with reflections, dB | Signal loss, dB | Gain in energy flux density loss, times |
|---|---|---|---|---|---|
| Coniferous trees | 20 | 0.028 | -59.67 | 19.57 | 90.52 |
| Mixed tree forest | 15 | 0.028 | -59.65 | 19.55 | 90.15 |
| Park land | 4 | 0.02 | -56.12 | 16.02 | 40.72 |

The analysis of table data shows, that vegetation use lowers energy flux density by about two orders of magnitude. At that, the most rational choice is green belt zones of high-rise coniferous and mixed tree. Park land with stocky vegetation is less effective to lower energy flux density level in residential areas. This conclusion is proved true by recommendations of monograph [16]. It is said that long attenuation of electromagnetic wave propagating through vegetation is between 0.15 and 0.26 dB/m depending on polarization type in frequency band of about 1 GHz.

Therefore, vegetation resolves electromagnetic ecology issues in residential areas.

## 4. Conclusions

The carried out research in additional actions efficiency (i.e. building stand-alone high-rises in mobile communications radio paths, using vegetation near the reception point) in resolving the issues of electromagnetic compatibility and ecology has shown the following:

- Planting vegetation near the receiver in the direction of jamming device provides signal loss of about 10 dB greater than the stand-alone high-rise on the radio path.
- Coniferous trees provide maximum gain in signal loss. The difference to the next best choice of mixed trees planting is 7.03 dB. Stocky vegetation has no significant influence on signal loss.
- Vegetation also provides energy flux density loss of about two orders of magnitude.

- It is necessary to consider nominal tree height when choosing vegetation type to resolve electromagnetic ecology issues. Park zones with stocky vegetation are less effective for energy flux density decrease in residential areas.

**References**

[1] Webster F. Theories of the Information Society 2014 4rd edition. Oxford: Routledge ISBN 9780415718790
[2] Recommendation ITU-R P.452-16 Prediction procedure for the evaluation of interference between stations on the surface of the Earth at frequencies above about 0.1 GHz Web-site: https://www.itu.int/rec/R-REC-P.452-16-201507-I/en (accessed February 11, 2017)
[3] Kharin A S, Kalugin V G, Sukhatsky S V Tvarovsky Yu V Shorin O A, Du K O (2010) *Telecommunications (*11) 8-12
[4] Ali S, Khusro S 2016 *Indian J. of Sci. and Techn.* **9** (19) doi: 10.17485/ijst/2016/v9i19/53088
[5] Zvezdina M Yu, Shokova Yu A, Kundrukova N I, Kutukova V D, Pozdnyakova A V 2017 *IOP Conf. Series: Earth and Environmental Science* **50** 012029 doi: 10.1088/1755-1315/50/1/012029
[6] Levitt B B, Lai H 2010 *Environ. Rev.* **18** 369-95 doi: 10.1139/A10-018
[7] Liu X, Burokur S N, Lustrac A de, Sabanowski G, Piau G P (2012) *PIERS C* **33** doi: 10.2528/PIERC12073007
[8] Gabriel'yan D D, Zvezdina M Yu, Sinyavsky G P 2007 *Uspekhi sovremennoi radioelektroniki* (*Achievements of Modern Radioelectronics*) (1) 48-65
[9] Gabriel'yan D D, Zvezdina M Y, Kostenko P I 2001 *J. Commun. Techn. and Electron.* **46** **(**8) 875-78
[10] Recommendation ITU-R P.526-13 Propagation by diffraction Web-site: https://www.itu.int/rec/R-REC-P.526/en (accessed February 11, 2017)
[11] Recommendation ITU-R P.1546-5 Method for point-to-area for terrestrial services in the frequency range 30 MHz to 3000 MHz Web-site: https://www.itu.int/rec/R-REC-P.1546/en (accessed February 11, 2017)
[12] ICNIRP (International Commission on Non-Ionizing Radiation Protection) 1998 Guidelines for limiting exposure to time-varying electric, magnetic and electromagnetic fields (up 300 GHz) *Health Phys.***74** 494-522
[13] IEEE (Institute of Electrical and Electronic Engineers) 2005 C95.1-2005 IEEE Standard for Safety Levels with respect to Human Exposure to Radio Frequency Electromagnetic Fields, 3 kHz to 300 GHz (New York IEEE) Web-site: https://standards.ieee.org/findstds/standard/C95.1-2005.html (accessed February 11, 2017)
[14] Ministry of Health of the Russian Federation 2003 *SanPin 2.1.8/2.2.4.1190-03 Hygienic requirements on land-based mobile communication location and operation* (Moscow)
[15] Ministry of Health of the Russian Federation 2002 *MG 4.3-1167-02 Methodological guidelines for energy flux density estimation for electromagnetic field for location of radio facilities with operating band of 300 MHz – 300 GHz* (Moscow)
[16] Spodobayev Yu M, Kubanov D P 2000 Electromagnetic ecology basics. Moscow: Radio i svyaz'.
[17] WHO Library Cataloguing-in-Publication Data 2004 Establishing a dialogue on risk from electromagnetic fields. http://www.who.int/peh-emf/publications/en/russian_risk.pdf
[18] Zvezdina M Y, Shokova Yu A, Shokov A V 2015 *Proc. Int. Conf. on Application of Information and Communication (Rostov-on-Don)* 399-403 doi: 10.1109/ICAICT.2015.7338588
[19] Zvezdina M Yu, Shokova Yu A, Shokov A V, Dymchenko A A, Parhomenko P A 2015 *Int. J. Theor. & Appl. Sci.* **03** (23) 10-17 doi: 10.15863/TAS.2015.03.23.3
[20] Zvezdina M Yu, Shokova Yu A, Shokov A V, Andronov M A, Tyul'pin V I 2015 *Int. J. Theor. & Appl. Sci.* **08** (28) 56-64 doi: 10.15863/TAS.2015.08.28.8

# Risk Assessment of Oil Pipeline Accidents in Special Climatic Conditions

**A N Vtorushina[a], Y V Anishchenko, E D Nikonova**

Ecology and Human Safety Department, Tomsk Polytechnic University
Lenin av. 30, Tomsk, Russia, 634050

E-mail: [a] anl@tpu.ru

**Abstract**. The present study identifies the main accidents' factors and causes for oil pipeline located in Siberia and operated in special climatic conditions. Various types of pipeline accident scenarios were modeled. It is showed that the most dangerous scenarios are oil spills fire and oil vapor explosion due to the loss of piping integrity (rupture) of the pipeline's section, laying on marshlands and oil spill on the water surface due to the loss of piping integrity (puncture). The most probable scenario is oil spills fire due to the loss of piping integrity (puncture) of the pipeline's section, laying on dry lands and marshlands. To estimate the scenarios «event tree analysis» is used. Also such risk indexes as individual, societal, public and potential risks were determined.

## 1. Introduction

The first paragraph Pipeline system is a source of increased danger due to a large amount of welded and flanged joints, shut-off and control valves, harsh environment and significant amounts of substances transported by them. Fluid transmission risks are associated with constant dynamic loads on the system and instability of the process. Pressure drops, dynamic and static loads create good conditions for the deformation aging of metal. From this point of view, the oil transportation by pipeline is the greatest danger, since this substance is characterized by the formation of two-phase flow, the existing of pulsation flow, the formation of shock waves and discharge zones. Non-stationary processes can lead to vibration of communications and equipment, leakage of pipelines up to their complete catastrophic destruction. A large number of connecting fittings creates an additional danger of piping integrity's loss.

Nowadays, the oil industry gives great attention to providing the reliability and safety during the construction and operation of main pipelines. Despite this fact, the emergency situations associated with oil transportation by pipelines occur very often [1-4]. In general, such accidents include high material costs for the oil pipeline's operating companies and significant damage to the environment, people, and property in the vicinity of the failure pipelines. Therefore, the decreasing of the probability of emergency situations on the main oil pipelines is one of the immediate issues in this field. The problem is especially important for pipelines, which are located and operated in special climatic conditions. The main factors, which contribute to the emergency occurrence, are the following: the usage of significant amounts of flammable and explosive substances in operating procedure; the existence of mechanical impurities in oil, causing abrasion of equipment and pipelines; operating procedure under pressure; harsh climatic conditions and others.

Preaccident analysis is an important aspect of emergency oil spills problem. The main causes of oil pipelines accidents are pipeline corrosion (including internal and external), construction defects, mechanical damage (including excavation work), diversions, unauthorized junctions, third party activity, oil pipelines' operational imperfection, manufacturing defects of pipes and equipment, natural disasters [1, 3].

During the oil transportation, the loss of piping integrity may occur, which would lead to release of hazardous substances into environment. The main result of such accident is the transported oil leakage, which creates toxic pollution of the environment (water, air and soil). Moreover, as a result of spilled oil evaporation the cloud of fuel-air mixture is formed. The presence of ignition source within the cloud may lead to its inflammation. The highest risk of accidents on pipelines is connected with longitudinal damages. They can occur during the formation of corrosive gas pocket or guillotine ruptures both on the base part of pipe and in the zone of welding joints. Therefore, the loss of piping integrity is considered as the most frequent initiating event, which causes the hazardous substance's release into environment [1, 5-8]. The loss of piping integrity mode influences the types of leakage. If discharge area is small, relatively longtime liquid outflow through a hole is observed. On the other hand, if pipeline integrity damage is significant, considerable volumes of hazardous substance reach environment immediately.

The most probable oil pipeline accidents are characterized by considerable volumes of oil blowout from different parts of the pipeline. If the considerable volumes of oil are released, the most hazardous accidents occur, since it is volumes of hazardous substances that cause the largest damage to the environment, human, infrastructure facilities. Therefore, the possible pipeline accident analysis includes the amount of substance assessment, which could be involved in the accident. Also the accident consequences and probability of their occurrence should be determined.

To analyze the accident the different scenarios of its development are prepared. They include the sequence of logically connected individual events, which are caused by a specific initial event. Some aspects should be taken into account as part of possible pipeline accidents determination of scenarios. For example, the characteristics of the transported oil, known oil pipeline accidents analysis, descriptions of technical security solutions, preaccident analysis, and others.

Typical scenarios of pipeline destruction may include:

- the formation of a persistent environment pollution due to oil blowout from the damaged pipeline;
- oil pool fire, causing the thermal impact on the environment, material objects and people;
- the explosion of fuel-air mixture;
- the distribution of explosive fuel-air mixture clouds with the wind, and possible explosion.

This work is devoted to compiling the possible scenarios of accidents during oil pipeline operation in special climatic conditions with the subsequent risk assessment.

## 2. Object of research

This paper presents the examination of the oil pipeline system located in the Siberian region. As the object of research we selected a pipeline section that passes through territory characterized by different conditions. Part of the pipeline is lined through dry land and 42% of the pipeline route includes marshes. The pipeline has 30 intersections with water bodies, the largest of which is a navigable river. Therefore, the analysis of scenarios of these pipeline accidents should take into account some factors of natural hazards, which could influence the loss of piping integrity or enlarge the level of it. For example, these factors are average annual air temperature gradients, high level of snow cover in the winter, the pipeline's passing through water bodies.

## 3. Methods of research

Nowadays there are several approaches to risk assessment: phenomenological, deterministic, expert and probabilistic.

Phenomenological approach is based on emergencies possibility determination using the analysis results that represent conditions connected with the nature laws. This approach is the easiest one to use and gives reliable results, if operating processes have sufficient margin to limiting levels. However, it is unreliable, when abrupt change of substances or systems occurs. This approach is preferable to compare the safety margins of different potentially dangerous objects and is useless in analysis of emergencies, which depends on reliability of different parts of the object or its safety facilities.

Deterministic approach provides an analysis of the sequence of the accident development from the initial event through a series of suggested failures. Emergency process is studied and predicted with the help of mathematical modeling, construction of simulation models and complex calculations. Disadvantages of this approach include the gaps in some important events in accident development; difficulty in adequate mathematical models construction; complexity and high cost of experimental studies to validate computer models.

Expert approach is based on obtaining quantitative risk assessments by processing of experts' opinions. The main disadvantage of this approach is the necessity to analyze the objectivity and reliability of experts' opinions.

The probabilistic approach involves the probability assessment of the accident. This approach allows analyzing the chain of events and equipment failures and estimating the total probability of accident. Main disadvantage of this approach is related to the lack of statistics on equipment failure. In addition the usage of simplified design schemes reduces the reliability of the risk assessment for serious accidents. However, this approach currently is considered as one of the most perspective ones.

In this paper probabilistic approach was used. Its application can be explained by capacity to model all possible pipeline accident scenarios. Different techniques (statistical data, event tree analysis, fault tree analysis, etc.) allows to identify and quantify all scenarios, as well as to determine the protective actions if emergency occurs.

To estimate the risk of oil pipeline failure, different risk assessment methods are used, for example event tree analysis [1,5,7], fault tree analysis [5,9,10], bowtie [11] and others [3,5,8,12]. For emergency modeling the methods of direct analysis of the events are applied. All events that may occur after the accident are connected with cause-effect relationships, depending on the operation or failure of the protection system components. In this paper, to identify cause-and-effect relationships of emergency the method of "tree of events" is used, which allows to provide qualitative and (or) quantitative risk assessment. The advantage of this method is the possibility to analyze the initiating events that can lead to various effects (thermal radiation, overpressure, emissions, toxic effects, and others). Each branch of event tree includes a separate effect, which is the result of specific functional relationships. The method helps identify accident scenarios with different effects from various initiating events; to determine the relationship of systems' failure with the consequences of accidents; to determine the accident sequences, which contribute most to the risk due to their high probability.

## 4. Results and discussion

Accident consequences depend on the sizes of the emergency port. To evaluate the risk of oil pipeline accidents two types of scenarios, connected with oil spill due to loss of piping integrity, are chosen.

For analyzed sections of the research object the following scenarios were identified:

- Type 1. The oil spill due to rupture of the pipeline - guillotine rupture (the typical size of the hole (the ratio of the hole's length to the nominal diameter (D) of pipeline) is 1.5D).
- Type 2. The oil spill due to puncture of the pipeline - gas pocket (the typical size of the hole is 0.3D).

The consequences of the oil spill will be different. It depends on the failure mode (rupture, puncture), season (summer, winter), local conditions (dry land, marsh, water objects). To estimate the maximum risk there is a good reason to consider the occurrence of accident in summer, since during this time air temperature is the highest and the occurrence of adverse factors is more possible. The accident analysis in winter is not discussed in this work due to less considerable consequences. However, the pipeline maintenance in winter is the essential factor, which increases the possibility of accident occurrence.

There are several adverse factors, which influence environment, people and property. In case of rupture of the pipeline they are the thermal radiation intensity of oil spill fire and blast overpressure of oil vapors and toxic action of oil. If there is puncture of the pipeline, the main adverse factors consist of the thermal radiation intensity of oil spill fire and toxic action of oil.

To estimate the risk indexes, the following data should be determined: hazardous substance spill area and the amount of the hazardous substance involved in the scenario. In the calculation some certain assumptions were made: calculations are performed for routine maintenance conditions; the most unfavorable conditions of the accident with a maximum quantity of dangerous substances are considered; the combustion occurs over all oil spill area and others.

The research object is divided into 5 sections. For all sections of the pipeline (dry land, marsh, water bodies) volume, mass and the oil spill area after rupture or puncture of pipeline have been calculated. To evaluate the maximum risk of oil pipeline accident areas with maximum emissions of hazardous substances (oil) were considered. The section No1 includes the part of pipeline, which is laid on dry land and marsh. As a result of rupture of pipeline oil spill area is equal to 5259 $m^3$, mass of oil blowout - 4313 tons. The section No2 is laid on dry land and marsh with crossing a small river and stream. The puncture of pipeline led to formation of oil spill area, which size is 198 $m^3$, and mass of oil blowout - 162 tons. The maximum volume of oil spill from the puncture is predicted to be the same for sections, which are laid on dry land, marsh and water objects. Further calculations will be made for the section No2 as the most dangerous because of the maximal length (13473 m). The unsensed emergency for a long period (1 day or more) is more possible to occur on this section. In addition to the mentioned above sections, the sections No3, 4, 5 are the most dangerous. There is the maximal number of intersections with water bodies (two rivers and five streams) on the section No3. It is expected the maximum oil spill area on the water surface would be up to 1599207 $m^2$ in the case of oil rupture. The sections No4 and No5 cross the navigable river (the largest watercourse). In this case, oil spills will cause the maximum environmental damage.

Statistical data and results of expert method were used to determine the probability of certain events for each scenario. The prediction of accidental leakages frequency is conducted in the simplified way by using average accidents data.

Any scenario begins with the initiating event (the loss of piping integrity) that can occur with some frequency. The frequencies of basic events leading to the pipeline accident are listed in Table 1.

**Table 1.** The frequencies of basic events.

| Failure mode | The probability of failure, 1/year |
|---|---|
| Rupture (section No 1) | $1.03\ 10^{-4}$ |
| Puncture (section No 2) | $2.70\ 10^{-3}$ |

Two types of scenarios can be represented in the form of «event trees» (Figure 1 and Figure 2).

The frequency of the accident scenario is product of the probability of equipment failure and scenario probability. As a result of the accident scenarios assessment, risk indexes were obtained and the zones of the adverse factors' action were defined. According to the calculations, the most dangerous are the following scenarios:

Scenario 1. This scenario describes oil spills fire due to the loss of piping integrity (rupture) of the section No1, laying on marshlands. The largest oil spill area (99454 $m^2$) and maximum property loss would be expected if the accident is developed by this scenario. The probability of output event is equal to $1.54\ 10^{-5}$ 1/year. The exposure probability of human by thermal radiation of 17 $kW/m^2$ at the distance of 135 m from the center of the oil spill is 99%. Lethal injuries of employees are not predicted, the expected number of victims is 3 people.

Scenario 2. This scenario describes oil vapor explosion due to the loss of piping integrity (rupture) of the section No1, laying on marshlands. The probability of output event is equal to 4.11 $10^{-6}$ 1/year. The largest area of the adverse factors (bursting radius is up to 1763 m) and the highest number of victims would be expected if the accident is developed by this scenario. The exposure probability of human by blast overpressure of 100 kPa at the distance of 157 m from the center of explosive cloud is equal to 99 %. The expected number of lethal injured is 2 people.

Scenario 3. This scenario describes oil spill on the water surface due to the loss of piping integrity (puncture) of the section No2. Oil spill area is predicted to be up to 66000 $м^2$ on water surface, which cause the maximal ecological damage. The probability of output event is equal to $5.25*10^{-4}$ 1/year.

Scenario 4. This scenario describes oil spills fire due to the loss of piping integrity (puncture) of the section No2, laying on dry lands and marshlands. The probability of output event is equal to 2.42 $10^{-4}$ 1/year. In the case of pipeline laying on dry land, the exposure probability of human by thermal radiation of 17 $kW/m^2$ at the distance of 28 m from the center of the oil spill is 97%. In the case of pipeline laying on marshlands, the exposure probability of human by thermal radiation of 17 $kW/m^2$ at the distance of 30 m from the center of the oil spill is 97%. The expected number of lethal injured is 1 person; the expected number of victims is 3 people.

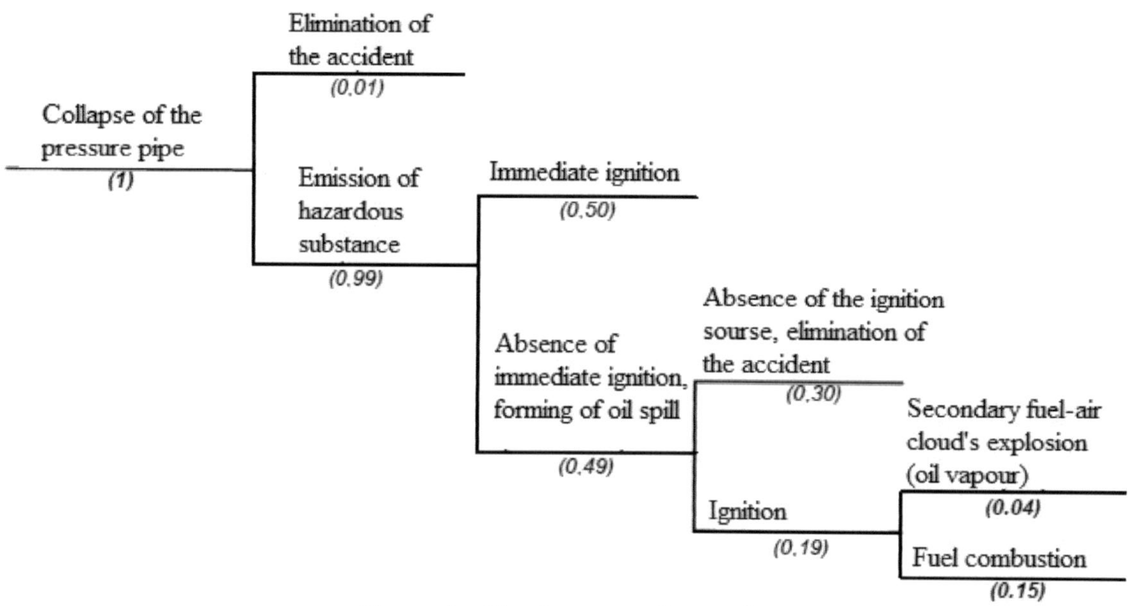


**Figure 1** – The first type of scenario

IOP Conf. Series: Earth and Environmental Science **66** (2017) 012006 doi:10.1088/1755-1315/66/1/012006

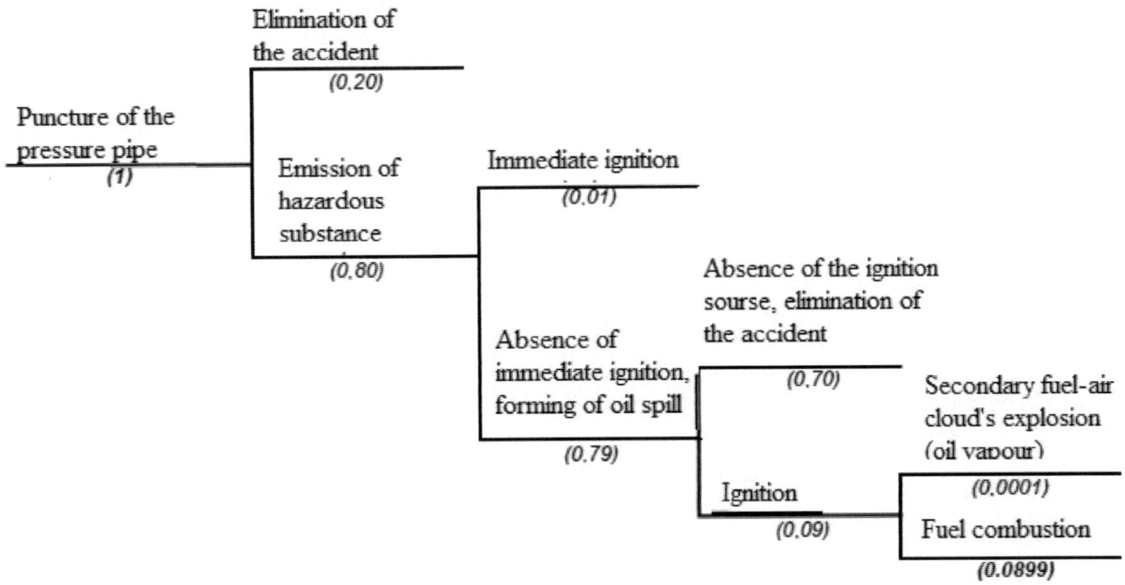


**Figure 2** – The second type of scenario

The main quantitative risk indexes are the individual risk, the potential risk of the territory, public risk and societal risk. Individual risk, which is estimated as the frequency of person's injuring under the influence of adverse factors of accident, is calculated for all pipeline's sections at the distance of 500 m and is equal to 3.95 $10^{-7}$ 1/year. The frequency of accident's adverse factors on the analyzed pipeline is independent from the fact of the people's presence in this area. It is assumed that conditional probability of a person's presence is 1, in other words a person stays at a territory point during the considered period of time). The results of potential risk evaluation for analyzed scenarios for oil pipeline are presented in Table 2.

**Table 2.** The results of potential risk evaluation.

| Section of pipeline | Failure mode | Potential risk, 1/year at the distance from possible accident, m | | |
|---|---|---|---|---|
| | | 600 | 700 | 800 |
| Section No 1 | Loss of piping integrity | 1.3 $10^{-6}$ | 6.2 $10^{-7}$ | 2.9 $10^{-7}$ |
| Section No 2 | | 0 | 0 | 0 |

Public risk is the expected number of affected people due to possible accidents within a certain period of time. This risk index depends on the hazards of production, the number of employees and other factors. It was estimated that the public risk for analyzed pipeline is 2.59 $10^{-4}$ person/year. Societal risk shows the number of victims and lethal injured, which depends on the frequency of accident. In this paper societal risk is estimated as probability of lethal injured due to fire, explosion of gas vapour mixture and is equal to more than 10 people in year. Consequently, the obtained results of individual, societal, public and potential risks can be compared with acceptable values, which mean that there is no need to risk reduction.

## 5. Conclusion

Risk assessment of oil pipeline accident involves the study of various types of scenarios, which are caused by natural and technogenic reasons. However, some factors may contribute to the accident occurrence namely the loss of piping integrity or increase the damage. These factors are average annual air temperature gradients, high level of snow cover in the winter, the pipeline's passing through water bodies. It is considered that the initiating event is the loss of piping integrity due to rupture or puncture of pipeline's section. The output event forms such adverse factors as thermal radiation (for oil spill fire) or blast overpressure (for oil vapor explosion). The results of this research include the following:

- The main reason of oil pipeline accident is depreciation and metal corrosion, which are caused as well as by local climate conditions.
- The most dangerous are the pipeline's sections, which have intersections with water bodies. The ecological damage is the main adverse affect for such accidents.
- The lethal injury for pipeline's servicers is possible during pipeline inspection or construction and assembling operations, if the time of their presence and accident occurrence will match.
- The obtained individual, societal, public and potential risks can be compared with acceptable values.

**References**

[1] Ramírez-Camacho J. G, Carbone F, Pastor E, Bubbico R and Casal J 2016 *Safety Science* **XX** Article in press
[2] Girgin S and Krausmann E 2016 *Journal of Loss Prevention in the Process Industries* **40** 578
[3] Barros da Cunha S 2016 *Journal of Loss Prevention in the Process Industries* **44** 282
[4] Siler-Evans K, Hanson A, SundayC, Leonard N and Tumminello M 2014 *International journal of critical infrastructure protection* **7** 257
[5] Shahriar A, Sadiq R and Tesfamariam S 2012 *Journal of Loss Prevention in the Process Industries* **25** 505
[6] Wang H and Duncan I.J 2014 *Journal of Loss Prevention in the Process Industries* **30** 177
[7] Bonvicini S, Antonioni G, Morra P and Cozzani V 2015 *Process Safety and Environmental Protection* **93** 31
[8] Lu L, Liang W, Zhang L, Zhang H, Lu Z and Shan J 2015 *Journal of Natural Gas Science and Engineering* **25** 124
[9] Alkhaledi K, Alrushaid S, Almansouri J and Alrashed A 2015 *Safety Science* **79** 184
[10] Liang W, Hu J, Zhang L, Guo C and Lin W 2012 *Engineering Applications of Artificial Intelligence* **25** 594
[11] A. de Ruijter and Guldenmund F 2016 *Safety Science* **88** 211
[12] Khan F, Rathnayaka S and Ahmed S 2015 *Process Safety and Environmental Protection* **98** 116

IOP Conf. Series: Earth and Environmental Science **66** (2017) 012007 doi:10.1088/1755-1315/66/1/012007

# Assessment of a Solar Cell Panel Spatial Arrangement Influence on Electricity Generation

**I A Anisimov[a], L N Burakova, A D Burakova[b], O D Burakova[c]**

Tyumen Industrial University
str. Volodarskogo 38, Tyumen, Russia, 625000

*E-mail:* [a]tkcc@list.ru, [b]burakova.1992@mail.ru, [c]burakova.96@bk.ru

**Abstract**. The research evaluates the impact of the spatial arrangement of solar cell panels on the amount of electricity generated (power generated by solar cell panel) in Tyumen. Dependences of the power generated by the solar panel on the time of day, air temperature, weather conditions and the spatial arrangement are studied. Formulas for the calculation of the solar cell panel inclination angle which provides electricity to urban infrastructure are offered. Based on the data in the future, changing of inclination angle of solar cell panel will be confirmed experimentally during the year in Tyumen, and recommendations for installing solar cell panels in urban infrastructure will be developed.

## 1. Introduction

At the present time technological and technical development level of fuel and energy production determines the structure of consumed energy resources, its environmental and development trends [1,2,3]. Technical development of industrial branches depends on the climatic conditions of the regions such as air temperature, solar radiation, barometric pressure and humidity [4,5,6,7,8,9,10,11,12,13]. We need to explore renewable, alternative and clean sources, as traditional forms of energy reserves are shrinking [14]. This trend of development is presented in the researches, which describe the use of stand-alone traffic lights, replacement of traditional forms of energy sources and so on [15,16].

In the modern world the proportion of energy that we get from alternative sources is 9.3%. The main part of it is made by means of hydroelectric power plants. It is 6.8%. The energy that we get from the sun is 0.33%. However, in 2050 the proportion of energy that we will get from solar power plants will exceed consumption of other types of fuel and the share of consumption will be 27% [17,18]. These data are presented in the International Energy Agency report. The development of solar energy industry is described by the cycle time of 6-7 years. The first leader in the development of solar energy industry was Japan. At the present time it is China.

In the Russian Federation, the development of solar energy began in 2013-2014. Now its share is 0.2%, which it is low in world production. At the present time in Russia, there are 23 solar power plants, which have a total capacity of 634 MW [19]. The development of solar energy industry depends on many factors, in particular, it is power plant geographical location [20].

IOP Conf. Series: Earth and Environmental Science **66** (2017) 012007 doi:10.1088/1755-1315/66/1/012007

The sun is "ecologically clean" source of energy, which is inexhaustible. It is one of these energy advantages that distinguish it from oil gas and other fuels. However, the disadvantages of solar energy industry are produced energy dependence on climatic conditions and time of day, also high investments and large occupied territory.

Solar radiation is converted into electric power by solar cell panels. Its advantages are long service life, which is 25 years or more and minimum maintenance during operation. However, the main drawback is the complex production and recycling, as well as solar cell panels contains toxic substances such as lead, cadmium, gallium, arsenic. Also one of disadvantages is a low coefficient of efficiency. The average value of this indicator is 12-25% in practice [21]. Its value depends on material used in the production of solar cell panels.

The main element is silicon. It does not take exposure to UV rays and it can only absorb infrared radiation. It reduces coefficient of efficiency. Therefore, one way of solar energy industry technical development is to create a multi-layer solar cell panel that can absorb all kinds of radiation. However, the amount of energy, which is obtained by means of the solar cell panel, can be increased not only by enhancing the efficiency and maximum spatial arrangement selection.

The potential amount of electricity generated by solar cell panels is estimated individually for each geographic location on Earth. Initial data for calculation is the amount of solar radiation and the nature of the cloud. Method does not consider the slope of surface, which absorbs solar radiation, and the presence of interfering objects. Therefore, the result is approximate.

The aim of this study is to increase the efficiency of solar cell panels in Tyumen by assessing the impact of spatial arrangement (geographic location), the climatic conditions and the presence of impeding objects on the amount of energy produced.

## 2. Equipment and devices used in studies

The amount of electricity, which is generated by the solar cell panel, is determined with a multimeter. The solar cell panel is a photoelectric converter HH MONO-60W, which is a silicon single crystal panel. Its power is 60W, and the current is equal to 3.5A. The peculiarity of these solar cell panels is a high coefficient of efficiency which can reach 17-22% at ideal conditions. The electrical network also includes the battery (its voltage is 12V and its condensance is 100A·h) and charge controller (its voltage is 12/24V).

## 3. Results and discussion

The amount of electricity that is generated by a solar cell panel depends on the amount of solar radiation coming onto its surface. Radiation from the sun spreads in all directions as electromagnetic waves. Passing through the layer of the atmosphere, the radiation is partly scattered, absorbed and reflected. This reduces the amount of solar radiation that reaches the earth's surface. Solar radiation depends on the geographical location and climatic conditions such as time of year, air temperature and humidity, cloudiness.

This paper describes the influence of solar cell panel spatial arrangement in different climatic conditions, as well as the existence of obstacles to change in the amount of electricity generated.

Solar radiation is distributed evenly on the surface, since the length of the path traveled by the rays in the atmosphere varies. Rays reach the shortest path the Earth's surface at the location of the sun at its zenith. This distribution of the rays occurs within latitudes 23° NL and 23° SL and is characterized by a small area of intensive energy transfer [22]. The angle of incidence decreases with increasing distance from this area, so the radiation propagates along a tangent line relative to the earth's surface and the area that receives the solar radiation increases. The amount of reflected energy is also rising. Dependence of the fraction of solar radiation weakening of the incidence angle is shown in Figure 1.

Reducing the angle of incidence of the rays increases the proportion of weakening rays therefore increasing the distance from the equator reduces the solar radiation that reaches the earth's surface. Dependence, which is represented in Figure 1, is one of the reasons for the low efficiency of solar cell panels in the Russian Federation in particular in Tyumen.

IOP Conf. Series: Earth and Environmental Science **66** (2017) 012007 doi:10.1088/1755-1315/66/1/012007

Geographical coordinates of Tyumen are 65° EL and 57° NL. In Tyumen solar radiation, which reaches the earth's surface, is shown in Figure 2.

Solar radiation of clear sky varies smoothly throughout the year. The form of this curve is close to the correct harmonic dependence [23]. This is because the radiation factors, which change periodically, and a set of local conditions. In the mid-latitude high and low total solar radiation corresponds to months with the minimum and maximum angle of sun declination (June, December). Thus, this work needs to be additional research that establishes the dependence of the angle of inclination of solar panels in Tyumen on the season, presence of impeding objects and weather conditions.

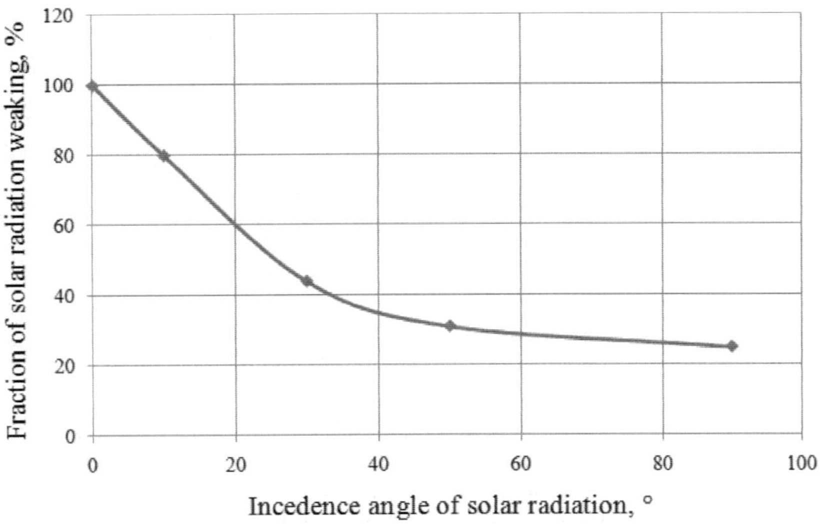


**Figure 1** – The dependence of the fraction of solar radiation weakening of the incidence angle

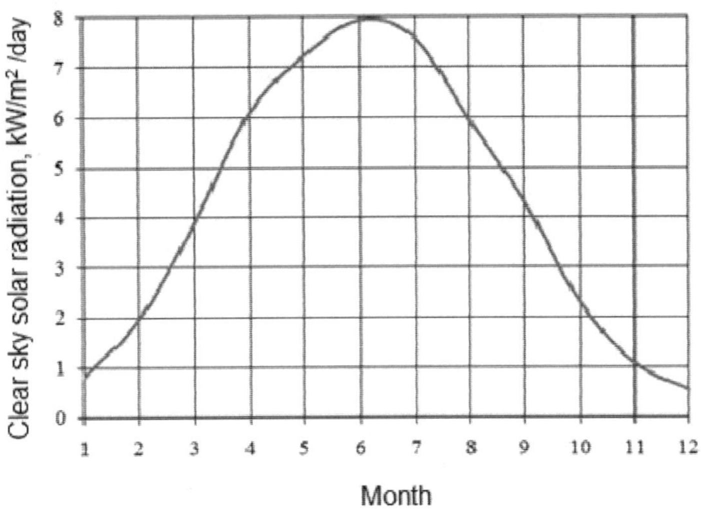


**Figure 2** – Changing of clear sky solar radiation in Tyumen during the year

Researches of Moldovan scientists describe a detailed methodology for calculating the inclination angle of the solar cell panel during the time of the year [24]. They share a total solar radiation into components such as direct, diffuse and back solar radiation from the surface. This approach allows you to get precise results. The study was conducted in Kishinev. Its geographical coordinates are 47° NL and 28° EL. The initial data of this calculation are the height and azimuth of the sun, the reflection coefficient of the ground, the amount of direct and diffuse radiation. The results are dependent changes in the inclination angle of solar cell panel of the month (Figure 3).

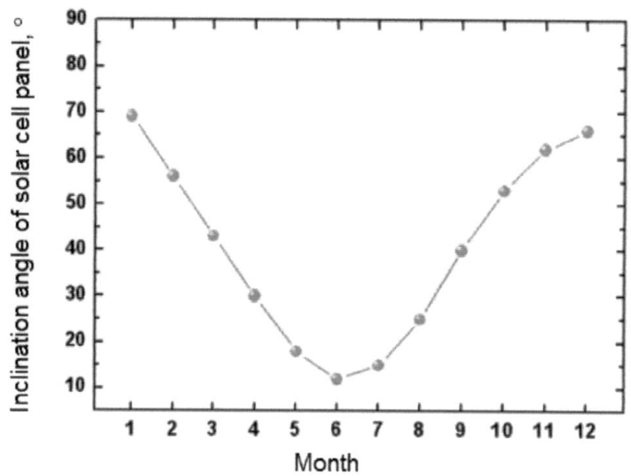


**Figure 3** – The dependence of inclination angel of solar cell panel on the month in Kishinev [24]

Calculations of optimum inclination angle of the solar cell panel in Tyumen were performed using a methodology, which is presented in the above study. The resulting dependence is shown in Figure 4.

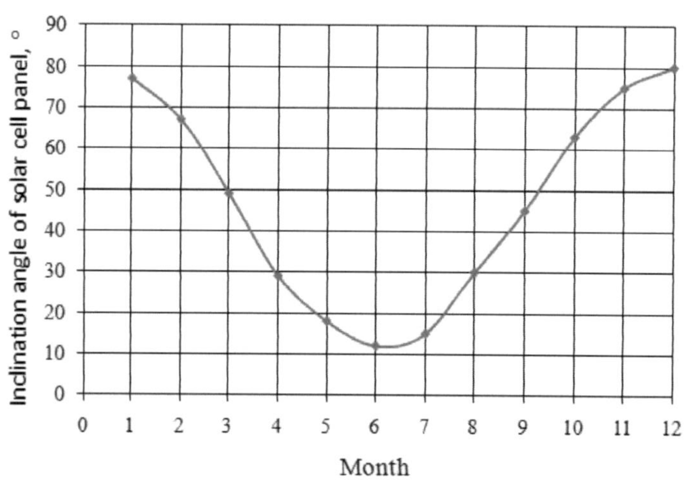


**Figure 4** – The dependence of inclination angel of solar cell panel on the month in Tyumen

The optimum inclination angle of solar cell panels, which is determined based on an analysis of theoretical data, ranges from 12 to 80 ° relative to the horizontal surface [25]. However, the theoretical calculation requires experimental confirmation.

The first stage of experimental research includes the study of the effect of air temperature, weather conditions, time of day and the presence of impeding objects on the amount of electricity that is produced by a solar cell panel (power generated by a solar panel). Influence of the time of day on the amount of electricity produced by a solar cell panel that has a similar relationship with the change in the clear sky solar radiation throughout the year. It is shown in Figure 5.

The maximum amount of electricity generated is observed from 14:30 to 15:30, as the transparency of the air changes continuously throughout the day and it depends on the content of dust in the atmosphere, water vapor and other impurities. At noon, the atmosphere heats up as much as possible, so the amount of updraft is increased, and the proportion of solar radiation penetration is reduced. Afternoon heat emission into the atmosphere is reduced, so the amount of solar radiation, which reaches the surface of the solar cell panel, increases. This explains the increase in the amount of generated electricity.

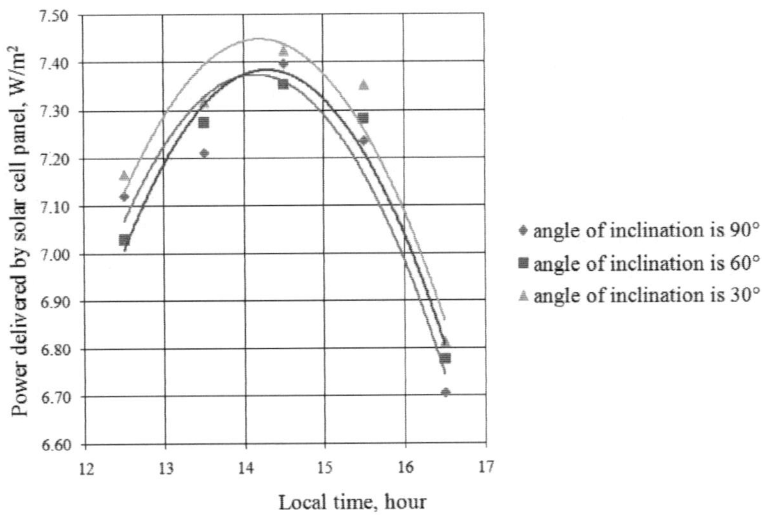


**Figure 5** – The dependence of the amount of electricity produced by a solar panel that is the time of day

In the atmosphere the amount of ascending currents depends on the ambient temperature. Rising temperatures increases the intensity of the mass heat exchange of the earth's surface and the air. It increases the amount of water vapor in the air, suspended solids, which prevent the spread of solar radiation. Thus, the air temperature increase reduces the amount of power that is generated by the solar cell panel.

This study examines the impact of rainfall on the amount of power produced by the solar cell panel. This is shown in Figure 6.

The presence of water droplets in the air is an obstacle to the spread sun rays, so an increasing of rainfall decreases amount of solar radiation, hence it decreases the amount of electricity generated [26].

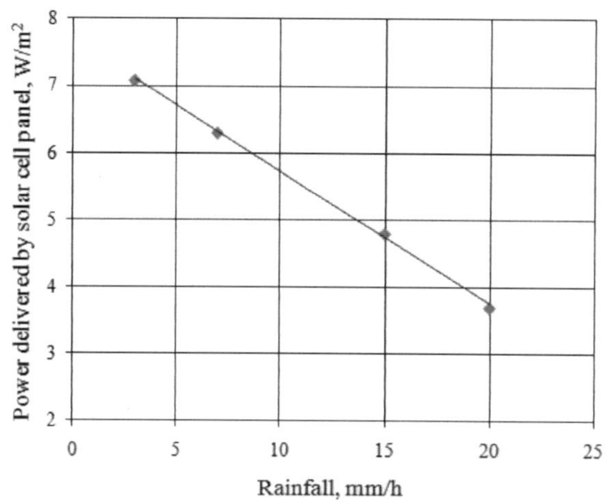


**Figure 6** – The dependence of the amount of electricity generated by the solar panel on rainfall

At the present time the increase in construction areas is an additional obstacle to the spread of direct sunlight [27]. The shaded areas are the areas of the earth's surfaces, which are mainly limited coverage or highlights daylight only for three hours. Shaded areas are determined by the duration of sunlight illumination and its intensity. In this case, the efficiency of the solar cell panels, which serve urban infrastructure, can be increased by performing the calculations and analysis scheme, which is presented in Figure 7.

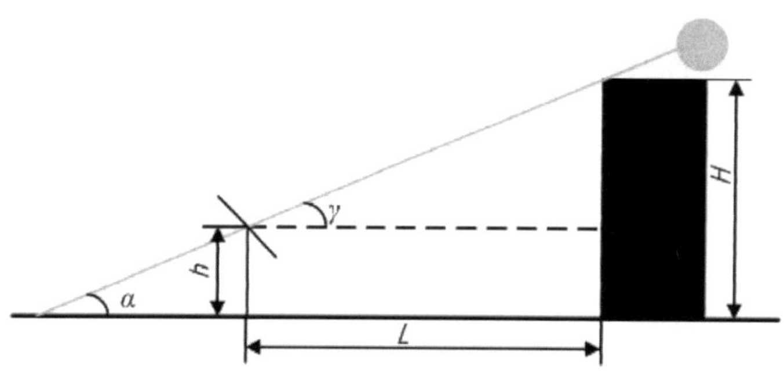


**Figure 7** – Scheme of the spatial arrangement of the solar cell panel which serves the city infrastructure

The main condition, which ensures the supply of solar radiation on the panel surface, can be represented by the formulas:

$$\alpha > \gamma, \tag{1}$$

$$\gamma = arctg(H - h/L), \tag{2}$$

Where $H$ is height of the building, m; $L$ is distance to the nearest high-rise building, m; $h$ is support height, m; $\alpha$ is the angular height of the sun, °; $\gamma$ is the incidence angle of the rays on a solar panel, °.

This inequality will improve the efficiency of the solar panel, so as it will increase the surface area, which will be illuminated by direct sunlight.

## 4. Conclusions

The first stage of the research establishes that the amount of solar radiation which enters the solar panel surface, as well as the amount of electricity generated depends on the time of day, air temperature, rainfall and the presence of impeding objects.

The increase of rainfall reduces the power generated by a solar panel. High-rise buildings are also an obstacle to the spread of direct sunlight. However, the optimum inclination angle of the solar cell panel can be calculated by analyzing the spatial position of the solar cell panels relative to the horizon and the surrounding high-rise buildings.

At the present time the changing of the inclination angle of the solar cell panel theoretically installed in Tyumen during the year. A further stage in this research will be obtained depending on the experimental confirmation, as well as it is developing recommendations for the installation of solar panels in urban infrastructure.

**References**

[1] Genon G., Panepinto D., Viggiano F. *Sustainability in energy production* // International Journal of Energy Production and Management 2016 Vol.1 p.16-32

[2] Erik Jarlsby *The global energy challenge: still fuel for progress?* // International Journal of Energy Production and Management 2016 Vol.1 p.33-49

[3] Giuseppe Emmi, Chiara Tisato, Angelo Zarrella, Michele De Carli. *Multi-source heat pump coupled with a photovoltaic thermal (ptv) hybrid solar collectors technology: a case study in residential application* // International Journal of Energy Production and Management 2016 Vol.1 p.382-392

[4] Chainikov D.A., Chikishev E.M., Gavaev A.S. *Influence of ambient temperature on the $CO_2$ emitted with exhaust gases of gasoline vehicles* // IOP Conference Series: Materials Science and Engineering 2016 Vol.142, Article number 012109. p. 1-6

[5] Ertman J.A., Ertman S.A., Anisimov I.A., Chainikov D.A., Chikishev Y.M. *Complex characteristic estimation of vehicle adaptability for low temperature operation conditions* // Research Journal of Pharmaceutical, Biological and Chemical Science 2015 No.1. p.1761-1770

[6] Anisimov I.A., Ivanov A.S., Chikishev E.M., Chainikov D.A., Reznik L.G. *Assessment of gas cylinder vehicles adaptability for operation at low ambient temperature conditions* // WIT Transactions on Ecology and the Enviroment 2014 Vol.190 p.658-695

[7] Anisimov I. A., Ivanov A.S., Chikishev E.M., Chainikov D.A., Reznik L.G., Gavaev A.S. *Assessment of adaptability of natural gass vehicles by the constructive analogy method* // International Journal of Sustainable Development and Planning 2017 Vol.12 p.1006-1017

[8] Chikishev E.M., Ivanon A.S., Anisimov I.A., Chainikov D.A. *Prospects of and problems in using natural gas for motor transport in Russia* // Innovative Technologies in Engineering VII International Scientific Practical Conference 2016 Vol. 142, Article number 12110. p.1-8

[9] Ivanov A.S., Chikishev E.M. *Level recession of emissions release by motor-and-tractor diesel engines through the application of water-fuel emulsions* // IOP Conference series: earth and environmental science 2017 Vol.50, Article number 012005. p. 1-7

[10] Ertman S.A., Ertman J.A., Zakharov D.A. *Raising of operating a motor vehicle effects on environment in winter* // Innovative Technologies in Engineering VII International Scientific Practical Conference 2016 Vol. 142, Article number 12119. p.1-8

[11] Chikishev E., Chainikov A., Anisimov I. *Increasing the use of natural gas on motor transport by an efficient location of the NGV RS (on the example of Tyumen)* // IOP earth and environmental science 2017 Vol.50, Article number 012004. p. 1-7

[12] Chikishev E., Chikisheva A., Anisimov I., Chainikov D. *Natural gas use on minibuses, engaged in the carriage of paasengers and baggage on the regular routes, as a measure for decrease in harnmful environment effects* // IOP earth and environmental science 2017 Vol.50, Article number 012008. p. 1-7

[13] Anisimov I., Magaril E., Magaril R., Chikishev E., Chainikov D., Gavaev A., Ertman S., Ertman Yu., Ivanov A. *Improving vehicle adaptability to the operating conditions of "Smart" cities in the northern regions* // E3S WEB of Conferences 2016 Vol.6, Article number 02003. p.1-6

[14] Hussain H. Al-Kayiem Hybrid *Techniques to enhance solar thermal: the way forward* // International Journal of Energy Production and Management 2016 Vol.1 p. 50 – 60

[15] Vitali D., Garbuglia F., D'Alessandro V., Ricci R. *The renewable energy in a led standalone streetlight* // International Journal of Energy Production and Management 2017 Vol.2 p. 118 – 128

[16] Rada E.C. *Environmental pollution from waste and biomass energy generation* // International Journal of Energy Production and Management 2017 Vol.2 p. 1 – 7

[17] *Energy: the answer to – and the cause of – some urgent problems* (International Energy Agency, World Energy Outlook 2014, Vol.12)

[18] Gudmestad Ove T., Kjell Traa. *Sustainable use and production of energy in the 21st century* // International Journal of Energy Production and Management 2016 Vol.1 p. 1 – 15

[19] *Catalog of Russian power plants.* Energy Base http://energybase.ru/power-plant (accessed February 15 2017)

[20] Ragazzi M., Ionescu G., Cioranu S.I. *Assessment of environmental impact from renewable and non-renewable energy source* // International Journal of Energy Production and Management 2017 Vol.2 p.8-16

[21] Hermann Scheer The solar economy: Renewable Energy for a sustainable global future Earthscan: Business & Economics 2004 - 347 p.

[22] Alisov B.P. *Climatologiya* [Climatology] / B.P. Alisov, B.V. Poltaraus – Moscow: Publishing House of Moscow University 1974 300 p.

[23] Bogoslovski V.I. *Kondicionirovanie vozduha I holodosnabdgenie* [Air conditioning and refrigeration] / V.I. Bogoslovski, O.Y. Kokorina, L.V. Petrov – Moscow: Stroiizdat 1985 367p.

[24] Aculinin A., Smicov V. *Evaluation of opportunities of solar exegetics on the basis of accurate ground-based measurements of solar radiation* // Problems of the regional energetic - 2016 No.3 (32) p.29-39

[25] *Power Surface meteorology and Solar Energy.* NASA https://asdc-arcgis.larc.nasa.gov/sse/ (accessed February 17 2017)

[26] Kolebaje O.T., Ikusika A., Akinyemi P. *Estimating solar radiation in Ikeja and Port Harcourt via correlation with relative humidity and temperature* // International Journal of Energy Production and Management 2016 Vol.1 p.253-262

[27] *Boeri A.*, Gianfrate V., Longo D. *Green buildings and design for adaptation: strategies for renovation of the built environment* // International Journal of Energy Production and Management 2016 Vol.1 p. 172 – 191

IOP Conf. Series: Earth and Environmental Science **66** (2017) 012008 doi:10.1088/1755-1315/66/1/012008

# Development of innovative architecture of the organizational and economic mechanism for the nature protection management

**V G Mikhailov[1,a], T V Kiseleva[2,b], V A Karasev[1,c], G S Mikhailov[1,d], V A Skukin[1,e]**

[1]T. F. Gorbachev Kuzbass State Technical University,
ul. Vesennyaya, 28, Kemerovo, 650000, Russian
[2] Siberian State Industrial University,
ul. Kirova, 42, Novokuznetsk, 654007, Russian

*E-mail:* [a]mvg.eohp@kuzstu.ru, [b]kis@siu.sibsiu.ru, [c]kva.gf@yandex.ru,
[d]mgs.kuzgtu@yandex.ru, [e]bos.skukin@mail.ru

**Abstract.** The problems of the efficient functioning of environmental and economic systems of various levels on the basis of the adequate organizational and economic management mechanism are considered in the article. The purpose of the study is the analysis and development of theoretical provisions for the formation of a modern, innovative organizational and economic mechanism of the nature protection management. The compliance matrix of the innovative elements presented in the structure of the organizational and economic mechanism of the nature protection management is developed. The main result of the study is the improvement of the existing management mechanism to minimize the negative impact on the environment, including through the incentive system, and to improve the financial performance of the economic entity. The practical component of the study conducted can be recommended to municipal, regional and federal authorities, as well as the industrial enterprises, to support the adoption of the effective, environmentally sound management decisions that are consistent with the global concept of sustainable development.

## 1. Introduction

The development of the real sector of the economy determines the high level of anthropogenic impact on the environment, stipulating the need to solve the problem of resource and nature conservation minimizing the financial costs [1]. The nature protection efficiency increase depends on the process of managing the territorially distributed environmental and economic system with its unique features. Of decisive importance in this case is the formation of the organizational and economic mechanism for the nature protection management, which corresponds to high modern requirements caused by the international integration of the enterprises and the environmental legislation enforcement [2]. This situation determines the relevance and practical significance of this study.

## 2. Material and methods

The object of the study is the geographically distributed environmental and economic system management. The subject of the study is the organizational and economic mechanism of the nature protection management. The study is based on an analysis of domestic and foreign literary sources on this issue. The particular attention is paid to the study of the innovative elements of the organizational and economic mechanism of the nature protection management, which have practical significance. The verification of the adequacy of theoretical developments is based on the factual data of the enterprises from environmental reporting forms. The elements of system analysis and the results obtained by the specialists in the management of environmental and economic systems are also used in the work.

## 3. Results and discussion

In the conditions of the environmentally stressed region which has the highly developed industry, the process of the effective environmental and economic system management, the simplified scheme of which is presented in Figure 1, seems to be of relevance [3 - 6].

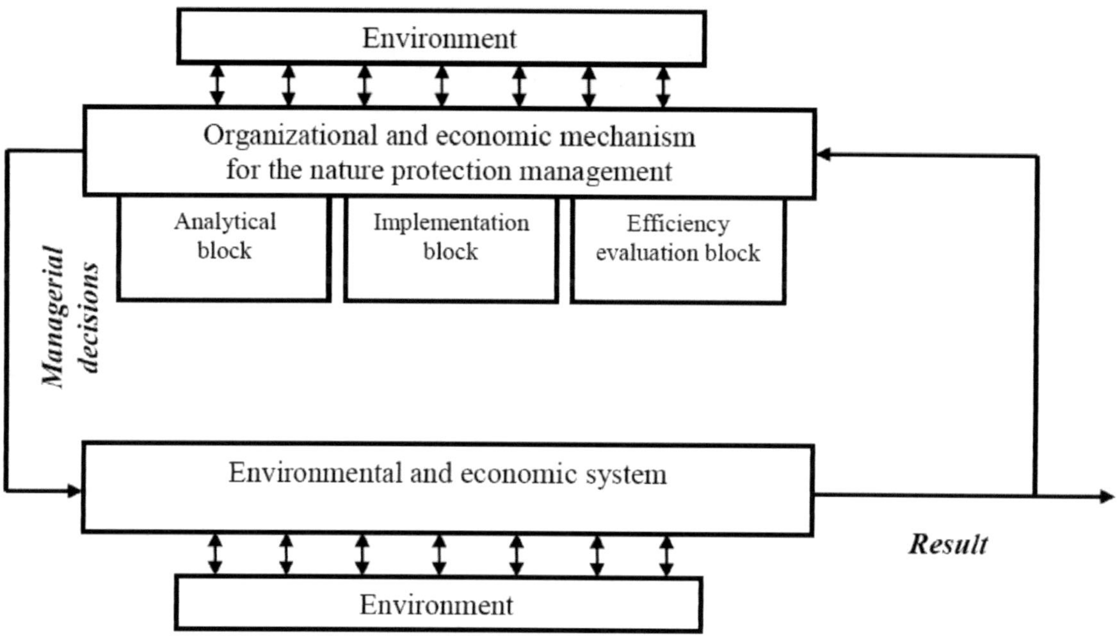


Figure 1 - Simplified scheme of the environmental and economic system management mechanism

Figure 1 shows that the key element of the presented scheme of the mechanism for managing the geographically distributed environmental and economic system is the organizational and economic mechanism for the nature protection management; in which three main elements being identified:

- Analytical block;
- Implementation block;
- Efficiency evaluation block.

In order to identify the developed innovative elements of the organizational and economic mechanism for the nature protection management, three main blocks have been identified, for which the compliance matrix presented in Table 1 has been developed.

Table 1 – Compliance matrix of the developed innovative elements in the structure of the organizational and economic mechanism for the nature protection management

| Innovative elements | Analytical unit | Implementation block | Efficiency evaluation block |
|---|---|---|---|
| Reduced integral indicator of the aquatic ecosystem state | + | | |
| The system of enterprises fining for providing the unreliable information on the level of negative impact on the environment | | + | |
| System of the universal environmental and economic indicators of the enterprise | + | | + |
| System of environmental and economic indicators of production and consumption wastes | + | | + |
| Diversification of the company's production program according to the environmental restrictions | | + | |

Table 1 shows that the greatest number of the developed elements belong to the analytical block, which additionally reflects its significance. To determine the practical feasibility of using the proposed innovative elements, their detailed study is necessary [7, 8].

Estimation of the quality of the aquatic ecosystem can be carried out using an integral indicator, the main disadvantage of which is the laboriousness due to the need to use a multitude of input data. The carried out sensitivity analysis allows to conclude that a number of components of the complex indicator yields little information because of the use of strongly correlated components. Therefore, some of them can be discarded in order to simplify the integral indicator and to insure its more effective practical application. As a result, the integral indicator was reduced [9].

The implementation block (Figure 1, Table 1) is represented by the development of the system of enterprises fining for providing the unreliable information on the level of negative impact on the environment [9]. The different levels of penalties for providing the inaccurate data with the differentiation according to hazard classes and depending on the range of deviations between the results of sampling of pollutants by an officially authorized state environmental management authority and the enterprise are shown below. In formulas (1) - (4), the interest means an additional level of penalization with respect to its base value for pollutants of different hazard classes.

$$F_{ADD\_I} = \begin{cases} 10\ \%\ F_{BASE}\ , if\ \Delta A^E > (1-5)\ \% \\ 15\ \%\ F_{BASE}\ , if\ \Delta A^E > (5-10)\ \% \\ 30\ \%\ F_{BASE}\ , if\ \Delta A^E\ is\ over\ 10\ \% \end{cases} \quad (1)$$

$$F_{ADD\_II} = \begin{cases} 8\ \%\ F_{BASE}\ , if\ \Delta A^E > (2-10)\ \% \\ 15\ \%\ F_{BASE}\ , if\ \Delta A^E > (10-20)\ \% \\ 20\ \%\ F_{BASE}\ , if\ \Delta A^E\ is\ over\ 20\ \% \end{cases} \quad (2)$$

$$F_{ADD\_III} = \begin{cases} 6\ \%\ F_{BASE}\ , if\ \Delta A^E > (3-10)\ \% \\ 10\ \%\ F_{BASE}\ , if\ \Delta A^E > (10-20)\ \% \\ 15\ \%\ F_{BASE}\ , if\ \Delta A^E\ is\ over\ 20\ \% \end{cases} \quad (3)$$

$$F_{ADD\_IV} = \begin{cases} 4\ \%\ F_{BASE}\ , if\ \Delta A^E > (4-10)\ \% \\ 5\ \%\ F_{BASE}\ , if\ \Delta A^E > (10-20)\ \%\ , \\ 10\ \%\ F_{BASE}\ , if\ \Delta A^E\ is\ over\ 20\ \% \end{cases} \tag{4}$$

where $F_{ADD}$ - additional amount of fines for providing the false information; $F_{BASE}$ - the base amount of fines for the pollution of the environment element;

$$\Delta A^E = \left|\frac{C_{SEM} - C_{IE}}{C_{SEM}}\right| 100\ \%\ , \tag{5}$$

where $\Delta A^E$ - is the amount of excess concentration of pollutants according to the officially authorized state environmental management authority, %; $C_{SEM}$ - is the actual concentration of pollutants according to the sampling data of an officially authorized state environmental management body; $C_{IE}$ - actual concentration of polluting substances according to sampling data of an industrial enterprise.

The conducted calculations showed that the proposed incentive system allows increasing the amount of fines by an average of 18 %, which should motivate the economic entities to provide the most reliable information in connection with the increase in costs included in the cost of production or attributed to the net profit of the enterprise.

The most important aspect of the analytical block is the development of the system of universal environmental and economic indicators of the enterprise performance, which can also be used to form the complex assessment of its environmental safety [10, 11]. If we consider the current activities of the enterprise, the developed schemes for the production greening, including the environmental and economic evaluation of the anthropogenic impact on the environment can be used to prepare the adoption of the environmentally economic decision [12]. The authors developed the system of environmental and economic indicators, taking into account the features of the studied enterprise [6, 13]. The final innovative element of this block is the system of environmental and economic indicators of production and consumption wastes [14, 15].

The implementation block of the organizational and economic mechanism for the nature protection management is also shown by the diversification of the enterprise's production program according to the environmental restrictions schematically depicted in Figure 2, where it is evident that after the first scenario of the environmentally safe production program, the main technical and economic indicator (profit) is compared with various environmental and economic components (ΔP – the increase of payment for the negative impact on the environment, ED – the economic damage from the environmental pollution, F - fines). The decision on the final inclusion in the production program is made on the basis of the Best Available Techniques (BAT) criterion, which is caused by the modern environmental requirements [16].

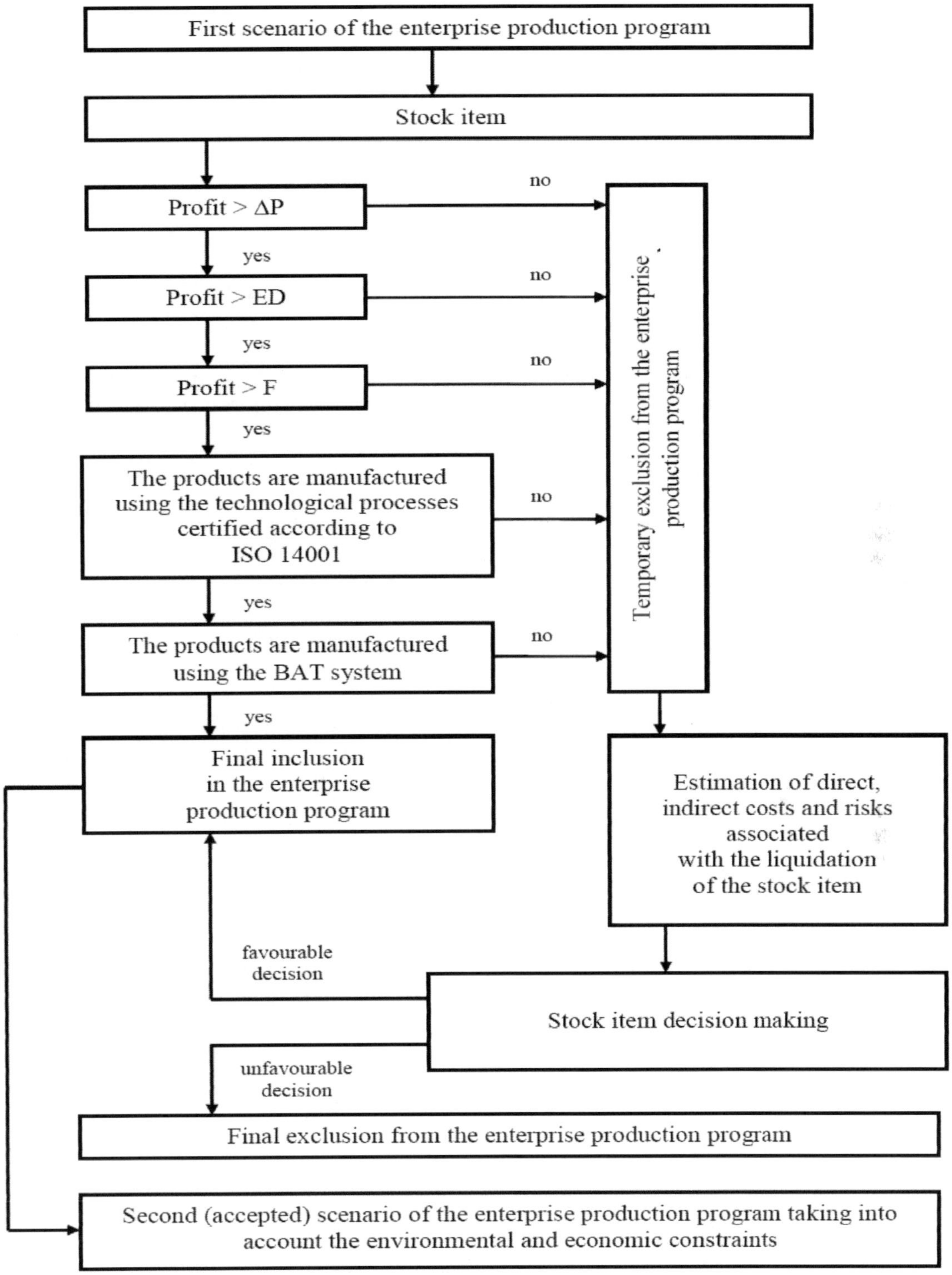


Figure 2 - Scheme for the formation of the production program of the enterprise, taking into account the environmental and economic constraints

## 4. Conclusions

The study of the development of innovative architecture of the organizational and economic mechanism for the nature protection management made it possible to draw the following conclusions:

- the necessity of improving the organizational and economic mechanism for the nature protection management taking into account the external requirements and the environmental legislation enforcement is substantiated;
- the simplified scheme for the environmental and economic system management mechanism with the differentiation of the blocks of the organizational and economic mechanism for the nature protection management is developed;
- the compliance matrix for the developed innovative elements in the structure of the organizational and economic mechanism for the nature protection management is formed;
- the analysis of various approaches to the assessment and interpretation of the environmental and economic indicators of the enterprise, including the study of the negative impact on all elements of the environment is made;
- the adequacy test of the developed innovative elements of the organizational and economic mechanism for the nature protection management for the purpose of its further practical implementation is carried out.

## References

[1] Zolotukhin V.M., Gogolin V.A., Yazevich M.Yu., Baumgarten M.I., Dyagileva A.V. Environmental management: the ideology of natural resource rational use // IOP Conference Series: Earth and Environmental Science. – 2017 – Vol. 50, Article number 012027. – p. 1-7.

[2] Zenkov I.V. *Obzor zarubezhnykh issledovaniy v oblasti ekologii gornodobyvayushchego proizvodstva // Gornyi Zhurnal* [Overview of foreign mining research // Mining Journal] 2016. No. 10. – p. 96-99.

[3] Burkov V.N., Burkova I.V. The network programming method in control over target programs // Automation and Remote Control. – 2014 – T. 75. No. 3 – p. 470-480.

[4] Burkov V.N., Burkova I.V. The principle of coordinated planning control of social and ecological-economic systems // Game Theory and Applications. – 2015 – T.17. – p. 17-36.

[5] Tretyakova E.A. Assessing sustainability of development of ecological and economic systems: a dynamic method // Studies on Russian economic development. – 2014 – Vol. 25. No 4. – p. 423-430.

[6] Kiseleva T.V., Mikhailov V.G., Karasev V.A. Management of local economic and ecological system of coal processing company // IOP Conference Series: Earth and Environmental Science. – 2016 – Vol. 45, Article number 012013. – p. 1-8.

[7] Bubnova M.B., Ozaryan Y.A. Integrated assessment of the environmental impact of mining // Journal of Mining Science. – 2016 – Vol. 52. No. 2. – p. 401-409.

[8] Agienko M.I., Bondareva E.P., Chistyakova G.V., Zhironkina O.V., Kalinina O.I. Eco-analytical methodology in environmental problems monitoring // IOP Conference Series: Earth and Environmental Science. – 2017 – Vol. 50, Article number 012022. – p. 1-9.

[9] Kiseleva T.V., Kulakov S.M., Mikhailov V.G., Mikhailov G.S. *Povyshenie effektivnosti upravleniya vodookhrannoy deyatelnostyu regiona na osnove postroeniya pereschetnoy modeli po kanalu "prirashchenie shtrafov - prirashchenie indeksa zagryazneniya" // Sistemy upravleniya i informatsionnye tekhnologii* [Management efficiency increase by region water-security activity on the basis of construction recounting models on the

channel an increment of penalties – an increment of index pollution // Control systems and information technology] 2005. Vol. 2 (19). – p. 84-87

[10] Dolzhenko E.N., Monich A.I., Kudryakov A.G., Sazykin V.G. *Model ekspertnoy otsenki v ekologicheskom menedzhmente predpriyatiya // Mezhdunarodnyy nauchno-issledovatelskiy zhurnal* [Model of expert evaluation in the environmental management of the enterprise // International research journal] 2016. Vol. 5-1 (47). – p. 75-78.

[11] Savon D.Yu., Tibilov D.P. *Upravlenie investitsionnoy deyatelnostyu predpriyatiya v oblasti okhrany okruzhayushchey sredy i ekologicheskoy bezopasnosti na otkhodoobrazuyushchikh proizvodstvakh ugolnoy otrasli // Gornyi Zhurnal* [Management of the company's investment activities in the field of environmental protection and safety at the waste-generating enterprises of the coal industry // Mining Journal] 2014. No. 12. – p. 31-35.

[12] Epifantseva E.I. *Razrabotka teoreticheskoy kontseptsii upravleniya prirodookhrannymi raskhodami promyshlennogo predpriyatiya // Vestnik Tambovskogo gosudarstvennogo tekhnicheskogo universiteta* [Development of theoretical concept of control for environmental expenditures of industrial enterprises // Transactions of the TSTU] 2003. Vol. 9. No 3. – p. 538-543.

[13] Mikhailov V.G., Koryakov A.G., Mikhailov G.S. Ecological risk management in coal mining and processing // Journal of Mining Science. – 2015 – Vol. 51. No.5. – p. 930-936.

[14] Mikhailov V.G., Golofastova N.N., Galanina T.V., Koroleva T.G., Mikhailova Ya.S. Environmental-Economic Assessment of Generation, Flow and Efficiency of use of Production and Consumption Waste // IOP Conference Series: Earth and Environmental Science. – 2017 – Vol. 50, Article number 012038. – p. 1-8

[15] Efimov V.I., Sidorov R.V., Korchagina T.V. *Obrazovanie otkhodov ot predpriyatiy ugolnoy otrasli na territorii Kemerovskoy oblasti // Ugol* [Generation of coal mining production wastes in the Kemerovo Region territory // Ugol] 2015. No. 12. – p. 73-76.

[16] Kiseleva T.V., Mikhailov V.G. *Modelirovanie diversifitsirovannoy proizvodstvennoy programmy promyshlennogo predpriyatiya s uchetom ekologicheskikh ogranicheniy // Trudy IV Vserossiyskoy nauchno-prakticheskoy konferentsii s mezhdunarodnym uchastiem "Modelirovanie i naukoemkie informatsionnye tekhnologii v tekhnicheskikh i sotsial'no-ekonomicheskikh sistemakh". Chast 2.* (Novokuznetsk: Publisher of the Siberian State Industrial University) [Modeling of the diversified production program of an industrial enterprise taking into account the environmental and economic constraints // Proceedings of the IV All-Russian Scientific Conference (with the international participation) "Modeling and knowledge-intensive information technologies in technical and socio-economic systems". Part 2. 2016. - pp. 27-31.

IOP Conf. Series: Earth and Environmental Science **66** (2017) 012009 doi:10.1088/1755-1315/66/1/012009

# Developing an Ecological Passport for an Open-Pit Dump Truck to Reduce Negative Effect on Environment

**V Yu Koptev[a], A V Kopteva[b]**

Saint-Petersburg Mining University,
21-th line 2, Saint-Petersburg, Russia, 199106,

*E-mail:* [a]Kvuy@mail.ru, [b]Alexandrakopteva@gmail.com

**Abstract**. Expanding the open-pit dump truck usage areas and the need to transport more and more minerals results in producing more and more powerful open-pit dump trucks, and this all is about environmental problems and potential health risks for the personnel. Harmful gas concentrations in working areas became threatening enough to have the work in some areas completely halted, until the contents of harmful substances in the air, as well as visibility on the roads, get back to norm. The article represents the new methodology for assessing comparatively the efficiency of modern transportation systems with performance and ecology characteristics taken into account, by developing an ecological passport for machines, facilitating design improvements and reducing pollution during operation.

## 1. Introduction

Implementing the recommendations by the UN Second International Conference (1992, Rio de Janeiro) on environmental challenges necessitates improving control over using and protecting natural resources, based on wide scale implementation of advanced methods for controlling their state, and developing ecological regulations and requirements when using water and soil resources. In connection therewith, the need emerges to develop new methodologies for assessing risks of adverse effect by technical facilities on various natural environment components.

The new concept resides on the fundamental statement that not only the existential threat for the humankind comes with the direct pollution of environment, but it also results from breaking the natural biological balance on Earth. To preserve it the way it is, one must not overcome the allowed disturbance threshold. It is necessary to develop the system for defining the effect on biological systems, based upon assessing qualitatively the changes that take place within those systems.

State of art methodologies used to design transport systems in mining enterprises do not account for ecological aspects of open-pit dump truck operation, neither do they allow comparing or selecting the optimal layouts of open-pit dump operation and transportation facilities, nor would they specify particular open-pit dump truck models (by structure, residual lifespan, cost). Analyzing the efficiency of the open pit atmosphere normalization is not a strong side of the state of art approach, too. What we have is the situation, when insufficient funds result in delayed equipment upgrade at enterprises, while the average wear-out of open-pit dump trucks at many facilities approaches 80% and more; open pit mining is to great extent monopolized in many lands, including in Russia, by products from several

IOP Conf. Series: Earth and Environmental Science **66** (2017) 012009 doi:10.1088/1755-1315/66/1/012009

manufacturers, like Caterpillar, Unit Rig and some others, with their competitors getting pushed away from the market, for example, this specifically relates to BelAZ.

## 2. Equipment and devices used in studies

Ecology is a weak point of any open-pit vehicle. Diesel engines produce harmful emissions in large amounts (all 17250 units of BelAZ open-pit dump trucks in CIS burn around 2.1 million tons of diesel fuel annually), resulting in degradation of ecology in mining work areas. Fines that businesses are charged for breaking environmental regulations became more costly recently. Therefore, open-pit dump truck manufacturers do their best to make their vehicles more fuel-saving, i.e. environmentally friendly.

When it comes to considering the operation of an open-pit dump truck, there are things that are usually neglected, although in reality those things are essential, too. This specifically relates to carter gases, oil leaks via failed seals, soil and water pollution with oil products, tire particles, processing sites being polluted with oil materials when fueling vehicles, servicing vehicles; powders from the load transported by vehicles, other open-pit dump truck parts wear-out particles, asphalt particles; noise, electromagnetic and electrostatic disturbances. The latter specifically relates to large heavy-duty open-pit dump trucks with electric mechanical transmission. In addition, a serious environmental challenge is the disposal of old broken cars, large car bodies and other metal structures. Large territories become cluttered, and environmental issues emerge in connection therewith.

Any open-pit dump truck, however perfect its design might be, would hardly perform well, unless optimized for working with open-pit excavators. The department of mining machines of the Mining University developed a math model for selecting optimal parameters for load/unload complex, with ecological factors taken into account [1, 2]. The methodology for defining an optimal excavator-vehicle complex for particular open pit mining site is simple to use, tested many times, and designers and workers successfully implemented it. This methodology still needs to be improved, as new machinery performance criteria start working under new conditions.

## 3. Results and discussion

System approach to assessing vehicle's ecological characteristics assumes analyzing all types and sources of environment pollution factors when operating a vehicle (fig.1); then it also specifies the contribution of each particular type and source of pollution in overall adverse effect on environment; third – a car is considered a complicated dynamic system with characteristics, including those related to ecology, being modified in time, as the resource is getting spent out. Therefore, the total ecological characteristics of a car [1, 3, 5] certainly includes sanitary characteristics that describe harmful effect on an operator by noise, vibrations, gas and dust getting into the vehicle cabin, and also electrostatic and electromagnetic fields, especially when operating vehicles with electrical mechanical transmission.

Therefore, one should talk apart the complex ecological characteristic of a new vehicle:

$$EX_{\kappa} = f\ (K_1, K_2, \dots, K_i), \tag{1}$$

where $K_1$, $K_2$, …, $K_i$ refer to design and technology parameters of individual subsystems and units in a vehicle. And the ecological characteristic of a vehicle in operation:

$$EXe = f\ (K`_1, K`_2, \dots, K`_i), \tag{2}$$

where $K'_1$, $K'_2$, …, $K'_i$ refer to design and technology parameters of a vehicle depending on its operating conditions and the lifespan part already spent.

For monitoring changes in ecological characteristics of a vehicle in operation, an open-pit dump truck ecological passport is proposed, which comprises the information about the natural resources used during the operation of the vehicle (natural, secondary etc.), and also the information, necessary to determine the effect on environment by years of its operation, after service and major repair work [3,4].

The passport is based on regulatory norms (GOST, industrial standards etc.), production indicators (official reference data), estimated values for maximum emissions allowed, maximum concentrations and conditions in connection therewith, and also the data of the state statistical register, other regulatory and technological documents. The overall structure of the passport is presented herewith (fig.2). The passport is filled by manufacturer to be attached to the technical documentation kit for an open-pit dump truck. Signed by the chief engineer of the enterprise, the passport is to be approved by local nature protection committee.

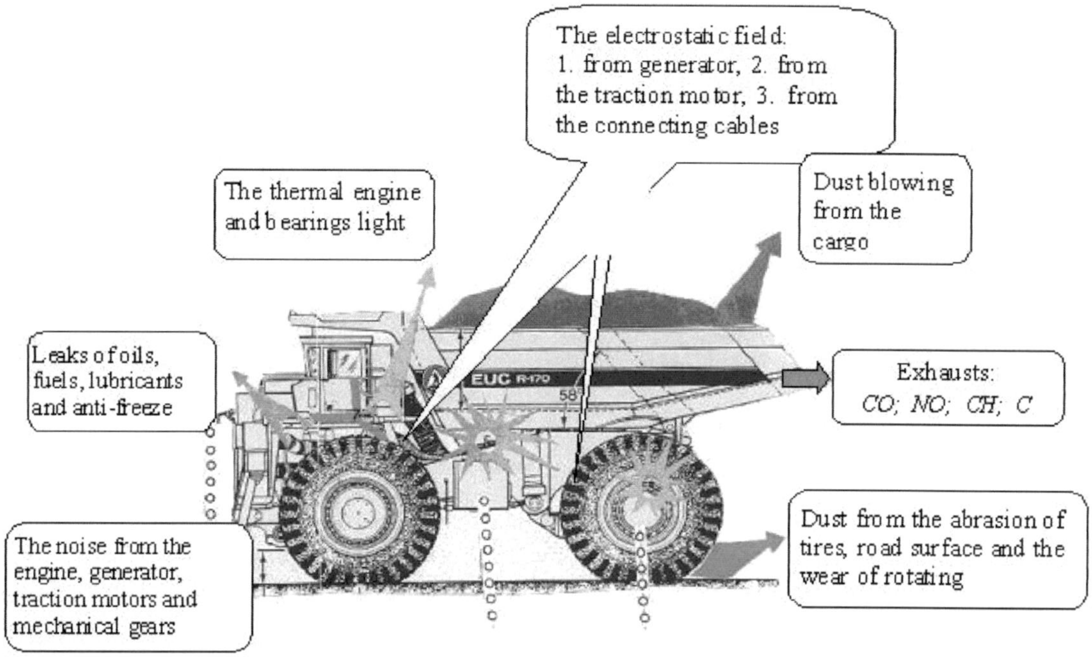


**Figure 1** – All types and sources of environment pollution factors when operating a vehicle

Solving tasks by the vehicle passport accounts for experience of leaders like USA, Germany etc., the ecological strategy in which focuses not only on pollution itself, but even more on negative influence on environment, assuming that financial methods might be more efficient than administrative ones. The qualitative evaluation of damage from influence of different nature (harmful gases, waste, radiation etc.) resides on the financial criterion, i.e. paying the risk of exceeding maximum pollution levels allowed.

The vehicle passport solves the following tasks:

- defining the list of natural resources used when operating open-pit dump trucks;
- defining the list and the amount of maximum emissions allowed, as estimated based on regulatory norms and other documents;
- defining the amount of harmful substances and adverse effects exceeding the maximum concentration allowed, expressed via indicators; performing quantitative evaluation of harmful substances and the effect in connection therewith;
- evaluating quantitatively the ecological condition of an open-pit dump truck in general and by individual types of influence to form the vehicle park structure in a manner that would minimize environmental pollution.

It is proposed to develop the open-pit dump truck ecological passport system for all types of mining and processing diesel equipment of both open pit and drilling kinds. The presence of such document in vehicle facilities of mining facilities, open pit facilities, etc. facilitates more accurate comparison of the regulated amount and structure of emissions at enterprises with the real one. In addition the vehicle park structure may be optimized for lowering environmental pollution, thus cutting costs on fines. Open-pit dump truck manufacturers become motivated for improving their product parts even further.

Onboard computers in modern open-pit dump trucks allow remote monitoring by relevant officials. The fig.6 illustrates such a system, developed on the basis of OLKON OJSC (former Olenegorsk mining and processing plant). The forms, reports and logs outputted correspond to standards accepted at the enterprise that issued the documents [6-9].

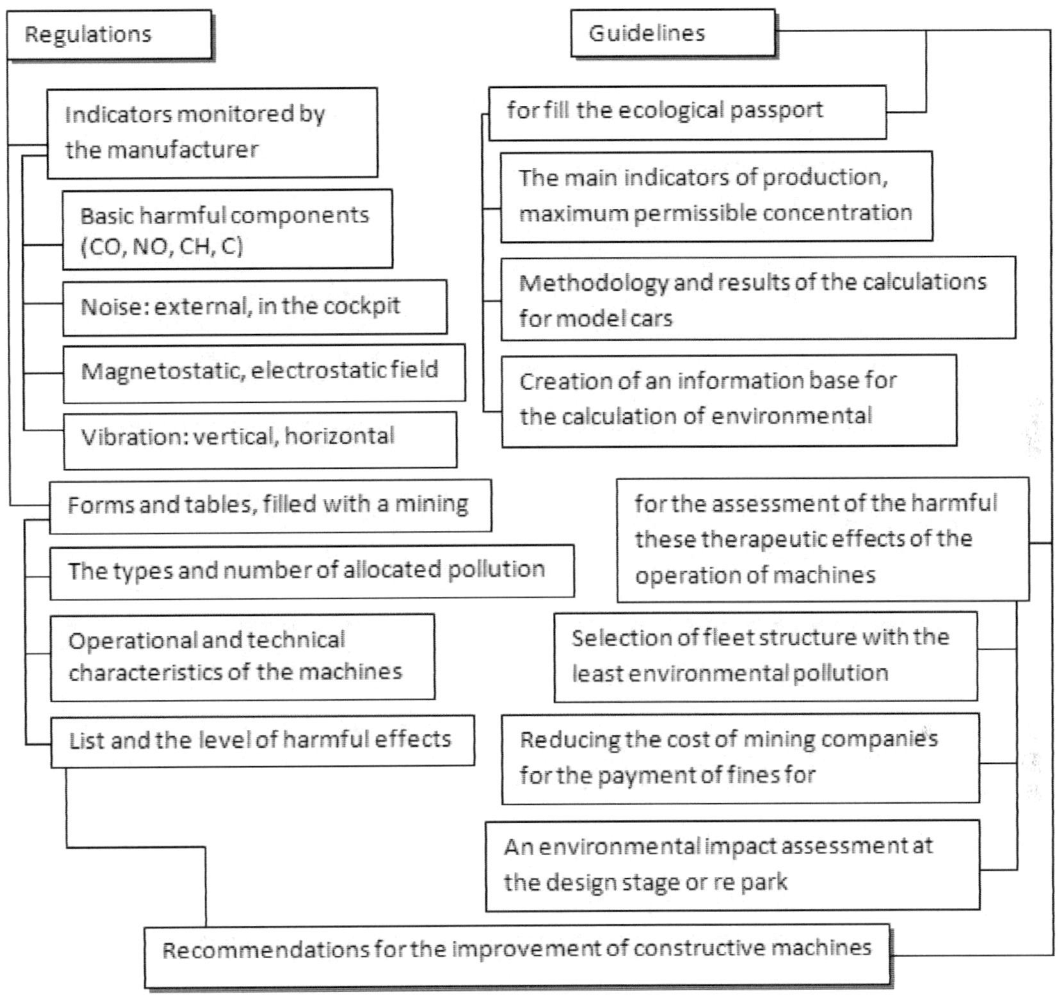


**Figure 2** – The overall structure of the passport

## 4. Conclusions

The proposed model suggests the selection of optimal transportation system to reside on minimum specific costs, accounting for ecology factors. The algorithm and the structure of economical and mathematical model are built in accordance with module principle, allowing inputting other calculation units for other types of transport without making serious changes. Module principle allows easy transformation into the model for combined transport. Fig.3 illustrates that ecological characteristic calculations may be set as reference for comparing various machine models to perform studies on disclosing the way for reducing harmful emissions.

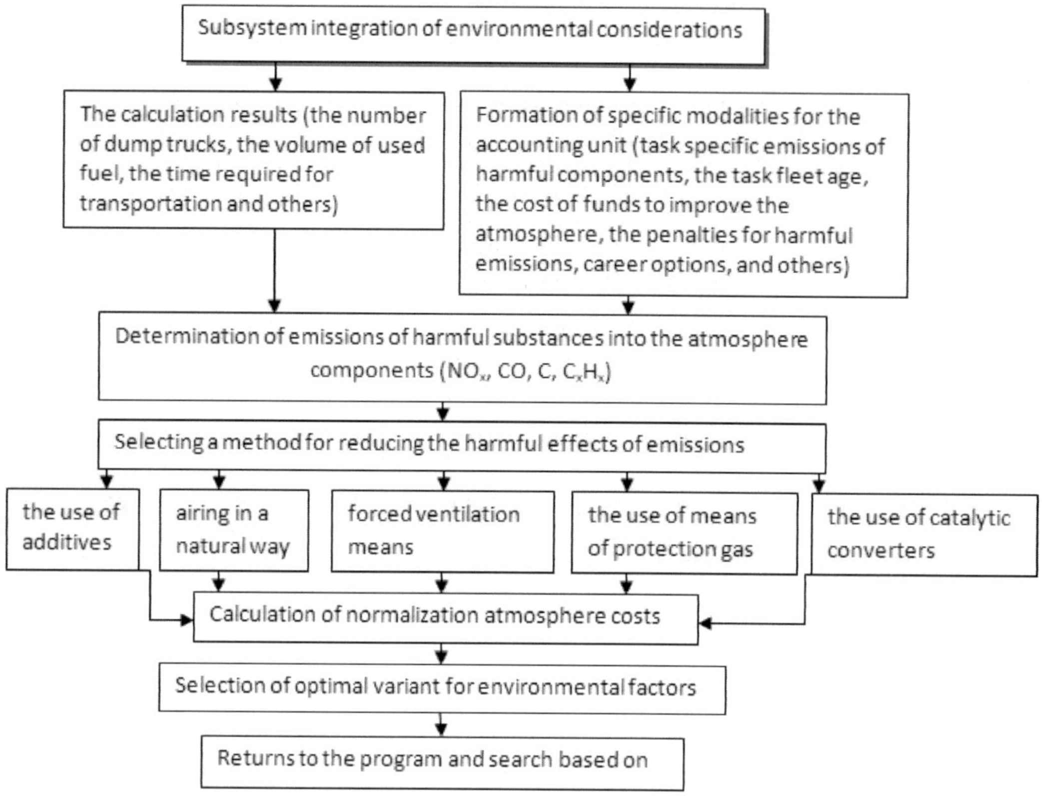


**Figure 3** – Subsystem integration of environmental considerations

**References**

[1] Koptev V. Y u. *Mining Business Transportation System Structure Optimization* // International Journal of Applied Engineering Research. 2016. Vol. 11, No. 11 p. 7402-7405.
[2] Koptev V. Yu. *Justification of mining enterprises transport machines selection* [Mining equipment] - St.Petersburg: Publ. House "Slavutich". 2012. p. 58-61
[3] Kuleshov A.A. *Principles of ecological passport dump* [Interuniversity thematic collection of works SPbASU]. – Saint-Petersburg: SPbASU, 1995. pp. 27-32.
[4] Kuleshov A.A. *Assessment of the environmental performance of mining dump trucks.* [a collection of works MANEB].-Moscow: MABEB, 2004. 160 p.
[5] Radkevich, J.M., Belyankina, O.V. *The mathematical expression for the basic target functions of mining vehicles and machines* // Mining informational and analytical bulletin (scientific and technical journal). 2006. No.2 p. 262-268.
[6] Radkevich, J.M., Skhirtladze, A.G. *Metrology, standardization and certification.* - M.: Moscow State Mining University. 2003. 340 p.
[7] Polovinkin, A.I. *Engineering creativitybases.* [Study guide for students of technical colleges] - Moscow: Engineering. 1998. 451 p.
[8] Anistratov K. Yu., Stremilov V. Yu., Teterin M. V. *Open pit technological transport vehicle enterprise functioning economic and mathematical model.* - Moscow: Gemos Limited Mining Industry. 2007. No. 2. p. 16-18
[9] Anistratov K. Yu. *The dump trucks park structure justification for vehicle load classes 40-100 tons.*- Moscow: Gemos Limited Mining Industry. 2015. No.5. p. 26-29
[10] Lishchuk A. N. *Mining technique. Business Directory* . S.Petersburg: Lamos. Vol.1. 2012.

p.40–43.

[11] Keerthika R., Jagadeeswari M. *Coal Conveyor Belt Fault Detection and Control in Thermal power plant using PLC and SCADA//* Int.l J. of Advanced Research in Comp. Engineering & Technology (IJARCET). 2015. Vol. 4. p. 1649-52.

# Models and Methods for Adaptive Management of Individual and Team-Based Training Using a Simulator

**L S Lisitsyna [1,a], N P Smetyuh [1,b], S P Golikov [2,c]**

[1] *ITMO University*
Kronverksky prospekt 49, St. Petersburg, Russia, 197101
[2] *Kerch State Maritime Technological University*
Ordjonikidze 82, Kerch, Russia, 298309

*E-mail:* [a]lisizina@mail.ifmo.ru, [b]nadya.s.2011@yandex.ua, [c]golosaa@mail.ru

**Abstract**. Research of adaptive individual and team-based training has been analyzed and helped find out that both in Russia and abroad, individual and team-based training and retraining of AASTM operators usually includes: production training, training of general computer and office equipment skills, simulator training including virtual simulators which use computers to simulate real-world manufacturing situation, and, as a rule, the evaluation of AASTM operators' knowledge determined by completeness and adequacy of their actions under the simulated conditions. Such approach to training and re-training of AASTM operators stipulates only technical training of operators and testing their knowledge based on assessing their actions in a simulated environment.

## 1. Introduction

In fact, this approach leaves out any identification of their knowledge and skills of simulating organizational and manufacturing situations and taking efficient managerial decisions; it doesn't enable any identification and assessment of their knowledge on the basis of multi-informational and least loss-making methods and information technologies. Hence the problem is to research and develop a methodology for systemic identification of professional problem-focused knowledge acquired by employees operating adaptive automated systems of training management (AASTM operators), which shall also further the theory and practice of the intelligence-related aspects thereof. A robot solves this problem as a general problem of matching mathematical knowledge models as prototypes to the formalized system of knowledge an AASTM operator as a system image [1-3].

Virtual operator training based on an AASTM simulator makes it less time-consuming and resource-intensive to obtain and master the required skills. This is achieved by eliminating the negative consequences of the risk factors inherent in the in-situ education and training of water transport specialists. This enables training of water transport specialists (WTS) without any significant financial losses. The fundament of the process is a direct linking of theoretical knowledge in automated industrial fishing and fishing ship navigation to the practical requirements. Before a WTS can use a simulator, they should complete theoretical learning to have a perfect comprehension of training tasks and objectives; they should also be able to use the best automated fishing practices of crews navigating the Azov-Black Sea Basin. To solve this problem, we have, as a part of our research, designed a scientific problem to create methods for adaptive management of individual and team-based training using a simulator.

The aim of the thesis is to research and develop methods for adaptive management of individual and team-based training using a simulator so as to facilitate the training of the Azov–Black Sea fishing fleet crews.

## 2. Equipment and devices used in studies

To solve the problems stated in the thesis, we have employed both systematic and situational approach as well as: control theory methods to state the decision making problem; similarity theory to construct a formal model of the system; probability theory apparatus, fuzzy and rough sets to enable the information system to process the inputs; situational control theory methods to construct a joint activity model; algebraic rings and graph theory to formalize the network model for knowledge representation and the tree structure consistency model; machine learning methods to solve the problem of data storage narrowing; adaptive planning methods to solve the activity planning problem.

The research object is the process of adaptive management of WTS training involving a professional maritime industrial bioresource extraction simulator [3-5].

The research subject matter is the methods for adaptive management of individual and team-based training using a simulator.

## 3. Results and discussion

The monitoring subsystem (WTM-SITT) provides a flow of observed system parameters of the joint activity model as well as a total of event flows for WTM-SITT; its operation is based on sensor measurements. A dynamic maritime object has a number of data sources of various accuracy, which provide navigation data (radar, hydroacoustic, sound ranging equipment, video cameras, etc.). In most cases these data are duplicated, for example, by the automated identification system, the radar equipment, and by visual observation. Thus, observing the navigational situation results in excessive data, which makes data processing more time-consuming and may cause delays in identifying a hazardous situation.

Consequently, the monitoring system signals the emergence of significant events, which enables the WTM-SITT and the joint activity model to solve their specific problems only most of the time. At certain moments of time $t_k$, measurements are made, and a vector of motion parameters $x_k^i$ is recorded for each maritime dynamic object (MDO) $A_i$:

$$x_k^i = \left\langle K_i, V_i, X_i, Y_i, \phi_i, \alpha_{ji}, l_{ij} \right\rangle_{t_k}, \quad (1)$$

where $X_i, Y_i$ is the MDO $A_i$ spatial coordinates;

$K_i, V_i$ are the direction and speed of the MDO $A_i$;

$\phi_i$ is the MDO angular velocity $A_i$;

$\alpha_{ji}$ is the bearing to $A_i$ from $A_j$;

$l_{ij}$ is the distance from $A_i$ to $A_j$.

As a result of monitoring, events like "change of direction $A_i$", "change of speed $A_i$" are recorded, with the significant events selected from the flow:

- MDO direction change;
- MDO speed change;
- MDO entering the interaction area;
- MDO leaving the interaction area.

As monitoring is performed at a specific discretization (as a rule, every 30 seconds), WTM-SITT generates event flows providing the basis for a further analysis of situation dynamics to forecast possible WTS actions and, consequently, possible changes in MDO routes [6-8]. Meanwhile, observation of how the parameters of motion $A_i$ alter enables: case-based determination of the scenario $\Sigma_{Ai}$ this MDO is to run; suggesting the possible plan $\Pi_{Ai}$ and forecasting its route $\wp(A_i)$. Using possible activities (plans) of observed situation disturbances, one can build control procedures $\Omega_i$ including a control action scenario to compensate them by maneuvering.

Key classification characteristics of a ship (MDO) are presented Table 1. Classifying based on $K^2$ makes sense only for moving targets, whereas classification on the basis of $K^3$ and $K^4$ is relevant for approaching targets as outgoing and equally spaced targets are non-hazardous by definition. Classification based on $K^5$ is made for a set of hazardous and potentially hazardous targets, and the target priority value $P_{0i}$ (corresponds to $\prec_{\Sigma}$) is generated on the basis of a fuzzy controller. Classification based on $K^6$ is made from the standpoint of possible changes in the MDO motion parameters: if no change in $(K_0, V_0)$ renders the object hazardous, it is classified as non-restrictive, and vice-a-versa.

Classification based on the spatial position $K^7$ is of particular importance as it is connected with clustering in case search in WTM-SITT with the following main criteria of context similarity: number of disturbances; spatial position of disturbances; motion parameters of disturbances. To simplify the search of similar situations, the classification $K^7$ is performed according to the chart in Figure 1. Controlled space is divided into sections numbered within the quadrants, each equal to $30^\circ$ as well as into circle areas *A*, *B*, *C* and *D* separated by the assumed borders of the areas $D_z$, $D_s$ and $D_u$ according to $K^4$. MDO location is assigned to a certain sector and is specified as its identifier.

**Table 1** - Classification characteristics to classify disturbances

| Classification | Characteristic | Value | Class |
|---|---|---|---|
| | | $V_i = 0$ | Immobile |
| $K^1$ | $V_i$, $\dot{V}_i$, $\dot{K}_i$ | $(V_i > 0) \wedge (\dot{V}_i = 0) \wedge (\dot{K}_i = 0)$ | Mobile non-maneuvering |
| | | $(V_i > 0) \wedge ((\dot{V}_i > 0) \vee (\dot{K}_i > 0))$ | Mobile maneuvering |
| | | $\dot{D}_{0i} > 0$ | Outgoing |
| $K^2$ | $\dot{D}_{0i}$ | $\dot{D}_{0i} = 0$ | Equally spaced |
| | | $\dot{D}_{0i} < 0$ | Approaching |
| | | $(D_i \le D_z) \wedge (0 < T_i \le T_z)$ | Hazardous |
| $K^3$ | $D_i$, $T_i$ | $(D_i \le D_z) \wedge (T_i > T_z)$ | Potentially hazardous |
| | | $(D_i > D_z) \wedge (T_i > T_z)$ | Non-hazardous |
| | | $T_i > T_z$ | MRA of unrestricted motion |
| $K^4$ | $T_i$ | $T_s < T_i \le T_z$ | MRA of a timely maneuver |
| | | $T_u < T_i \le T_s$ | NRA of a delayed maneuver |
| | | $0 < T_i \le T_u$ | MRA of an extreme maneuver |
| | | $P_{0i} = 0$ | Of equal priority |
| $K^5$ | $P_{0i}$ | $P_{0i} = 1$ | Passive |
| | | $P_{0i} = -1$ | Active |

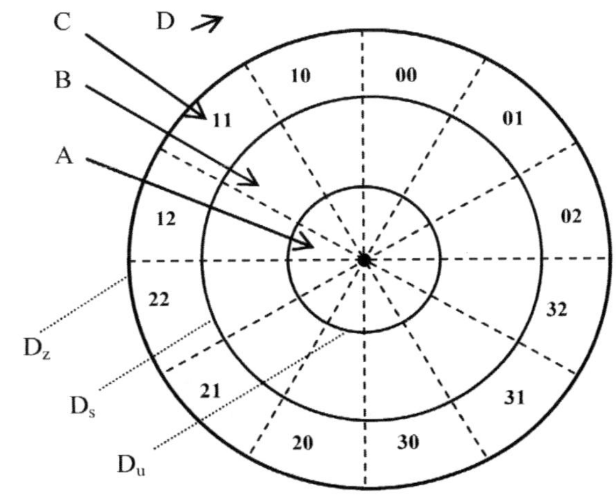


**Figure 1** – Chart for spatial position classification of MDOs

To classify targets in compliance with $K^3$, $K^4$, $K^6$, $K^7$, it is necessary to estimate the values of safe area borders, particularly $T_z$, $T_s$, $T_u$, as well as the values $D_z$, $D_s$, and $D_u$ corresponding thereto at a constant speed of the MDO. Safe area borders are determined as approximated interval estimates set by the border areas of $BND_{H1}$, $BND_{H2}$, $BND_{H3}$ in terms of the rough set theory:

$$POS_{H3} = NEG_{H2} \cup BND_{H2} \cup POS_{H2},$$
$$POS_{H2} = NEG_{H1} \cup BND_{H1} \cup POS_{H1}. \quad (2)$$

The interaction space $C'$ is divided into areas (see Figure 2) by $K^4$ using the approximate estimates of safe borders $\tilde{D}_z$, $\tilde{D}_s$, and $\tilde{D}_u$.

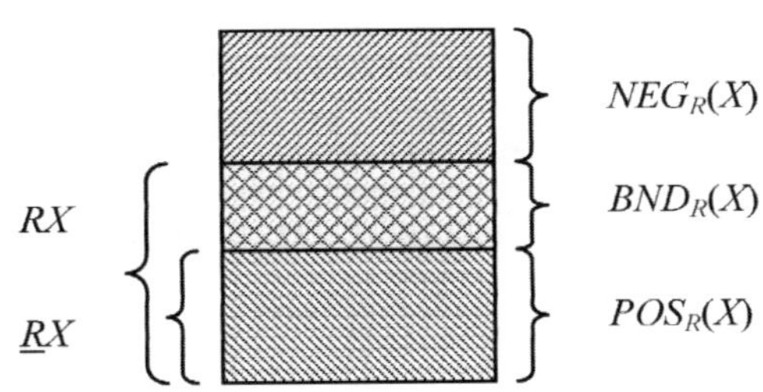


**Figure 2** - Rough set area determination

Being aware of the specified estimates and assuming the MDO speed is constant within the current time period, it is possible to determine the values $\tilde{T}_z$, $\tilde{T}_s$, and $\tilde{T}_u$. As there are no formal methods for getting reliable estimates of safe area borders, while it is obvious such estimates recur in similar situations for the same-type disturbance classes, WTMITT uses the following principles to train finding the approximate estimates of safe area borders [9-11].

1) to represent the estimates of safe area borders, a formal rough set apparatus is used, while an estimate for each safe area border is determined by the interval:

$$\tilde{D}_z = \left[\tilde{D}_{z_{\min}};\tilde{D}_{z_{\max}}\right],\ \tilde{D}_s = \left[\tilde{D}_{s_{\min}};\tilde{D}_{s_{\max}}\right],\ \tilde{D}_u = \left[\tilde{D}_{u_{\min}};\tilde{D}_{u_{\max}}\right]. \quad (3)$$

2) an approximate estimate of a border for each area can "contract" in case of navigation data are specified and ship approaching occurs, which makes disturbance classification more accurate.

3) in a real-world setting, estimates $\tilde{D}_z$, $\tilde{D}_s$, and $\tilde{D}_u$ can be obtained by the ASTM using a case-based analysis, see Figure 3.

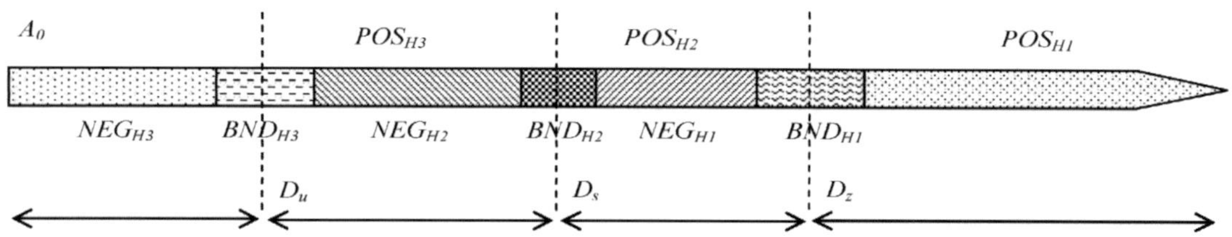


**Figure 3** – Safe border values determination

Appropriate cases are searched made according to the next neighbor method $K_{NN}$ based on the function that assesses the similarity of the case to the currently observed situation. Calculations are made using the coordinate-based method of comparing the situation and the case in terms of context parameters.

For $t_k$, the current navigation situation is a record:

$$s_k = \left\langle t_k, n, x_k^W, \left(x_k^i, x_C^i, \left\{x_k^{ji}\right\}_{j=1}^{n-1}\right)_{i=1}^{n} \right\rangle, \quad (4)$$

where $t_k$ is the current moment of time;

$x_k^W$ is the vector of external environment parameters at the moment $t_k$;

$x_k^i$ is the vector of MDO $A_i$ motion parameters at the moment $t_k$;

$x_C^i$ is the vector of MDO $A_i$ static characteristics;

$x_k^{ji}$ is the vector of relative motion parameters of $A_i$ and $A_j$ at the moment $t_k$

$n$ is the number of MDOs interacting at the moment $t_k$.

## 4. Conclusions

Preliminary simulator-based training of a ship driver helps re-train the skills of developing tactics for navigating in the expected maneuvering area, learn the specifics of movement, and make a rational choice of motion control methods. Preliminary fishing training and the necessity to plan curved sections enables the acquisition of knowledge to solve tactic and technological maneuvering problems while also reducing the risk of facing unfavorable trawl casting.

Overview of the existing SCM and the proposed formalized models have enables the researches to design a new structure of the system for automated control of a ship during trawl casting. It was made possible by creating of an integrated hierarchic three-level system for maneuvering control as well as taking into account the specific nature of ship control at during trawl casting. The paper contains the list of measures that shall be taken to implement the PID.

## References

[1] Zhilenkov A, Chernyi S 2015 Investigation performance of marine equipment with specialized information technology, Procedia Engineering 100 1247–1252

[2] Chernyi S and Zhilenkov A 2015 Modeling of complex structures for the ship's power complex using XILINX system. Transport and Telecommunication 16 (1) 73–82

[3] Chernyi S 2016 Use of Information Intelligent Components for the Analysis of Complex Processes of Marine Energy Systems. Transport and Telecommunication Journal, 17 (3) 202–211

[4] Chernyi S 2016 Analysis of the energy reliability component for offshore drilling platforms within the Black Sea. Neftyanoe Khozyaystvo - Oil Industry 2 106-110

[5] Chernyi S., Use of Information Intelligent Components for the Analysis of Complex Processes of Marine Energy Systems, Transport and Telecommunication Journal, vol. 17, no. 3, 2016. DOI: 10.1515/ttj-2016-0018

[6] Rutkowska D., Pilinsky M., Rutkowski L. Neural networks, genetic algorithms and fuzzy systems, Hotline Telecom, 452, 2004.

[7] Budnik V., Chernyi S. 2016 Future Development of the World Ocean Mining for the Industry, Procedia Engineering, vol. 150, pp. 2150-2156

[8] Computational Intelligence and Feature Selection: Rough and Fuzzy Approaches, Kybernetes, vol. 38, no. 34, 2009.

[9] Chaudhuri A., Stenger H. Survey sampling theory and methods, New York: Chapman &Hall, 416, 2005.

[10] Yue H., Xing K., Hu H., Wu W. and Su H. 2016 Petri-net-based robust supervisory control of automated manufacturing systems, Control Engineering Practice, vol. 54, pp. 176-189.

[11] Nyrkov A., Chernyi S., Zhilenkov A., Sokolov S. 2016 The use of Fuzzy Neural Structures to Increase the Reliability of Drilling Platforms, Proceedings of the 26th International DAAAM Symposium 2016, pp. 0672-0677.

IOP Conf. Series: Earth and Environmental Science **66** (2017) 012011 doi:10.1088/1755-1315/66/1/012011

# The Survey of Sorption Ion-Exchange Properties of Paleozoic Natural Minerals in the Static Conditions

**T V Germanova[a], V A Nefedov, I R Valieva[b]**

Tyumen industrial University,
St. Volodarsky, 38, Tyumen, *Russia,* 625000

*E-mail:* [a]ecogtv@mail.ru, [b]irina.valieva@list.ru

**Abstract**. An experimental and exploring process` results of iron`s ion, manganese and ammonium sodium extraction from aqueous solution by some Ural`s natural minerals in a static regime at certain physicochemical conditions: by the zeolite, montmorillonite and vermiculite in the static conditions. Values of maximum sorption capacity are estimated on a gram of natural sorbents for the present conditions. Comparison of the survey`s results shows that effective sorptions from exploring natural minerals can be used as a clinoptilolite and zeolite-mineral montmorillonite type.

## 1. Introduction

The research of the zeolite`s ion exchange properties is necessary for possible application the natural minerals in the field of environmental protection.

Surface waters and underground horizons in the Western Siberian region are characterized as an increased pollution by different substances of natural and technogenic origin, which include heavy metal cations, iron, ammonia nitrogen and others. Consequently, it is essential to make a reliable barrier from a wide range of pollution in the process of cleaning natural and waste water. Filtration is traditionally one of the latest technological parts for the cleaning natural and waste waters. For this purpose it widely used in filter layers of granular natural materials Nowadays zeolite breeds from different deposits are investigated as a download filter [1, 2]. In the subpolar and polar Ural are opened deposits of zeolite breeds, it has an industrial sense.

## 2. Equipment and devices used in studies

### 2.1 Sorption and ion-exchange properties of natural minerals

The work`s purpose is to explore the sorption activity of natural Paleozoic minerals in Ural: zeolite-monmorillonite breeds, clinoptilolite, habasit and vermiculite concerning the ions of external solutions. The researchers noted that the exchanging complex`s nature of a given material`s sorption ability influences on it [8, 9]. One of the most important aims in the kinetics of ion exchange is to find out the mechanism of the sorption process and the ability to rule its speed. During the survey of the new natural material in order to the comparing mark with known natural sorbents, a method is brought as the foundation which is used by explorers of the University of mineralogy, geochemistry and crystal chemistry of rare elements [3, 5].

This work is performed within the scientific-applied themes to the investigation of physicochemical properties of Paleozoic zeolites in Ural. The survey is leaded in the laboratories of the Industrial University of Tyumen on the basis of the Central laboratory of Tyumen [1, 2].

**3. Results and discussion**

For the evaluation of ion-exchange and sorption properties of zeolite-montmorillonite rocks, clinoptilolite, habazite, vermiculite, experiments were conducted on the sorption at static conditions, ions of iron, manganese and ammonium nitrogen. The solutions were prepared with the following parameters (mg/dm3): $Ca^{2+}$=40.0; $Mg^{2+=}$13.4; $K^{+}$=1.5 (approximate average composition of the surface waters of the West Siberian artesian basin). Then in the prepared solution was injected salt of the studied pollutants, the list of which was chosen taking into account the specifics of natural water pollutants and existing research findings in the literature [1-5]. In a flask were poured 10 g of sorbent and poured the prepared solution containing a salt of the investigated metal. The weight ratio of sorbent/solution 1:20. The contact time of sorbent with solution averaged about 14 hours to determine the stable adsorption. In the waste liquid were determined the contents of the investigated metal. Then the solution was replaced and the experiments were continued until complete saturation of the sorbent.

As the granular filters used powders fractions of 0.07-0.3 mm zeolite-montmorillonite rocks, clinoptilolite, chabazite, vermiculite, and to map the investigated organic sorbent-carbon (AG-3). The results are shown in the graphs (Fig. 1-3).

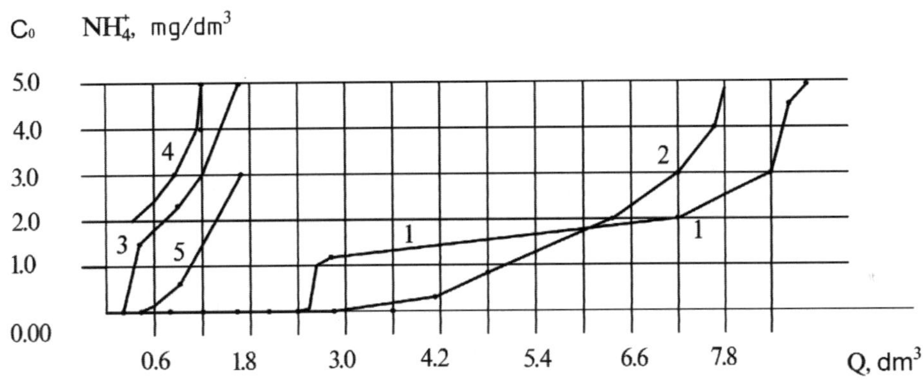


**Figure 1** – Sorption of ammonium nitrogen ($NH_4^+$) inorganic and organic sorbents in static conditions (initial content of the element in a solution of 5 mg/dm$^3$, T= 200$^0$C, the fraction 0.07-0.3 mm):1 – zeolite-montmorillonite sorbent; 2 – clinoptilolite; 3 – habazite; 4 – vermiculite; 5 – carbon (AG-3); $C_0$ $NH^{4+}$ - content of ions of ammonium in the solution after contact with the adsorbent (contact time 14 hours); Q – volume of preconditioning solution with the sorbent, dm$^3$.

A competitive movement of ions to mineral occurs in a complicated system of the mineral and the sorbing ions. All happened processes make essential influence on a character of diffusion transfer and common speed of exchange.

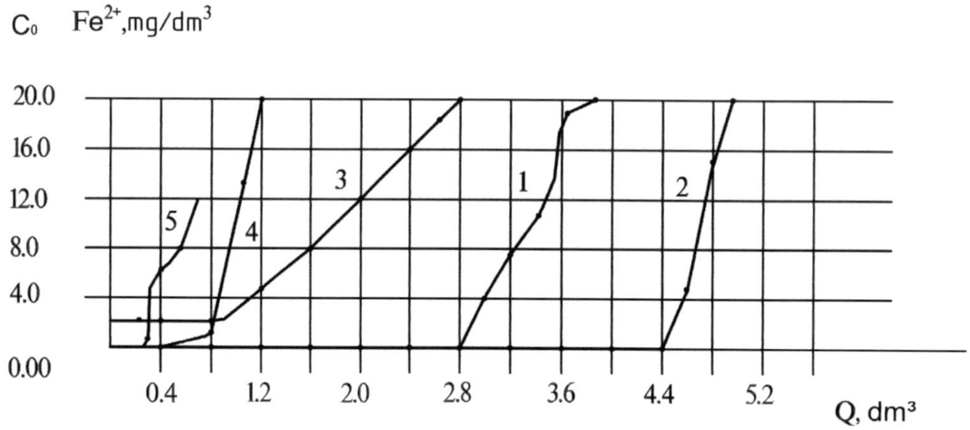


***Figure 2*** - Sorption of iron ions ($Fe^{2+}$) under static conditions (the content of $Fe^{2+}$ 20 ions of 20 mg/dm$^3$, T= 200$^0$C, the fraction 0.07-0.3 mm): 1 – zeolite-montmorillonite sorbent; 2 – clinoptilolite; 3 – habazite; 4 – vermiculite; 5 – carbon (AG-3); $C_{0Fe}{}^{2+}$ - the content of iron ions ($Pb^{2+}$) in solution after contact with the adsorbent (contact time 14 hours); Q – volume of preconditioning solution with the sorbent, dm$^3$.

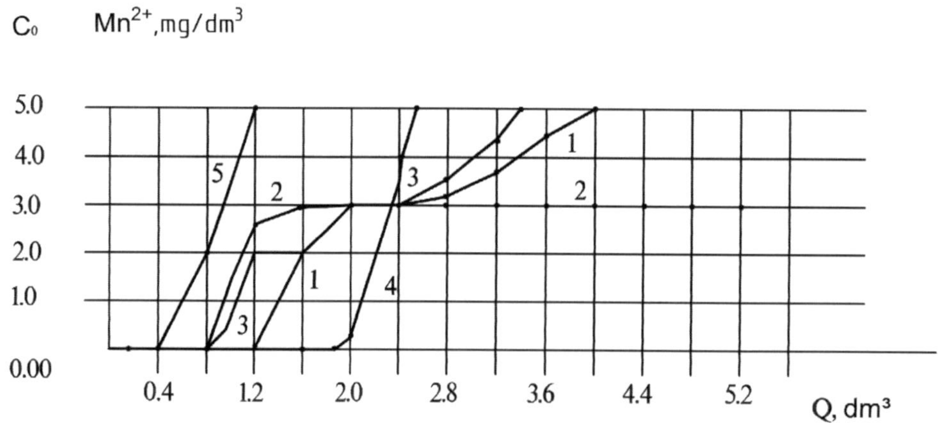


***Figure 3*** - Sorption of manganese ions ($Mn^{2+}$) in static conditions (initial content of the element in a solution of 5 mg/dm$^3$, T= 200$^0$C, the fraction 0.07-0.3 mm): 1 – zeolite-montmorillonite sorbent; 2 – clinoptilolite; 3 – habazite; 4 – vermiculite; 5 – carbon (AG-3); $C_{0Mn}{}^{2+}$ - the content of manganese ions ($Mn^{2+}$) in solution after contact with the adsorbent (contact time 14 hours); Q – volume of preconditioning solution with the sorbent, dm$^3$.

Analysis of the own and previous results of exchange of the one-bivalent natural zeolites [1-3] cations shows that the speed of the ion processes increases with the concentration of the solution, also by temperature and decrease of the counter-ion size.

The main parameters of ion-exchange capacity are the sorption and technological properties. The exchange capacity is necessary at study of kinetic and equilibrium values of ion-exchange [8]. Distinguish the maximum exchange capacity, which is corresponded to the full substitution of one ion to another and the exchange capacity which is realized at the certain physicochemical conditions. The last one is especially important by the choice and calculation of resource technological equipment.

Calculated at the certain specific physical and chemical conditions of the exchange capacity indicates the sorption capacity of the zeolite-montmorillonite mineral, clinoptilolite, chabazite and vermiculite. Zeolite-montmorillonite and clinoptilolite species are highly sorbent for extraction of iron from liquids to 9.2 mg per 1 gram of sorbent (on this indicator has no competitors among the studied sorbates); extracts from the solution (1 gram of sorbent) of 1.5 mg of manganese; 4.0 mg ammonium nitrogen [6].

Vermiculite absorbs from the solution effective only to the manganese exchange capacity (by 1 gram of sorbent) of 1.13 mg of manganese. Thus, the investigated minerals - zeolites of the Paleozoic Urals characteristics are known (younger) zeolites of the Far East, Carpathians, Caucasus [7].

## 4. Conclusions

Presented dependencies experimentally demonstrate the existence of the optimum time`s contact, quantity of the sorbent and the initial concentration of the removed substance, for the maximum sorption`s attainment. High effectiveness was being achieved at the removing iron from water and the ammonium nitrogen, manganese zeolite-montmorillonit and clinoptilolit. It testifies to these natural materials can be used at water processing, purification of the turnaround and waste waters.

## References

[1] Germanova T.V., Valieva I.R.application of zeolites and zeolite-montmorillonitovyh of breeds of Ural Mountains for effective sewage treatment from аммнийного nitrogen and accompanying катионов//News of the Samara centre of science of the Russian Academy of Sciences. 2014. T. 16. N 1-7. P. 1828-1.

[2] Valieva I.R., Germanova T.V.ionic an exchange of natural Paleozoic zeolites under various physical and chemical conditions. Natural and engineering science. 2015. N 6 (84). P. 114-119.

[3] Edy Wibowo, Mamat Rokhmata, Sutisnaa, Khairurrijala, Mikrajuddin Abdullah. Reducing the salinity of sea water using natural zeolite (Clinoptilolite): adsorption isotherms, thermodynamics and kinetics. Desalination Volume 409, May 1, 2017, p. 146-156.

[4] H. Y. GU, J. W. K. Wong, R. D. Tyagi. Bioleaching of heavy metals from sewage sludge for land use. Modern developments in biotechnology and Bioengineering, Solid Waste Management 2017, p. 241-265.

[5] Kun-Yi Andrew Lin, Ah, Zhang-Xuan business woman, Andrew P. Jochems. Adsorption behavior methylimidazolium ionic liquids with Y-type zeolite in water: kinetics, isotherms, thermodynamics and interference. Journal of molecular liquids. Volume 232, April 2017, p. 269-276.

[6] Tinpan Man,Hong Chen,Junzhong Lin, Zhang Lin, Junliang Sun. Zeolite synthesized from alkaline helped pre-activated halloysite for effective removal of heavy metals in polluted river water and industrial wastewater. Journal of the science of the environment.

[7] Watts, H. Zhou, H. Liu, P. W. Chay, T. Kuznicki, S. M. Kuznicki. Natural zeolite clinoptilolite-phosphate composite membranes for desalination by pervaporation. Journal of membrane science, Volume 470, 15 November 2014, p. 431-438.

[8] Ayten Ates, Christopher Hardacre. The effect of various treatment conditions on natural zeolites: ion exchange, acidic, thermal and steam treatment Article Journal of Colloid and Interface Science, Volume 372 , Issue 1, 15 April 2012, Pages 130-140.

[9] Carlo Perego, Roberto Bagatin, Marco Taliabu, Rodolfo Vignola. Zeolites and related mesoporous materials for multi-purpose environmental solutions. Microporous and mesoporous materials, volume 166, January 15, 2013, p. 37-49.

IOP Conf. Series: Earth and Environmental Science **66** (2017) 012012 doi:10.1088/1755-1315/66/1/012012

# Ecological and Toxicological Characteristics of Metalworking Fluids Used in Finishing Processing in Russian Federation

S N Grigoriev[1,a], N M Bobrovskij[2,b], P A Melnikov[2,c], I N Bobrovskij[2,d], O O Levitskih [2,e]

[1]Moscow State Technological University Stankin,
Vadkovskij per.1, Moscow, Russian Federation
[2]Togliatti State University,
Belorusskaya st. 14, Togliatti, Russian Federation

*E-mail:* [a]Rector@stankin.ru, [b]bobrnm@yandex.ru, [c]topavel@mail.ru, [d]bobri@yandex.ru, [e]loo-05@mail.ru

**Abstract.** Nowadays, metalworking fluids (MWF) in the design of technological processes in most cases are considered as mandatory persistant components despite the constant improvement of the technology of machining, tools and equipment. Three main functions of MWF: cooling, lubrication, waste chips removal – seems to be the essential condition for stable process. In most cases, cooling reduces wear of tool and improves the quality of the processed surface. The cooling characteristics of the MWF affect not only the heat capacity and thermal conductivity, but metal surfaces wettability and vaporization. If processing speed and temperature of the fluid are high then it may not be in direct contact with the surface of the tool due to low wettability or vapor blankets. Improvement of machining process with applying the MWF is accompanied with negative factors. Due to the high temperatures in the treatment area it is exposed to MWF vaporization. This article presents estimation of the applicable in Russian Federation MWF: fire risk, toxicological and environmental hazards.

## 1. MWF fire risk

Modern MWF contains components that have a limit of inflammability. For example, the mechanical production assembly of JSC AvtoVAZ contains 356 units of metalworking machinery, which use about 2 thousand tons of emulsive MWF where 60% are irrecoverable loss (Table 1). MWF vapors accumulate and increase fire risk due to inadequate ventilation and air filtration (Fig. 1).

In this way fire occured in engine plant "KAMAZ" in April 14, 1993. The fire engulfed the whole enterprise in minutes, completely destroyed not only the production facility, but also almost all of the equipment. Firemans managed to eliminate fire only five days later, the latest were extinguished the burning oil pockets in the basement of the plant. According to experts, spontaneous ignition of MWF used in machining equipment was one of the most probable causes of fire.

Automatic machining line of main brake cylinder casing mechanical in assembly production of JSC AvtoVAZ caught fire in July 20, 2002. Due to the strong fire plant was given the increased number of the fire complexity, the evacuation of workers was carried out. The fire was located and extinguished two hours later by efforts of 200 people and 23 fire units. According to experts, the cause of the fire was also coolant spontaneous ignition.

Table 1 – Annual need MWF mechanical assembly production of JSC AvtoVAZ

| № | Brand name | Limit of inflammability, °C | Modulus of flow (kg) |
|---|---|---|---|
| 1 | BASIS OF LOW VISCOSITY MWF TU 38.401 188-98 | 80°C | 66767.000 |
| 2 | MWF LIQUID MP-10K TU 0253-015-27833685-2001 | 175 °C | 149028.570 |
| 3 | BASIS OF WORK LIQUID RJ-8U TU 0258-01-70015191 1-99 | 125 °C | 1034117.600 |
| 4 | LUBRICOOLANT LZ MWF1T TU 38.101.85 | 180 °C | 14252.500 |
| 5 | ROSSOIL MP TU 0258-012-06377289-98 | 158 °C | 152117.970 |
| 6 | ROSSOIL-1MIO GREASE TU 0258-013-06377289-98 | 160 °C | 252076.300 |
| 7 | ROSSOIL 23M ADDITIVE TU 0257-015-06377289-99 | 160 °C | 25693.100 |
| 8 | ROSSOIL-26MO ADDITIVE TU 0258-014-06377289-98 | 156 °C | 140641.770 |
| 9 | GREASE ROSSOIL-SHOK TU 0258-001-06377289-94 | 169 °C | 3902.439 |
| 10 | GREASE ROSSOIL-305 TU 0258-006-06377289-96 | 90 °C | 38311.000 |
| 11 | TECHNOLOGICAL GREASE ROSSOIL-503 TU 0258-018-06377289-2000 | 135 °C | 1160.200 |
| 12 | INDUSTRIAL OIL I-12A GOST 20799 | 105 °C | 180529.000 |
| 13 | MWF MP-7 TU 0258-154-06377289-01 | 180 °C | 83970.000 |
| **TOTAL** | | | **2142567.449** |

**Figure 1** – Trace of fire in industrial ventilation system

## 2. Toxicological characteristic of MWF

Modern MWF represent complex multicomponent system containing additives for different purposes, some of which may be toxic for workers. Chemical compounds that do not possess toxic characteristics may acquire them as a result of the interaction or synergy of various chemical constituents of MWF, workpiece and tool materials.

Principle routes through which MWF gets into human organism are represented on Fig. 2 [1,2,3].

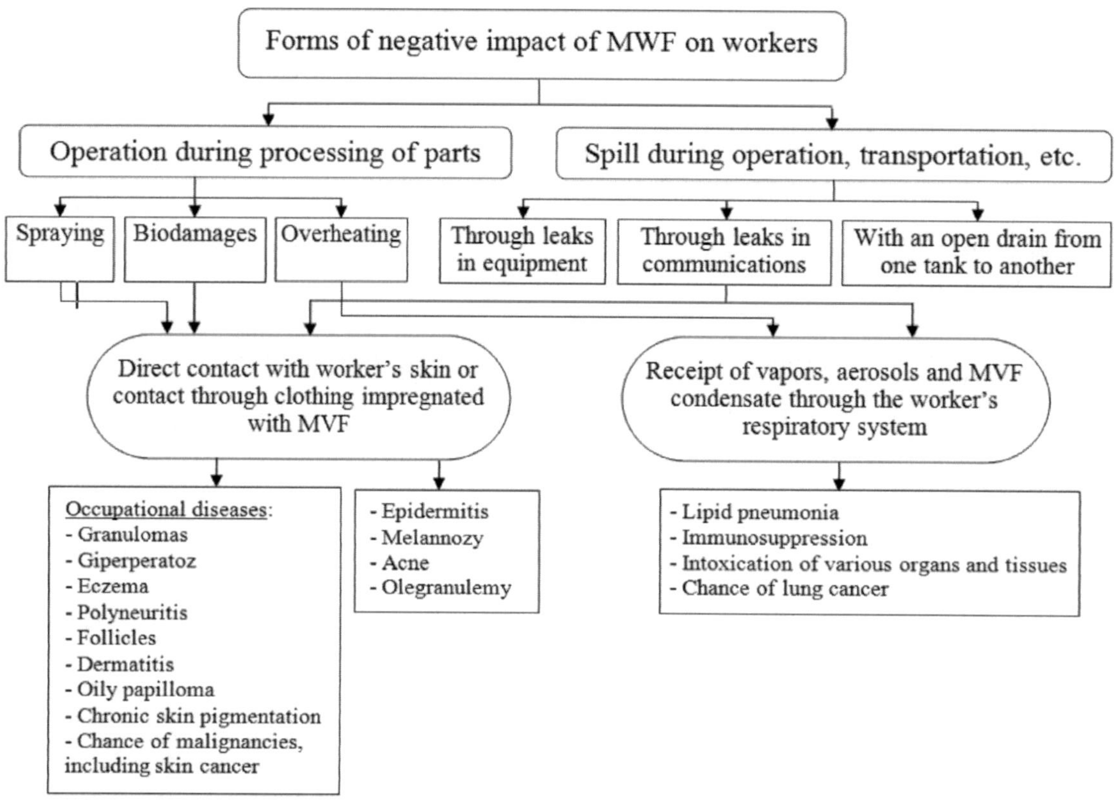


**Figure 2** – Principle ways of harmful substances in the human body and their occupational diseases

Several studies show that the average lifetime of MWF varies from two weeks to a month and a half. The MWF which consist of industrial oil, alkali, glycol, asidol, and a number of other substances presents significant danger. Moreover, the used MWF in 15-30 times more toxic than new, it is shown in [4].

Fragment of a generalized classification of MWF on toxicological characteristics which most commonly used in Russian enterprises presented in Table 2.

MWF can expose with prolonged contact with the skin and the flow of vapor (fog) through the respiratory tract. It depends on the extent and nature of cleaning MWF and various protective and ventilation devices, as well as human environment (temperature, humidity etc.).

Smoke emission and oil mist occur during processing with mineral oils and oil MWF at elevated cutting modes (in the processing of hard-machinable materials under intense spraying liquid). Flow of inhaled aerosol and volatile oil products of MWF with thermo-destruction (carbon monoxide, hydrogen chloride, hydrocarbons, sulfurous anhydride, aldehydes) may exceed the norm in 2…3 times (maximum permissible concentration (MPC) oil MWF aerosol in the air of the working area is 5 mg/m$^3$, hydrocarbon vapors – 300 mg/m$^3$). List of chemicals subject to hygiene supervision in the work area when operating the MWF may be found in sanitary regulations of Russian Federation [5].

Long work in such conditions leads to irritation of the mucous membranes of the upper respiratory tract and even the development of lipoid pneumonia, reduced total immunobiological reactivity, the change of the nervous system [6].

Table 2 – Classification of MWF on toxicological characteristics, danger class 4

| № | Brand name | Limit of inflammability, °C | Toxico-hygienic characteristic |
|---|---|---|---|
| 1 | AVITOL-1<br>TU 38.5901193-90 | - | provides irritating action on the skin, has cumulative properties |
| 2 | AKVOL-18<br>TU 38.5901134-90 | - | slight effect on skin and mucous membranes of the eyes, causing sensitizing action and has no cumulative properties |
| 3 | AKVOL-6<br>TU 38.101875-87 | - | - |
| 4 | AKVOL-15<br>TU 38.5901162-89 | - | No irritating action on the mucous membranes of eyes.<br>Has no locally-irritating, skin resorbtive and allergic properties. |
| 5 | AKVOL-18<br>TU 38.5901134-90 | - | Has no skin resorbtive and cumulative effect, slight effect on skin and mucous membranes of the eyes |
| 6 | SINTAL-2<br>TU 38.1011022 | - | Has no action on the mucous membranes of eyes and no sensitizing effect,<br>has skin resorbtive effect. |
| 7 | SINHO-2M<br>TU 38.1011060-86 | - | No effect on skin and mucous membranes of the eyes, has no skin resorbtive effect, no sensitizing effect. |
| 8 | SN-C<br>TU 38.41097-81 | 170 | slight effect on skin and mucous membranes of the eyes |
| 9 | UVEROL<br>TU 38.5901183-90 | 139 | slight effect on skin and mucous membranes of the eyes |
| 10 | UKRINOL-5/5<br>TU 38. 101389-79 | 150 | no effect on mucous membranes of the eyes, no sensitizing effect |
| 11 | UKRINOL-5OU<br>TU 101978-84 | - | slight effect on skin and mucous membranes of the eyes, no cumulative and sensitizing effect |
| 12 | UKRINOL -202 | 75 | Has no skin resorbtive and sensitizing effect |

With prolonged contact with oil MWF workers can get professional skin lesions (dermatosis): folliculitis decalvans, occupational hyperkeratosis, papillary tumor, pigmentation, xerodermia and desquamation [7]. Occurrence and progression of disease are determined by individual predisposition and presence of skin micro trauma (scratch, abrasion, irritation overalls). The most frequent localization of dermatosis is on the hands, especially on the right.

Emulsion MWF containing special additives may cause damage to the fat grease skin redness and appearance of nodules, vesicles, crusts (dermatosis). Synthetic MWF containing some kinds of soaps can cause maceration of the skin (stratum corneum, softening the appearance of cracks and abrasions). The water miscible MWF causes sharp skin dryness, peeling, moderate painful cracks.

Intensive evaporation of water during water base MWF operation leads to an increase in the concentration of vapor in the working area, as well as the concentration of additives in the MWF, exceeding the hygienic standards. With prolonged exposure of emulsions with sodium nitrite content over 0.2% workers have headaches, fatigue, loss of appetite, poor sleep, pain in the extremities as a result of the violation of gas exchange in the body, abnormal hemoglobin in blood. Thereethanolamine have an allergic effect on the organism and causes dermatitis. The emergence of contact dermatitis is also promoted by chromium, nickel, cobalt, which possible in MWF when processing alloyed steels and alloys [8, 9].

An important factor in the sanitary-hygienic condition of the MWF is the degree of microorganisms pollution rate. Microflora of MWF can cause secondary infection of follicles and micro trauma exacerbates skin maceration; as a result of vital activity of microflora in the MWF may occur also toxic substances. Bacterial and fungus microflora develops in all kinds of water mixable MWF when they are stored or used. Bacteria prevail in emulsion MWF, yeast and fungi - in synthetic MWF. Semi-synthetic liquid occupies an intermediate position. Different measures (cleaning, introduction of biocides, etc.) which carried out during the operation can significantly prolong the life of the MWF.

Modern MWF before introduction in industry receive comprehensive laboratory toxicological evaluation and sanitary-hygienic studies in production environments. Therefore, in accordance with the sanitary rules [5] to industrial application are suitable only MWF with permission of the health authorities. However, elimination of MWF adverse effects on the workers can be achieved only through the implementation of preventive measures and measures of individual protection.

## 3. MWF – ecological risks

Mechanical engineering industry ranked fourth among hydrosphere pollutants after steel, chemical and cellulose-paper industry. Out of the total drain, equal more 7 $km^3$/year, only slightly more than half passes through sewage treatment plants, which usually do not meet modern environmental requirements. Proportion of so-called greasy water (wasted MWF, leakage of lubricating systems and oil storage) is 40...60% of plant total drain. Their characteristics are presented in Table. 1. Spilling, splashing, losses from chips and cleaning material, simply MWF drain to sewage system lead to the contamination of soil, water and air. Thereby, permanently lost portion of the components (including water), which could be extracted and reused for making new batch of MWF or for other purposes.

MWF lead to pollution of surface and subsoil on industrial site of the enterprise and potentially has a number of dangerous and difficult liquidating consequences. MWF in the soil generates multiple types of geological pollution:

- soil;
- zone of suspended water;
- surface of underground water.

Soil pollution is formed primarily in the areas of the MWF opening – the primary contamination. Soil is the most heavily polluted by heavy and viscous oil products of MWF, incapable of deep penetration in subsoil. Secondary pollution occurs in some cases as a result of the rise in groundwater levels [10].

Under the influence of petroleum products increasing number of water-stable soil particles bigger than 10 mm, aggregation of soil particles occurs. Coarse particles content increases, and the content of agronomically valuable small particles decreases. Petroleum-rich soil loses the ability to absorb and retain moisture [5].

Recovery (self restoration) period of soil, polluted by MWF, ranges from 1-2 to 10-15 years.

In machining process with using MWF under the influence of high temperature in the treatment area may form aerodisperse systems based on aerosols MWF, that further enhances their migration ability [3]. Figure 3 shows the analysis of possible migration ways of MWF aerosols.

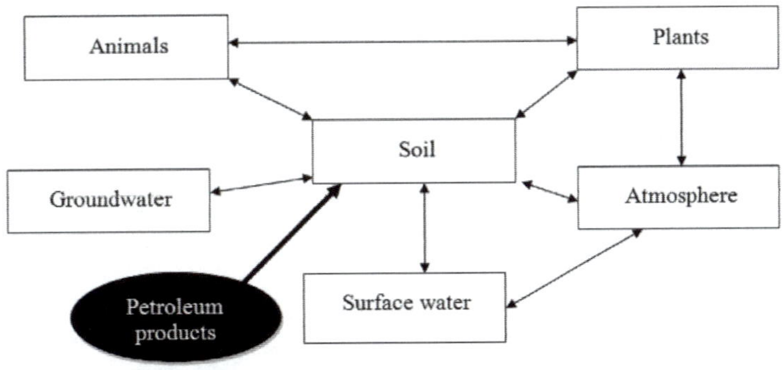


**Figure 3** – Ways of MWF migration leading to soil pollution

The presence of 2 g of petroleum products in 1 kg of soil make it uninhabitable for plants and soil microflora. Petroleum derivatives influence almost all groups of soil invertebrates. The large species (insects, worms) die quickest. Lowering of oxygen concentration in soil contributes to the development of anaerobic microorganisms and aerobic microflora braking. Initially, even the weak soil contamination by petroleum products leads to a decrease the number of soil microorganisms. Restore strength observed a few months after contamination, sometimes even further growth of microorganisms through the use of carbon-mineral oil as a nutrient. However, the intensive growth of microorganisms, with soluble compounds, strongly impoverishes the soil nitrogen and phosphorus compounds. Soil contamination by petroleum products creates a new environment with the appropriate number of organisms in soil. A common feature of all oil-contaminated soils - limited species and ecological diversity pedobionts [4].

## 4. Conclusions

On the basis of the foregoing, for environmental friendliness and the expected passing of economic effect in the case of "dry" (without traditional MWF) machining many companies are actively engaged in developing the theoretical foundations and practical ways to implement processing without the use of MWF.

The experience of global engineering shows, that increasing legislation and tax sanctions, aimed at protecting the environment, technology's that minimize the use of MWF, becoming more popular.

Currently used MWF have a negative impact not only on workers health, but also the ecosystem. Research of ways to optimize MWF work is relevant along with minimizing of their use in view of the increasing influence of thermal factors on machining.

One of the modern approaches is to replace the MWF by applying the special coatings or structured workpiece and tool surfaces.

## References

[1] Maximum permissible concentration (MPC) of harmful substances in the air of the working area. Hygienic standards. Information on: http://www.gosthelp.ru/text/GN22568698Predelnodopusti.html

[2] N.M. Bobrovskij, I.N. Bobrovskij, A.V. Ezhelev, P.A. Melnikov The surface-plastic deformation technology for parts processing without the use of lubricating and cooling technological means [Tehnologija obrabotki detalej poverhnostno-plasticheskim deformirovaniem bez primenenija smazyvajushhe-ohlazhdajushhih tehnologicheskih sredstv], Samara Scientific Center RAS, Samara, 2012. ISBN: 987-5-93424-598-7.

[3] General hygiene requirements for working zone air. Information on: http://www.generent.ru/files/1216182858.pdf

[4] A.V. Vasiliev, L.R. Hamidullova Analysis of negative influence of lubricating-cooling liquids to the man and to biosphere and of methods of it reduction, Vector of TGU Science. 4 (2011) 45-49.

[5] L.S. Janovskij, A.A. Harin, I.V. Shevchenko, V.P. Dmitrenko Aviation ecology. Impact of aviation fuels and lubricants on environment, INFRA-M Publishing House, Moscow, 2015. ISBN: 978-5-16-010830-8.

[6] P.A. Melnikov, N.M. Bobrovskyi, A.N. Popov, D.V. Gussarova Features education aerodisperse systems when using coolants in machine-building enterprises, Vector of TGU Science. 3 (2011) 87-91.

[7] E. Brinksmeier, D. Meyer, A.G. Huesmann-Cordes, C. Herrmann Metalworking fluids — Mechanisms and performance, CIRP Annals - Manufacturing Technology. 64 (2015) 605–628.

[8] K. Weinert, I. Inasaki, J.W. Sutherland, T. Wakabayashi Dry Machining and Minimum Quantity Lubrication. Annals of the CIRP — Manufacturing Tech-nologies 53 (2004) 511–537.

[9] E. Brinksmeier, C. Heinzel, M. Wittmann Friction, Cooling and Lubrication in Grinding. Annals of the CIRP — Manufacturing Technologies. 48 (1999) 581–598.

[10] N. Bay, A. Azushima, P. Groche, I. Ishibashi, M. Merklein, M. Morishita, T. Nakamura, S. Schmid, M. Yoshida Environmentally Benign Tribo-systems for Metal Forming. Annals of the CIRP — Manufacturing Technologies. 59 (2010) 760–780.

IOP Conf. Series: Earth and Environmental Science **66** (2017) 012013 doi:10.1088/1755-1315/66/1/012013

# Research of Tool Durability in Surface Plastic Deformation Processing by Burnishing of Steel Without Metalworking Fluids

**S N Grigoriev [1,a], N M Bobrovskij[2,b], P A Melnikov[2,c], I N Bobrovskij[2,d]**

[1]Moscow State Technological University Stankin
Vadkovskij per.1, Moscow, Russian Federation
[2]Togliatti State University
Belorusskaya st. 14, Togliatti, Russian Federation

*E-mail:* [a]Rector@stankin.ru, [b]bobrnm@yandex.ru, [c]topavel@mail.ru, [d]bobri@yandex.ru

**Abstract.** Modern vector of development of machining technologies aimed at the transition to environmentally safe technologies - "green" technologies. The concept of "green technology" includes a set of signs of knowledge intended for practical use ("technology"). One of the ways to improve the quality of production is the use of surface plastic deformation (SPD) processing methods. The advantage of the SPD is a capability to combine effects of finishing and strengthening treatment. The SPD processing can replace operations: fine turning, grinding or polishing. The SPD is a forceful contact impact of indentor on workpiece's surface in condition of their relative motion. It is difficult to implement the core technology of the SPD (burnishing, roller burnishing, etc.) while maintaining core technological advantages without the use of lubricating and cooling technology (metalworking fluids, MWF). The "green" SPD technology was developed by the authors for dry processing and has not such shortcomings. When processing with SPD without use of MWF requirements for tool's durability is most significant, especially in the conditions of mass production. It is important to determine the period of durability of tool at the design stage of the technological process with the purpose of wastage preventing. This paper represents the results of durability research of natural and synthetic diamonds (polycrystalline diamond - ASPK) as well as precision of polycrystalline superabrasive tools made of dense boron nitride (DBN) during SPD processing without application of MWF.

## 1. Introduction

The most important factor in making decisions about the advisability of the appointment of material as a tool for the SPD is a durability, as well as the optimal processing conditions.

Accelerated determination of tool's durability in burnishing processing was carried out closely to plant conditions. Previously, the authors investigated the tool durability for wide burnishing steel 45 samples of high-duty cast iron [1, 2] and this study addresses the processing of chromium-manganese steel 17HG (C: 0.16-0.21%; S: 0.015-0.030%; P: <0.035%; Mn: 0.90-1.10%; Cr: 0.70-0.90%; Si: 0.17-0.37%; Ni: <0.30%; Cu<0.30%), austenitic grain at most 5 micron, snowflake sensitive. This steel is intended for the production of small parts with cementing sections running on friction.

As a criterion for tool wear was used a surface's roughness tolerable limit Ra = 0.4 μm. This value is the limit of the technical requirements for surface: lip seal necks of crankshafts (Fig. 1).

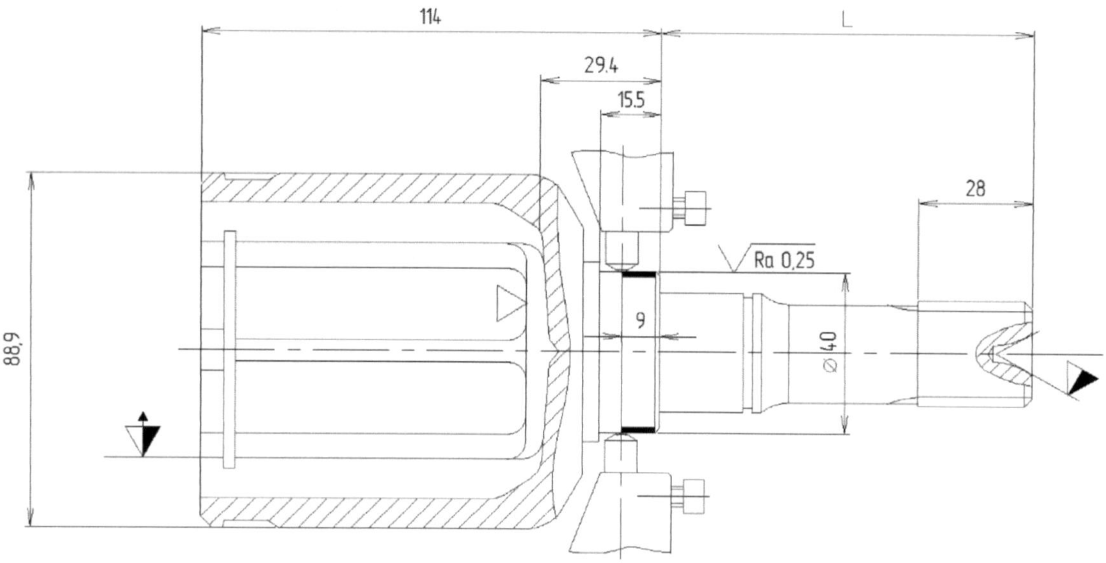


**Figure 1** – Processing scheme of crankshafts

## 2. Experiments

As a basis for experiments was used SPD machine for differential side joint housing.

Process-dependent parameters:

- work surface hardness HRC 58…62;
- basic surface roughness Ra = 0.6…0.8 μm;
- roughness after processing Ra = 0.2…0.25 μm;
- number of simultaneous working tools: 2;
- tool design parameters: radiused;
- tool radius: 2.5 mm;
- tool body roughness Ra = 0.06…0.07 μm;
- actuator 150…250 N;
- spindle rotational speed 540 rpm;
- feed motion 0.1 mm/rev (0.05 mm/rev on each tool) ;
- feed and force tool actuator: fluid;
- actuator system (feed and stroke): fluid;
- cycle: 32 sec;
- production level (duty factor 0.8): 90 pieces/hour

Machining conditions:

- speed (V): 67.9 m/min;

- feed (S); 0.1 mm/rev;
- revolutions (n): 540 rev/min;
- tool's force (P): 175 N.

**Figure 2** – Industrial machine with installed work piece

## 3. Object of research: tool materials

As tool materials were used: natural diamond, synthetic polycrystalline diamond (ASPK), and DBN - polycrystalline superhard materials (SHM) with high (up to 99.5%) contents of dense boron nitride (cubic - CBN and vurcit - VBN) with (particle size 10–100 μm) [3].

**Table 1** – Tool material's properties

| Tool material | Natural diamond (ND) | Synthetic polycrystalline diamond (ASPK) | Dense boron nitride (DBN) [4] |
|---|---|---|---|
| density, $kg/m^3 \cdot 10^3$ | 3.47 - 3.55 | 3.5 - 4.0 | 3.411 - 3.482 |
| microhardness, $MPa \cdot 10^2$ | 931.57 - 986.48 | 800 - 1000 | 900 - 950 |
| elastic module, GPa | 930 | 850 | 900 |
| ultimate compressive strength, GPa | 2.00 | 0.4 - 0.8 | - |
| thermostability, °C | 800 - 1000 | 600 - 700 | 1300 |
| crack growth resistance, MPa m(1/2) | 6.5 - 8.5 | - | - |
| Cost relation of prises,% | 200 - 300 | 100 | 150 |

## 4. Experiment results

First stage: synthetic polycrystalline diamond (ASPK), results presented on Fig 3. Tool durability 58 750 meters, that corresponds to, for example, 5200 pieces of differential side joint housing of LADA auto.

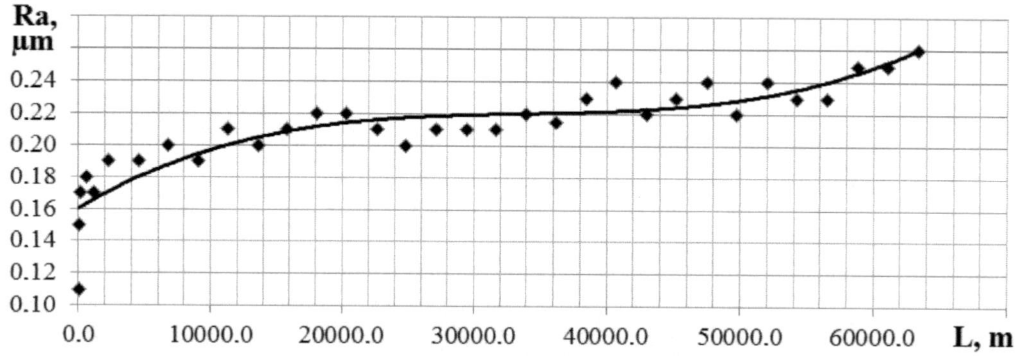


**Figure 3** – Dependence of traversed path of burnisher made of ASPK on processed surface roughness in burnishing processing without metalworking fluids of differential side joint housing gland necks from chromium-manganese steel (surface hardness HRC 63)

Second stage: natural diamond. Tool durability: 70 000 meters, that corresponds to, for example, 7000 pieces of differential side joint housing of LADA auto (Fig. 4).

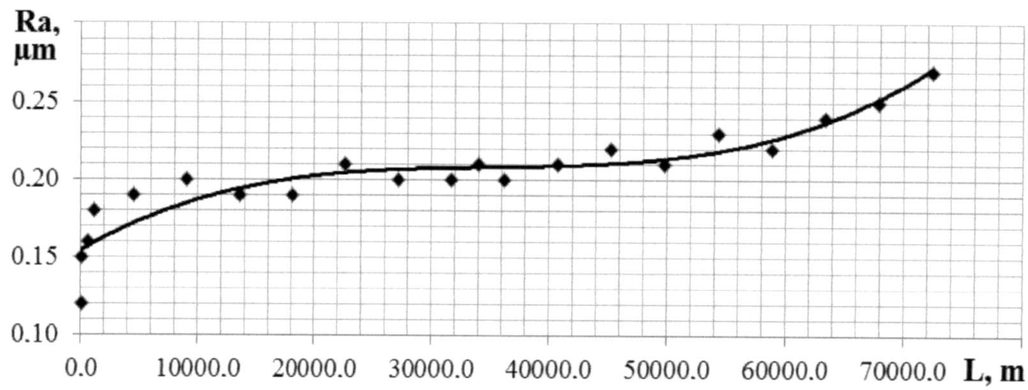


**Figure 4** – Dependence of traversed path of burnisher made of natural diamond on processed surface roughness in burnishing processing without metalworking fluids of differential side joint housing gland necks from chromium-manganese steel (surface hardness HRC 63)

Third stage: dense boron nitride (DBN) [5]. Tool durability in this case: less than 28000 meters, which corresponds to, for example, 2500 pieces of differential side joint housing of LADA auto (Fig. 5). In mass production with annual program more than 350 000 pieces per year the result is not satisfactory.

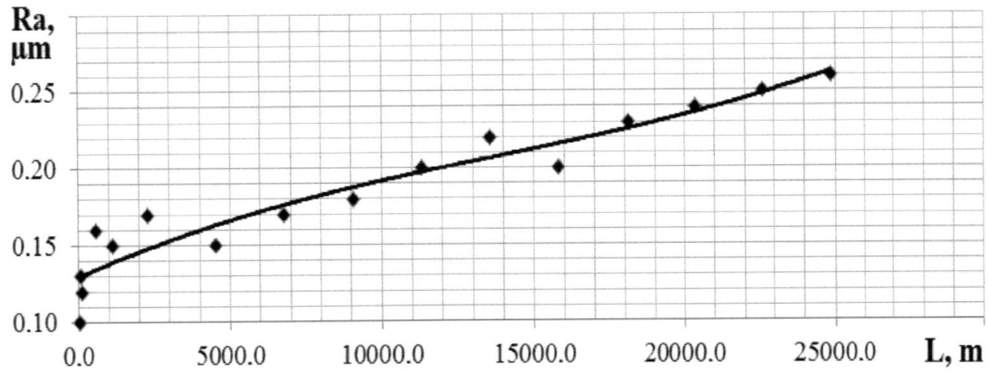


**Figure 5** – Dependence of traversed path of burnisher made of DBN on processed surface roughness in burnishing processing without metalworking fluids of differential side joint housing gland necks from chromium-manganese steel (surface hardness HRC 63)

## 5. Results and Discussion

Despite the fact that the durability of natural diamond tools for 30% bigger, than synthetic polycrystalline diamond, it is economically sound to use synthetic polycrystalline diamond because its price lesser than natural diamonds for 200-300%. It should be noted that natural diamonds are more prone to chip during operation and instability.

Tools made of dense boron nitride (DBN) showed the lowest durability. Taking an ASPK for base (100%), natural diamonds has shown 120% durability, and DBN has shown 47%.

Probably with different set of machining conditions (high work surface hardness 58…62 HRC is the main reason for DBN failure) DBN could be used as material for tool working part. Purpose of this study was to try different tool materials in constant machining conditions, which corresponds to plant conditions. Tool force (150…250 N) could not be lowered enough to make DBN work longer because than smoothing function of burnishing will not work properly [6] and designed roughness after processing (0.2…0.25 µm) will not be achieved [7]. Combination of significant hardness of work surface, absence of MWF (high heat stress), and designed roughness leads to exclusion of DBN under these conditions of work [8].

It is hard to compare this results to other papers because there are no similar studies data available on DBN [9] and durability test are rare because of their cost, as shows in newest survey article on burnishing where analysis of more than 60 sources with range for 13 years [10].

## 6. Conclusion

1. Burnishing without using of MWF is cost effective with using of synthetic polycrystalline diamond, despite of lesser thermostability than natural diamond; cost of unit surface area machining of tools with ASPK lesser than natural diamond for 20%.

2. DBN can be used as hard turning tool material, but to use it with SPD processes tools needs to have a coating. If it is not acceptable then adjusting machining conditions could be useful.

## References

[1] F.V. Grechnikov, A. Khaimovich The study of plastic deformation at high strain rates in upset forging of cylinders, Key Engineering Materials, 684 (2016) 74-79.

[2] I.N. Bobrovskij, S.N. Grigoriev, P.A. Melnikov, N.M. Bobrovskij Research Of Hardalloyed Burnishing Tool Durability With Coatings By Ion-Plasmous Sputtering Method, Applied Mechanics and Materials, 770 (2015) 274-278.

[3] Precision polycrystalline superabrasive cutting tools made of Dense Boron Nitride (DBN), information on: http://www.microtechnika.ru/modules/catalog/index.php?categoryid=15
[4] S.N. Dub, I.A. Petrusha Mechanical properties of polycrystalline cBN obtained from pyrolytic gBN by direct transformation technique, High Pres. Res., 26(2) (2006) 71–77.
[5] N.M. Bobrovskij, P.A. Melnikov, S.N. Grigoriev, I.N. Bobrovskij, Simulation of thermal fields using different types of wide burnishing, IOP Conf Series: Materials Science and Engineering, 91 (2015) 012034. Information on http://iopscienceioporg/article/101088/1757-899X/91/1/012034 DOI: 101088/1757-899X/91/1/012034.
[6] V.P. Kuznetsov, A.S. Skorobogatov, V.G. Gorgots, A.S. Yurovskikh, The analysis of speed increase perspectives of nanostructuring burnishing with heat removal from the tool, IOP Conference Series: Materials Science and Engineering, 124(1) (2016) 012127. Information on: https://www.scopus.com/inward/record.uri?eid=2-s2.0-84975745777&partnerID=40&md5=d172fb1942804f2147dedaf58b28ecaa DOI: 10.1088/1757-899X/124/1/012127.
[7] V.P. Kuznetsov, A.I. Dmitriev, G.S. Anisimova, Yu.V. Semenova, Optimization of nanostructuring burnishing technological parameters by Taguchi method, IOP Conference Series: Materials Science and Engineering, 124(1) (2016) 012022. Information on: https://www.scopus.com/inward/record.uri?eid=2-s2.0-84975706896&partnerID=40&md5=c9fb4cd34a1ffcdaa0d9521d84324eb6 DOI: 10.1088/1757-899X/124/1/012022.
[8] V.M. Bushlya, O.A. Gutnichenko, J.M. Zhou, J.-E. Ståhl, S. Gunnarsson Tool wear and tool life of PCBN, binderless cBN and wBN-cBN tools in continuous finish hard turning of cold work tool steel, Journal of Superhard Materials, 36(1) (2014) 49-60.
[9] L. Tanovic, P. Bojanic, R. Puzovic, S. Klimenko Polycrystalline Cubic Boron Nitride (PCBN) Tool Life and Wear in Turning of Amorphous-Crystalline Iron-Based Coatings, Journal of Mechanical Engineering, 57(12) (2011) 904-910.
[10] A. Verma, D. Mahato ROLLER BURNISHING-A Literature Review of Developments and Trends in Approach to Industrial Application, Asian Journal of Engineering and Technology, 1(3) (2013) 68-83. ISSN: 2321–2462.

IOP Conf. Series: Earth and Environmental Science **66** (2017) 012014 doi:10.1088/1755-1315/66/1/012014

# Technological Desition of Extraction of Melanin from the Waste of Production of Sunflower-Seed Oil

**Yu N Kartushina, E E Nefedieva, G A Sevriukova, N V Gracheva, V F Zheltobryukhov**

Volgograd State Technical University
pr. V.I. Lenina 28, Volgograd, 400005 Russia

*E-mail:* kartysina@rambler.ru

**Abstract**. The research was realized in the field of the technology for re-use of waste of sunflower-seed oil production. A technological scheme of production of melanin from sunflower husk as a waste was developed. Re-cycling will give the opportunity to reduce the amount of waste and to obtain an additional source of income.

## 1. Introduction

Attention is paid to the problems of chemical and biotechnological utilization of waste of agricultural and food production, especially to the treatment of plant biomass. It is the necessity to utilize big volume of wastes and to get useful secondary materials with new properties. Sunflower-seed oil production is based on a technology with large volume of wastes of 5th degree of dangerous such as shells of sunflower seeds (husk). It consists of 100% of plant residue, so there are no dangerous properties. However, it is a problem of utilization of the waste by burial or combustion. There is a possibility of littering of enterprise territory or ignition in case of insufficient storage and transportation of big volumes of husk.

The following methods such as use of the husk as a part of the forage or mat in animal industry as well as an addition to compost for the growing of mushrooms, and for improvement of mechanical properties of soil are the most prevalent. Husk is known to be used for the production of decorative tiles, as a raw material for hydrolysis production, for alternative fuel, as a sorbent for the treatment of contaminated soils, for nutrient medium, humectants material, and lignocellulose raw material [1, 2, 3].

Despite the wide variety of directions of use of the husk of sunflower the real potential of the use of that substance did not answer the volumes of enterprises and the dumps which are accumulated. That wastes are difficult for access for microbial destruction. They contaminate soil, they need additional squares for storage, so it seems an ecological problem for the environment. Problems listed above is a background for the elaboration of new approaches for the re-cycling of the named waste.

Sunflower husk (seed shells) is a perspective feedstock raw material used in the manufacture of antioxidant products [4, 5]. Antioxidant properties depends on the content of phenolic polymers both of low-molecular and high-molecular nature. Those compounds are melanins and humic substances.

Low-molecular phenolic substances have following unique properties: they are able to protect cells from primary reactive oxygen species as well as from secondary reactive oxygen species such as lipid peroxides. In addition, they are able to associate ferric and copper ions which activate peroxidation. However, their content in sunflower seed shells is low, that fact specifies the economic inexpediency of their separate extraction.

Melanins and humic substances are phytogenous or mycogenous high-molecular compounds. They have high stable paramagnetic centers as well as variety of functional groups. They are characterized by conjugate system of bonds in molecules. Their unique property is steady free radical state. Monomers of melanin are appeared as phenoxil or semi-quinone radicals. In oxidative reaction that natural polymers react not only in reduced hydroquinone form of phenols, but as a system of polyphenols and quinones which has semi-quinone radical as an obligatory intermediate component. Those compounds promote the high antioxidant activity of melanins. Also melanins are chelating agent against ferric and copper ions which initiate radical reactions.

Content of black melanin and humic melanin is up to 1.5% [1] and 12% [6] correspondingly.

Economical expedience of processing of this type of husk raw materials requires the development of methods for the simultaneous extraction of compounds belonging to different classes, such as melanins and humic substances, for their further use as antioxidants complex with high activity.

Now the most perspective method of the recycling of sunflower husk is extraction of melanin. Melanins which were got from the husk of sunflower are nitrogen-free compounds. They generate in the process of oxidation of pyrocathechin by enzymes. That melanins have high antioxidant and antiradical activity. That property gives the possibility of production of antioxidants, inhibitors of radical reactions, sorbents, biostimulants, and antiaging agent using melanins from the husk of sunflower [7].

## 2. Materials and methods

Chemical and element composition of husk was researched. Cross-sections of sunflower seed shells were prepared by standard anatomical method. Sections were impregnated by different colorants. So sudan was used to identify lipids, floroglucin was used to identify lignificated elements, and lugol solution was used to identify starch. Photographs were made by light microscope MX 300 with digital camera DCM 900 and computer program Scope Photo 3.0. Element compound was assessed by scanning electron microscope FEI Versa 3D LoVac with energy-dispersion detector EDAX Apollo X for micro-roentgen-spectroscopic analysis.

Biochemical investigations of final product were also carried out for assessing the rate of the lipid peroxidation. The substrate of the lipid peroxidation was suspension of chicken egg yolk. It contains two types of lipid-protein complexes which correspond to low and very low density lipoproteins of blood plasma. Out of all types of lipids, those lipoproteins are known to primarily affected by lipid peroxidation.

Described model system has the following advantages: commercially availableness, easy of lipoprotein extraction, stability during the storage and biochemical oxygen demand.

Chicken egg yolk was dried on filter paper and then it was solved in equal volume of phosphate buffer (40 мM $KH_2PO_4$ + 105 mmol KCl, pH 7.5). The obtained phospholipid suspension was diluted 25-fold with the same buffer before use. Suspension may be stored for a week at 4°C.

The tests were realized by the following way. To the sample (volume 0.5 ml), 1 ml of lipoprotein suspension and 3 ml of 1% solution of orthophosphoric acid (pH 2.0) were added. Lipid peroxidation in the samples was initiated by addition of 0,1 ml solution of ferric sulfate (30 mg of $FeSO_4 \cdot 7H_2O$ in 10 ml of distillated water). Control sample did not contain antioxidants. Intimate mixtures were incubated in water thermostat for 20 min at 37°C.

Rate of lipid peroxidation was assessed by the amount of accumulated TBA-products (malonic dialdehyde) which reacts with thiobarbituric acid – TBA. After the incubation in each sample 1 ml of 0.6% solution of TBA was added. Intimate mixtures were heated in boiling bain-marie for 20 min, periodically stirring till it became of pink color. Then 4 ml of n-butanol was added to the mixture, samples were centrifuged for 10 min at 3000 rpm. Optical density of the butanol phase was assessed at 540 nm by photoelectrocolorimeter. The path length of cell was 0.5 cm. Butanol was used as a control.

Concentration of accumulated TBA-products was maid taking into account the coefficient of molar extinction of malonic dialdehyde which was $1.56 \cdot 10^5$ моль·см$^{-1}$:

$$A = E_{op} \cdot 106 \cdot 4 \text{ ml} / 1.56 \cdot 10^5 \cdot 0.5 \text{ ml} = E_{op} \cdot 51.3 , \qquad (1)$$

if A is content of malonic dialdehyde (μmol/l or nmol/ml); 4 ml is volume of butanol phase; 0.5 ml is volume of sample; Еоп is optical density in samples and in control.

Antioxidant activity (AOA, %) was calculated using formula:

$$AOA (\%) = (E \text{ control} - E \text{ sample} / E \text{ control}) \times 100, \qquad (2)$$

if E control and E sample are optical densities in samples and in control.

## 3. Results and discussion

The phytomelan layer is situated in the husk of the sunflower; it protects the seed from parasites. Husk of cultivars of sunflower differs by their color. They can be white, grey or black.

Husk was colored by sudan and floroglucinol due to the presence of lipids (fig. 1a) and lignin (fig. 1b) correspondingly. There was no starch in husk (fig 1c) because they were not colored by lugol. The content of lignin is known to be 25.1% … 29.6%. Content of lignin depends on the oil content of the sunflower seeds; the more of oil the more of lignin.

a b c

**Figure 1** – The sections of husk impregnated by sudan, floroglucinol and lugol solutions

**Table 1 –** Element content of the husk of sunflower

| Color of husk | Element | | | | | |
|---|---|---|---|---|---|---|
| | C | O | Na | Mg | K | Ca |
| Black | 64.36±3.7 | 32.8±3.1 | 0.27±0.09 | 0.57±0.17 | 1.03±0.31 | 1.02±0.25 |
| White | 62.01±2.0 | 32.1±3.7 | – | 1.37±0.16 | 1.43±0.51 | 2.83±0.92 |

The table 1 demonstrates that organic substances consists of only carbon and oxygen (hydrogen was not assessed by this method). So there was no nitrogen in those substances. There were no sufficient differences in contain of carbon and oxygen, but in white husk the content of mineral elements was more than in black ones.

The research of components of husk evidences that the useful components can be extracted from the material.

Searches of the optimal way of extraction of melanin from the husk of sunflower were realized at the laboratory. There is a patent for invention [8].

Due to melanins and humic substances as natural high-molecular compounds are slow-diffusible substances via cell walls, their extraction is a time-consuming process that entails significant intensification. The extraction of melanins from the raw materials is carried out preferentially at an elevated temperature and under the external power influences [9-12]. Despite the effectiveness of those ways of intensification their application to industrial production is not practicable due to high power consumption. As concerns economic expediency a perspective way of intensification of the process of extraction of melanins is vibrating influence [6, 8, 13]. It is possible to extract melanins from husk under standard conditions.

Duration of the process of extraction in vibrating extractor allows increasing of hydrodynamic conditions in large voids of raw material and decreasing of the retention time. Initiated convection promotes the increase of the rate of mass carry at the stage of diffusion and convection extraction. Vibration reduces dead diffusion space in extractor and enhances the coefficients of mass exchange. Experimental and calculated data of extraction of antioxidant are shown in table 2.

Elaborated method provides the increase of output of melanoid antioxidant on 40-170% and 27-98-time decrease of power consumptions.

Use of alkali solution as an extractant increases the solubility of melanins and humic substances, so this effect would improve the quality of the antioxidant and extend the output of the final product. Also the disintegration of insoluble complex of biopolymers such as cellulose-lignin-hemicellulose arises from the treatment of raw material by alkali solution. Use of sodium hydroxide solution as an extractant allows realizing the process at room temperature18-22 °C without pre-heating.

**Table 2** – Dependence of output of the final product on the parameters of extraction

| Variant | Extractant | Extraction time, min | Mass ratio husk: extractant | Power consumption for extracton of 1 kg of raw material, kW·h | Output of the final product, % of the mass of raw material |
|---|---|---|---|---|---|
| in accordance to **Ошибка! Источник ссылки не найден.**] | water | 55 | 1:10 | 12.76 | 3.55 |
| in accordance to [14] | water | 30 | 1:10 | 6.96 | 2.62 |
| 1 | 0.5M NaOH | 30 | 1:7 | 0.13 | 7.04 |
| 2 | 0.1M NaOH | 45 | 1:7 | 0.20 | 5.05 |
| 3 | 0.5M NaOH | 45 | 1:7 | 0.20 | 8.67 |
| 4 | 0.1M NaOH | 60 | 1:7 | 0.26 | 5.98 |
| 5 | 0.5M NaOH | 60 | 1:7 | 0.26 | 9.67 |

Appropriate technology of production of melanin was elaborated and confirmed experimentally (fig. 2).

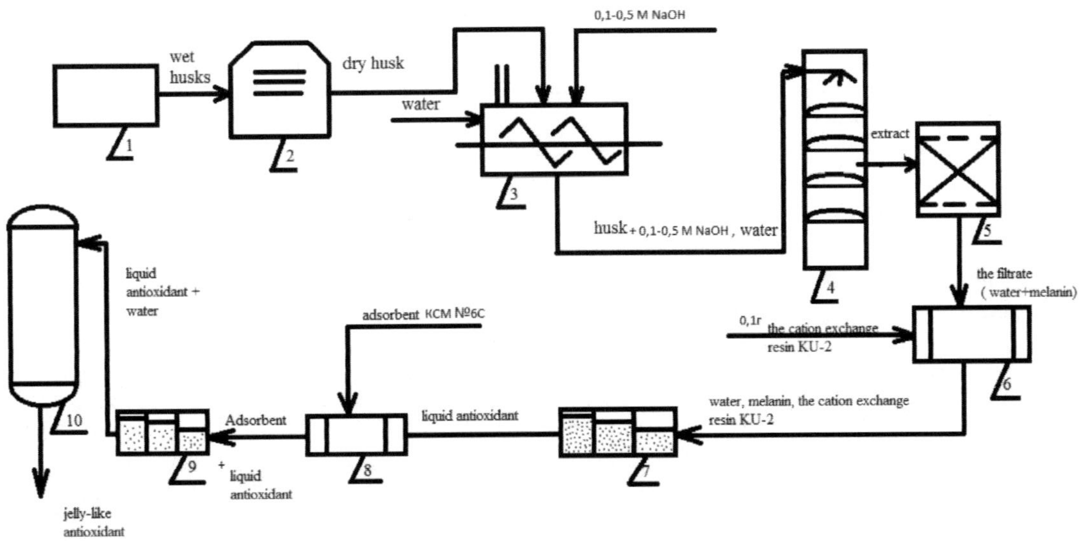


**Figure 2** – Technological scheme of the process of extraction of melanin from the husk of sunflower: 1 – bunker; 2 – lattice drying chamber; 3 – mixer; 4 – vibrating extractor; 5 – filter with setted bed; 6, 8 – tanks; 7, 9 – settlers; 10 – evaporation installation

Because of the necessity to extract some estimable components from solid phase of sunflower husk we used leaching as a main step of the technology. This process allows getting melanin which is useful for different branches of industry.

The process is realized in conditions of high temperature, heat supply and intensive mixing. We propose to use 0.1-0.5 M solution of sodium hydroxide. Taking in account all the peculiarities of the process we change an extractor. It is an apparatus equipped by oscillating drive component. The vibration process was used to intensify the process of extraction. The extractor has nickel coating to protect the apparatus in alkalified medium. Nickel coating has high hardness and adhesion to the base part especially after the thermal treatment [15]. Nickel coatings are used in different branches of industry both as a back layer and all by itself for protective, decorative and special purposes. That coatings are characterized by hardness, corrosion stability, good reflecting power, and resistivity $8.3 \cdot 10^{-2}$ Ohm·m [16]. Those coatings are applied for protective and decorative finish work of machine elements, instruments, apparatuses, and devises in industry. Also nickel coatings protect the equipment in severe corrosive conditions such as high temperature or special medium – alkalis and acids. It can be used as a back layer to coating of steel by other metals, it provides rigid adherence of coatings with the base. Nickel coating enhances the hardwearing of rubbing surfaces.

The process of leach needs some additional operations. Grinding husk is washed by water, then it follows to the bunker (1). Than it is dehumidified to free-flowing condition by hot air which has temperature 90-110oC in lattice drying chamber (2). The mass is mixed with 0.1-0.5 M solution of sodium hydroxide in mixer (3). The mass ratio of husk and extragent is 1 : (6÷7). The leach process is realized in vibrating extractor (oscillating drive component) (4).

After extraction the finished product is prepared by filtration of extract on the filter with scuttled bed (5). Filtrate and 100 mg of cation exchange resin KU-2 are mixed in a tank (6) under control of pH.

When pH runs to 7.0 the filtrate is separate from cation exchange resin in a settler (6). Then supernatant is baffled to tank (7), and adsorbent is added to the mixture. Adsorbent is precipitated in a settler (8). Dissolvent (water) is evaporated by boiling treatment in evaporation installation (10) [13].

Bag filter is recommended to provide high efficiency of filtration. Bladed rotary pump is preferable to realize continuous and uninterruptable flow of solutions.

Final product is dark-brown gel-like substance, it has sweet woody smell. It has antioxidant properties which connected with the presence of melanin. It enters into reactions with hydroperoxide, and the solution is discolored. The product reacts with potassium permanganate, the color changes into green, later the residue precipitates and the solution is discolored. The reaction of the final product with ferric chloride (III) is accompanied by precipitation of flaky residue and its following dissolution in the overbalance of reagent.

Final product contains polysaccharides as well as small amounts of hemicelluloses, soluble sugars and carbon acids besides the phenolic compounds which would increase its adsorbing activity.

Antioxidant activity of melanin was high (table 3, [6]).

**Table 3** – Antioxidant activity of product and residual quantity of TBA products

| Material | Concentration of antioxidants, μg/ml | Antioxidant activity, % | TBA products, μmol/l |
|---|---|---|---|
| Sunflower husk | 544.3 | 87.51±0.12 | 4.11±0.07 |
| | 216.5 | 71.09±0.19 | 9.40±0.10 |
| | 108.3 | 10.93±0.17 | 29.75±0.14 |
| Control | 0.0 | - | 32.44±0.12 |

**Conclusion.**

A technological scheme of production of melanin from sunflower husk as a waste was developed. Final product has antioxidant properties which connected with the presence of melanin. Final product melanin can be applied as a prophylactic mean and medicinal agent for treatment of human diseases. It is a substance for food and perfumery industry. The simplest variant is addition of melanin to sun-protection cream. Presence of melanin in food products provides their long storage. In the process of digestion melanin partially dissimilates by assistance of microflora of intestinal tract. It controls the peristalsis and affects the microflora of intestine. Melanin plays a role of enterosorbent, so it is an active antidote and it helps in the cases of acute poisoning. Melanin brings out toxicants on the incipient stage of contamination before their absorption to blood. Melanin is applied for treatment and prophylaxis of disease of liver, neurological diseases such as stress and chronic fatigue syndrome, and oncological diseases. Treatment of melanin as an oxidant has the rejuvenating effect.

**References**

[1] Kartushina Yu.N. Perspektivy ispolzovaniya otkhodov masloekstraktsionnogo proizvodstva (luzgi podsolnechnika) s tselyu polucheniya melaninov /Yu.N. Kartushina. N.V. Gracheva. M.A. Danilova //Ekologiya i bezopasnost v tekhnosfere: sovremennyye problemy i puti resheniya. Tomsk: NI TPU. 2014. p. 90-93.

[2] Gromyko N.V. Primeneniye podsolnechnoy luzgi v kachestve sorbenta dlya ochistki prirodnykh vod ot ionov tyazhelykh metallov /N.V. Gromyko //Innovatsionnaya nauka. № 1-3 (13). 2016. p. 41-42.

[3] Kalashnikova L. I. Issledovaniye tekhnologicheskikh svoystv rastitelnykh otkhodov kak alternativnogo ekologicheskogo topliva /L.I. Kalashnikova. A.A. Ovchinnikova. A.V. Aleksandrova. A.A. Kalashnikova //Vektor nauki Toliattinskogo gosudarstvennogo universiteta. 4. 2011. p. 32-34.

[4] De Leonardis A. A first pilot study to produce a food antioxidant from sunflower seed shells (Helianthus annuus) /A. De Leonardis, V. Macciola, N. Di Domenico //Eur J Lipid Sci Technol. 2005. 107:220-227.

[5] Szydłowska-Czerniak A. Optimization of extraction conditions of antioxidants from sunflower shells (Helianthus annuus L.) before and after enzymatic treatment /A. Szydłowska-Czerniak, K. Trokowski, E. Szłyk //Ind Crop Prod. 2011. 33: 123-131.

[6] Gracheva N.V. Sposob polucheniya melanina iz luzgi podsolnechnika i issledovaniye ego antioksi-dantnoy aktivnosti /N.V. Gracheva. V.F. Zheltobryukhov //Vestnik Kazanskogo tekhnologicheskogo universiteta. 2016. T. 19. 15. p. 154-157

[7] Gabruk N.G. IK - spektroskopiya v izuchenii sostava kompozitov. poluchennykh iz rastitelnogo i zhi-votnogo syria /N.G. Gabruk. I.I. Oleynikova. A.V. Metelev [dr.] //Nauchnyye vedomosti BelGU. Seriya Este-stvennyye nauki. 2011. 15. 16 (110). p. 95-98.

[8] Pat. 2578037 RF. MPK A61K36/28. B01D11/02. Sposob polucheniya melanoidnogo antioksidanta iz luzgi podsolnechnika /N.V. Gracheva. Yu.N. Kartushina. M.A. Danilova. A.B. Golovanchikov. V.F. Zheltobryukhov; VolgGTU. 2016.

[9] Gracheva, N.V.Studies of the intensification of the extraction of biologically active substances from chaga using direct current electric fields /N.V. Gracheva, A.B. Golovanchikov //Pharmaceutical Chemistry Journal. 2010. T. 44. 11. p. 608-610.

[10] Issledovaniye ekstraktov i melaninov griba inonotus obliquus (pers.) Pil.. poluchennykh posle obra-botki syria VChE-plazmoy / O.Yu. Kuznetsova, I.Sh. Abdullin, M.F. Shayekhov, G.K. Ziyatdinova, G.K. Bud-nikov //Uchenyye zapiski Kazanskogo universiteta. Seriya: Estestvennyye nauki. 2016. T. 158. 1. p. 23-33.

[11] Pat. 2406514 Rossiyskaya Federatsiya. MPK A61K36/06. B01D11/02. A61P39/06. Sposob polucheniya vodnykh ekstraktov chagi /M.A. Sysoyeva, E.V. Sysoyeva, A.V. Sysoyeva, V.S. Gamayurova; zayavitel i patentoobladatel Kazanskiy gos. tekhnologicheskiy un-t. № 2009113400/15; zayavl. 09.04.09; opubl. 20.12.10. Byul. 35. 5 p.

[12] Pat. 1805968 Rossiyskaya Federatsiya. MPK A61K35/78. Sposob polucheniya sredstva, obladayushchego protivoyazvennoy i adaptagennoy aktivnostyu /G.L. Ryzhova, S.S. Kravtsova, I.V. Bogdanova, A.Ya. Korbut, V.I. Grebnev, M.P. Garbusenko, V.T. Pashinskiy, N.V. Grisel; Tomskiy gos. un-t. Tomsk. opubl. 30.03.93. Byul. 12. 5 p.

[13] Kartushina Yu.N. Polucheniye melanina na osnove otkhodov masloekstraktsionnogo proizvodstva /Yu.N. Kartushina. M.A. Kirichenko. G.A. Sevryukova //Vestnik Kazanskogo tekhnologicheskogo uni-versiteta. 2016. T. 19. 16. p. 124-126

[14] Pat. 2281779 Rossiyskaya federatsiya. mpk7 A61K 36/28. Sposob polucheniya prirodnogo melanoidnogo antioksidanta/L.A. Zhorina, R.N. Kashevatskaya, A.L. Ivanov, V.L. Ivanov; zayavitel i patentoobladatel L.A. Zhorina, R.N. Kashevatskaya, A.L. Ivanov, V.L. Ivanov. №2004134636/15. zayavl. 26.11.2004; opubl. 10.06.2006. Byul. 23. 6 p.

[15] Dasoyan. M.A. Tekhnologiya elektrokhimicheskikh pokrytiy /M.A. Dasoyan. I.Ya. Palmskaya. E.V Sakharova. Lipetsk: Mashinostroyeniye. 1972. 464 p.

[16] Voldman. G.M. Osnovy ekstraktsionnykh i ionoobmennykh protsessov v gidrometallurgii /G.M. Voldman. M.: Metallurgiya. 1982. 376 p.

[17] Alekseyeva. T. N. Antimutagennyye svoystva rastitelnogo melaninovogo pigmenta /T.N. Alekseyeva. A.V. Oreshchenko. A.V. Kulakova [i dr.] //Khraneniye i pererabotka selkhozsyria. 2001. 5. p. 37-38.

# Physical and Theoretical Models of Heat Pollution Applied to Cramped Conditions Welding Taking into Account the Different Types of Heat

**Y I Bulygin[a], D A Koronchik[b], A N Legkonogikh[c], M G Zharkova[d], N N Azimova[e]**

Don State Technical University
Gagarin square 1, Rostov-on-Don, Russian Federation, 344000

*E-mail:* [a]bulyur_rostov@mail.ru, [b] koronchic@mail.ru, [c]lan-666-121285@mail.ru, [d]mariazharkova@mail.ru, [e]arkomaazimov@mail.ru

**Abstract**. The standard k-epsilon turbulence model, adapted for welding workshops, equipped with fixed workstations with sources of pollution took into account only the convective component of heat transfer, which is quite reasonable for large-volume rooms (with low density distribution of sources of pollution) especially the results of model calculations taking into account only the convective component correlated well with experimental data. For the purposes of this study, when we are dealing with a small confined space where necessary to take account of the body heated to a high temperature (for welding), located next to each other as additional sources of heat, it can no longer be neglected radiative heat exchange. In the task - to experimentally investigate the various types of heat transfer in a limited closed space for welding and behavior of a mathematical model, describing the contribution of the various components of the heat exchange, including radiation, influencing the formation of fields of concentration, temperature, air movement and thermal stress in the test environment. Conducted field experiments to model cubic body, allowing you to configure and debug the model of heat and mass transfer processes with the help of the developed approaches, comparing the measurement results of air flow velocity and temperature with the calculated data showed qualitative and quantitative agreement between process parameters, that is an indicator of the adequacy of heat and mass transfer model.

## 1. Introduction

Design and development of systems and means of protection of welder operator is detailed in the mathematical description unsteady heat and mass transfer processes of pollutants and excess heat from sources of pollution [1,2]. For one of studied objects of research include welding in the construction of pipelines. Welding ring pipe joints are carried out within the special devices (booths, tents) in order to prevent environmental influences (gusts of wind, rainfall, inflow of cold air) in the weld zone, i.e. processes take place in closed ventilated spaces.

## 2. Statement of research problem

Previously considered transient three-dimensional mathematical model of heat and mass transfer and heat of harmful impurities in the production environment with actively ventilated and sources of pollution and heat, which takes into account the turbulence of air-gas flows [3]. The standard k-epsilon turbulence model, adapted for welding workshops, equipped with fixed workplaces with sources of pollution, took into account only the convective component of heat transfer, which is quite reasonable for large-volume space (with low density distribution of pollution sources), especially since the results of model calculations taking into account only the convective component correlated well with experimental data.

## 3. The purpose and objectives of the study

However, for the purposes of this study, when we are dealing with a small confined space, where it is necessary to take into account the body heated to a high temperature (for welding), which are adjacent to each other as additional sources of heat, cannot be neglected radiative heat exchange. Therefore, it is reasonable to set purpose - experimental study of different types of heat transfer in a closed space is limited in welding and the behavior of a mathematical model describing the contribution of the various components of the heat exchange, including radiation, field influencing the formation of concentrations, temperatures, and air thermal stress mobility in the medium under study [4].

## 4. The results of experimental studies on physical models and discussing

This has been specially designed experimental setup and pick up the equipment for the purpose of studying the processes of heat diffusion in a fluid metering of heat in solids and radiant. To this end, it has been selected and assembled model object in the form of a ventilated cube with a simple geometric shape with an internal heat source (500W incandescent lamp) placed inside and the ability to create it in the ventilation air flows (carried outlet tube with a built-in fan). Field investigations of heat transfer processes carried out in the research laboratory of the department "BZhiZOS" DSTU.

The experimental setup is, in fact, the device is ventilated cube. The very size of the cube 0.5x0.5x0.5m was made of removable sheets of black metal thickness of 1 mm. Inside the cube in the center was secured incandescent lamp - a source of heat. On the front side (panel) cube holes were made for thermocouples through which the measurement of the air temperature inside the cube in its various locations. On the side of the cube in the middle there is a hole with a diameter of 50 mm, which is built into a plastic tube with a built-in fan it "ERA-5", providing extract hot air from the bottom. Fan through the adapter 250 mm was connected to a plastic tube with a diameter of 50 mm. When the fan is running, then the opposite wall cube is removed, providing a predetermined air capacity. General view of the installation shown in Figure 1.

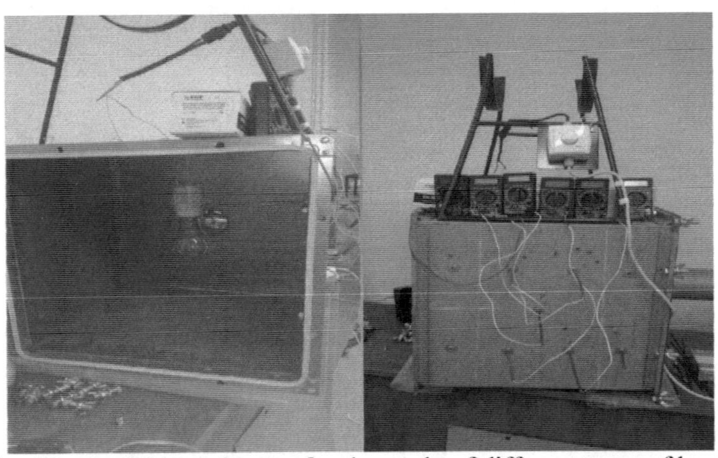

**Figure 1** – The experimental setup for the study of different types of heat transfer

As a source of heat than an incandescent lamp of 500 W was chosen randomly, since almost all supplied to the lamp energy is converted into radiant light. The losses due to conduction and convection are small. Such power lamp provides a high enough temperature to the glass bulb according to the source [5] 200-300 °C.

In addition, specially selected small internal volume of a cube V = 0,125 $m^3$ thermal stress within the object of the study was $Q= N / V = 500/0.125=4\ kW/m^3$, which is significantly higher hygiene standards for workplaces. As the driving force of air flow fan of the brand has been involved "ERA-5" with a maximum of 20 W, providing a predetermined maximum air flow 143 $m^3$ / h and discharge pressure 42.5 Pa.

The dynamics of the air temperature in the inner section of a closed cube with a device that simulates the propagation of heat welding gases from the heat source capacity of 0.5 kW.

Experiment 1. The ambient temperature in the room where he was set closed cube was T ambient air = 23.7 °C. Total test time was - 16 minutes. The initial temperature in the cube to turn on the lamp was 24 °C. The experimental results are shown in Fig. 3-4 and in Table 1.

As can be seen from the graphs in Figure 3 heat source heating mode, there is separation of temperature fields in the test section. If the maximum temperature near the source was 156 °C (measuring point 5) then measuring points remote from the source (point 1,3,6,8,9) it was in the range 45-60 °C. Furthermore, temperature measurements were performed on the cube of its faces surfaces cooled by ambient air. The table 1 shows the results of measuring the temperature in the cooling mode, the cube is not ventilated.

**Table 1** – The surface temperature of the cube faces, °C

| | Upper | Rear | Lateral (right) | Lateral (left) | Centre |
|---|---|---|---|---|---|
| Centre | 59 | 56 | 59 | 50 | 56 |
| Corner | 48 | 48 | 48 | 42 | 48 |

In the second phase of the study used a vented cubic meters with an internal source of heat (fig. 2).

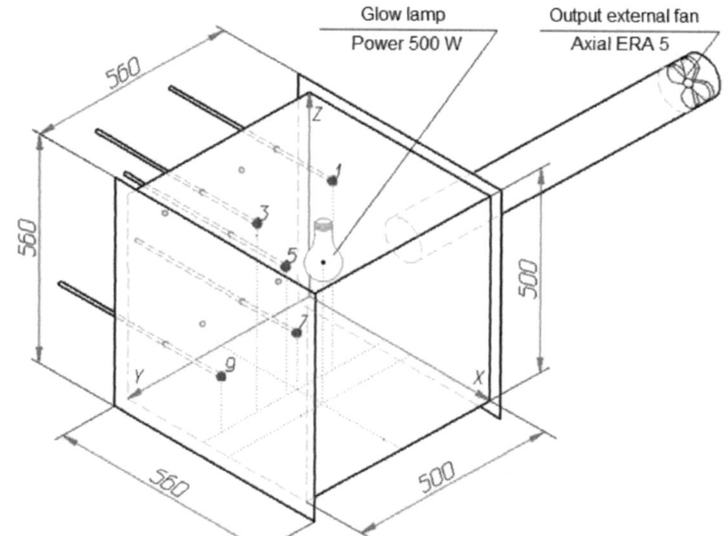


**Figure 2** – Location of measuring points in a cube

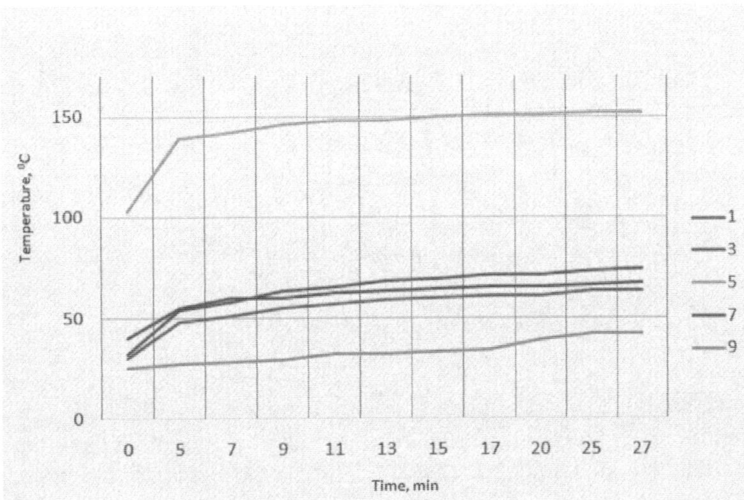


**Figure 3** – Changes in air temperature in the measurement points within the closed not ventilated cube

Experiment 2. Turn on the fan. Ambient air temperature is 23.7 ˚C. Air velocity in the pipe was 2.9 m/s. Using differential manometer Testo 521 using a Pitot tube air velocity of air from the fan side was measured v = 3.8 m/s at a temperature 29.5 ˚C, same air velocity at the entrance to the pipe v = 4.5 m/s, 30 ˚C temperature, air flow rate Q = 127m$^3$/h. For a given fan performance opposite side of the cube has been removed.

The experimental results are shown in Figure 4. It is seen from experiments that after the creation of conditions of aeration cube, the temperature in all points of the cross section of heat source decreases to 40-50 ˚C, and the difference between the maximum temperature at the point 5 and the minimum temperatures at 9 becomes further smaller than in the first experiment, about 60 ˚C.

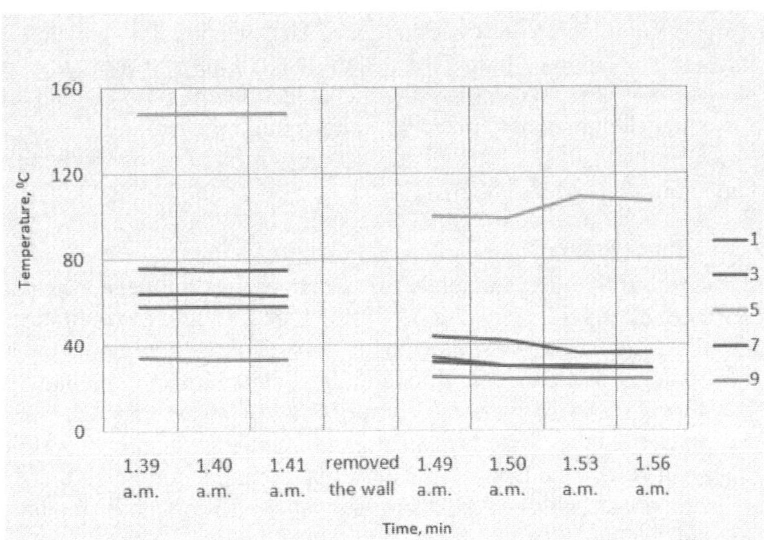


**Figure 4** – Changes in air temperature in the measurement points within the closed ventilated cube

Experiment 3. In order to clarify the boundary conditions for the mathematical model of heat and mass transfer were carried out additional studies of the temperature field on the glass bulb incandescent 500 W. Temperature measurement results at five points of heat source are presented in Figure 5.

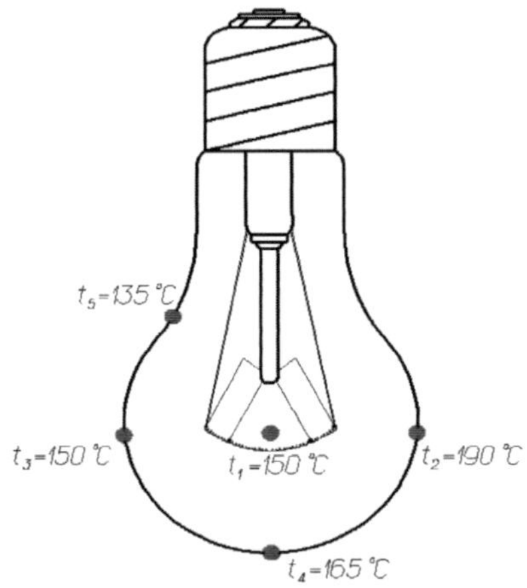


**Figure 5** – The temperature field in the glass bulb heat source

## 5. The results of theoretical research. Mathematical model.

In the numerical solution of the problem is carried out in a CAD-CAE Solid Works package, to this end, developed a method of construction in a software environment CAE model package ventilated cube with different power heat sources. Conducted field experiments to model cubic body and modeling of heat and mass transfer processes with the help of the developed approaches, comparison of air flow velocity and temperature measurement results with the calculated data showed qualitative and quantitative agreement between process parameters, that is an indicator of the adequacy of heat and mass transfer model.

The software for semi-enclosed and confined spaces where there is a welder working in cramped conditions, considering contamination and discharge of heat during the welding process, which allows to calculate the concentration of harmful substances field, temperature, air flow and thermal stress. The problems of convergence calculations and selection of a finite element mesh. Using software developed by the calculations of, possible to determine the optimum operating parameters of the mobile complex ventilation [6], ensuring safe sanitary conditions of work welders at work in cramped conditions and confined spaces [7-10].

## Conclusions

Developed and tested for adequacy physical and theoretical models of heat and mass transfer of pollutants in confined confined space taking into account active ventilation. The proposed mathematical model allows you to determine with a sufficient degree of accuracy values of temperatures, air mobility, and also concentrations and heat stresses in the medium under study. The model describes the contribution of various components of heat transfer, including radiation, influencing the formation of the investigated parameters in the working zone of welding. The obtained results allow us to proceed to the studies of heat and mass transfer of pollutants both in terrestrial ventilated chambers during construction of main gas and oil pipelines, and in the protective chambers when welding under water.

The work was supported by the Russian Federal Property Fund within the framework of a research project 16-3860055 mol_a_dk.

IOP Conf. Series: Earth and Environmental Science **66** (2017) 012015 doi:10.1088/1755-1315/66/1/012015

**References**

[1] Grishagin V.M. *Svarochny aerosol: obrazovaniye, issledovaniye, localizaciya, primineniya: monografiya* [Welding fumes: education, research, localization, use: monograph] / V.M. Grishagin. - Yurga Institute of Technology. - Tomsk: Publishing house of Tomsk Polytechnic University, 2011. - 213 p.

[2] Rykalin N.N. *Raschyoty teplovih processov pri svarke* [Calculations of thermal processes in welding] / N.N. Rykalin. M.: Engineering, 1951 296 p.

[3] Final element modeling of mass transfer of impurities in the production environment taking into account the turbulence of air flow / B.Ch. Meskhi [and others] // Don Herald. state. tehn. Univ. 2012 Vol. 12, №6 (67). p.10-16

[4] The methodology and principles to find solutions and safety of welders work in cramped conditions and on hard objects / D.A. Koronchik [and others] // Don Herald. state. tehn. Univ. 2016 Vol. 12, 3 (86). p.141-147.

[5] Taubkin S.I. Fire and explosion, especially their expertise. M.: 1999. 600 p.

[6] *Ventilyacionniy complex: patent 2600717 MPK: F24F7/00* [Vent Complex: Patent 2600717 IPC: F24F7 / 00 / B.Ch. Meskhi, Y.II. Bulygin, A.L. Gaidenko, D.A. Koronchik, O.V. Denisov, D.V. Rogozin, A.N. Sitnikov 2015 130 241/12; appl. 07/22/2015; 27.10.2016 bull. 30.

[7] Levchenko O.G., Metlitski V.A., *Sovremennie sredstva zashchity svarshchikov* [Modern protective remedies of welders] / O.G. Levchenko, V.A. Metlitski. K .: Ecotehnologiya, 2001. 84 p.

[8] Flow Simulation 2012 Technical Reference.

[9] http://www.solidworks.ru/

[10] G. Plymouth AB. *Chistiy svezhiy vozduh: Katalog.* [Clean fresh air: Catalogue. - Malmo (Sweden), 1988 30 p.

IOP Conf. Series: Earth and Environmental Science **66** (2017) 012016 doi:10.1088/1755-1315/66/1/012016

# Modelling of the combustion velocity in UIT-85 on sustainable alternative gas fuel

**N M Smolenskaya**[a], **N V Korneev**[b]

[a]Togliatti State University,
Belorusskaya st. 14, Togliatti, Russian Federation, 445667
[b]Volga region state university of service,
Gagarin st. 4, Togliatti, Russian Federation, 445017

E-mail: [a]Nata_smolenskaya@mail.ru, [b]niccyper@mail.ru

**Abstract.** The flame propagation velocity is one of the determining parameters characterizing the intensity of combustion process in the cylinder of an engine with spark ignition. Strengthening of requirements for toxicity and efficiency of the ICE contributes to gradual transition to sustainable alternative fuels, which include the mixture of natural gas with hydrogen. Currently, studies of conditions and regularities of combustion of this fuel to improve efficiency of its application are carried out in many countries.
Therefore, the work is devoted to modeling the average propagation velocities of natural gas flame front laced with hydrogen to 15% by weight of the fuel, and determining the possibility of assessing the heat release characteristics on the average velocities of the flame front propagation in the primary and secondary phases of combustion.
Experimental studies, conducted the on single cylinder universal installation UIT-85, showed the presence of relationship of the heat release characteristics with the parameters of the flame front propagation. Based on the analysis of experimental data, the empirical dependences for determination of average velocities of flame front propagation in the first and main phases of combustion, taking into account the change in various parameters of engine operation with spark ignition, were obtained.
The obtained results allow to determine the characteristics of heat dissipation and to assess the impact of addition of hydrogen to the natural gas combustion process, that is needed to identify ways of improvement of the combustion process efficiency, including when you change the throttling parameters.

## 1. Introduction

Hydrogen as an additive to natural gas, can significantly improve efficiency and reduce toxicity of exhaust gases of reciprocating ICEs. In particular, there is actual necessity for conducting studies when working at low frequencies of the crankshaft rotation, including the throttling modes, which have great influence on the economic and environmental performance of the engine, as the engine is operating in these modes for significant amount of time when driving in the megacity [1].

In this regard, determination of regularities connecting the average velocity of flame propagation and the dissipation characteristics is of high priority. That will allow at the stage of designing of engines, operating on gas fuels, to improve the efficiency of work in these modes.

Accordingly, the purpose of the conducted studies was to develop the computational method for determining dissipation characteristics in engines operating on gas fuels, taking into account the obtained dependences for calculation of the average velocities of the flame front propagation.

## 2. Experimental technique

Studies of the characteristics of the combustion process and their relationship with parameters of combustion process were conducted on the experimental stand with a single-cylinder installation UIT-85, in which the electric motor maintains the constant velocity, homogeneity of the fuel-air mixture (FAM) was provided by the structure of the heated inlet pipe.

Registration of the combustion front propagation inside the cylinder installation UIT-85 near the spark plugs, was carried out using the ionization sensor installed in the spark plugs adapter. The parameters of the flame front propagation in the combustion chamber and the electrical conductivity in the final phase of combustion were evaluated using the multielectrode ionization sensor, installed in the zone of the combustion chamber, most remote from the spark plug.

In the study the major variable factors were the excess air ratio, varying from 0.8 to the maximum for the extremum dilution (1.3 – 1.7) and composition of gaseous fuels, where the share of natural gas ranged from 85 to 100%, and proportion of hydrogen, from 0 to 15% (by fuel weight). Operation of the installation was carried out at the ignition advance angle, varying from 0 to 30° of the crankshaft position sensor and velocities of 600 and 900 $min^{-1}$. The experimental procedure included parallel registration by the means of the digital multichannel ADC L-783M, manufactured by the "L-Card" company, of signals from the following sensors: ionization, spark plugs, crankshaft position, and cylinder pressure of the engine, mass air flow, and a gas analyzer, that determines the composition of the mixture by the excess air ratio and toxicity. As a result of experiments the series of oscillograms in each mode of testing was obtained. For each mode, with mathematical analysis the averaged signal indicator pressure signal and the signal from the ionization sensors were obtained, by which the characteristics of heat dissipation and the average velocity of the flame front propagation were obtained.

Determination of the experimental data errors was carried out according to GOST R 8.736-2011. The signal amplitude error from the ionization sensors and the signal pressure mostly did not exceed 6% of the time from spark discharge to the occurrence of maximum pressure in the cylinder of the ICE and to an impulse of ion current at ionization sensors – 3%, excess air ratio – 3%; consumption of natural gas and hydrogen – 3%.

## 3. Results and Discussion

The analysis of the studies of combustion in reciprocating ICEs has shown that the main parameters characterizing the combustion are the velocity of flame propagation in the 1st, 2nd and main phases of combustion, as well as characteristics of heat dissipation. Existing formulas for the flame propagation velocity in turbulent flow, reflecting the process behavior in the best way is the Damkohler – Karlovic model [2], which is often used to calculate combustion in large-scale turbulence. This model is considered to be continuation of Schelkin's works [2], which, in turn, are often taken as the basis for calculating combustion in small-scale turbulence. In both models, both flow turbulence and laminar combustion rate are considered. Therefore, for modeling the flame front propagation velocity in the combustion chamber of reciprocating ICEs it is necessary to determine the laminar velocity under these conditions, as it is the base to determine the empirical dependences of average velocities of the 1st and main phases of combustion.

From the existing models for calculation of the laminar velocity of the flame front propagation we will choose the model of determining the laminar velocity of flame front propagation for the methane-air mixtures, presented in the Heywood's work [3]:

$$U_{L(\mathrm{CH4})} = U_{0L(\mathrm{CH4})} \cdot \left(\frac{\mathrm{T}_u'}{T_0}\right)^{\alpha_{t(\mathrm{CH4})}} \cdot \left(\frac{P_u}{P_0}\right)^{\beta_{p(\mathrm{CH4})}} \quad (1)$$

where $U_{0L(CH4)}$ – the normal rate of combustion at $T_0 = 298\,K$; $P_0 = 101325\,Pa$; $T_u'$ and $P_u$ – the temperature and pressure for which the calculation is carried out; and the coefficients $\alpha_{t(CH4)}$ and $\beta_{p(CH4)}$ – are functions of the mixture composition:

$$\alpha_{t(CH4)} = 2.18 - 0.8 \cdot (\alpha - 1)$$
$$\beta_{p(CH4)} = -0.16 + 0.22 \cdot (\alpha - 1)$$

The current pressure was determined from the indicator diagrams of the pressure and current temperature on the equation of the real gas state at the spark moment.

To determine the normal propagation velocity of the methane-air mixture flame front we will use the data given in [4], which presents the experimental dependence of the normal propagation velocity of the flame front from the volume content of methane in the FAM. As a result of recalculation of the volume content of methane in the mixture to the excess air ratio the graph of the normal propagation velocity of flame front from α [4] was obtained. Following approximation of this formula, the polynomial describing the dependence of the normal velocity of flame front of methane-air mixture from the air excess factor for $T_0 = 298$ K and $P_0 = 1$ atm was obtained:

$$U_{0L(CH4)} = -2.7537 \cdot \alpha^4 + 15.612 \cdot \alpha^3 - 32.609 \cdot \alpha^2 + 29.257 \cdot \alpha - 9.0757 \qquad (2)$$

Under the terms of the ongoing research, in ICEs s combustion of the natural gas laced with hydrogen to 15% occurs, therefore, it is necessary to determine the laminar propagation velocity of the flame front also for the hydrogen-air mixture. For this we will take the Ujima and Takeno model given in [5]:

$$U_{L(H2)} = U_{0L(H2)} \cdot \left(\frac{T_u'}{T_0}\right)^{\alpha_{t(H2)}} \cdot \left(\frac{P_u}{P_0}\right)^{\beta_{p(H2)}} \qquad (3)$$

where $T_0 = 291\,K$; $P_0 = 101325\,Pa$; $P_u$ and $T_u'$ – the pressure and temperature for which the calculation of the normal combustion velocity is conducted, corresponding to the pressure and temperature, defined for the hydrogen-air mixture; $\alpha_{t(H2)}$, $\beta_{p(H2)}$ and $U_{0L(H2)}$ are determined from the formulas:

$$\alpha_{t(H2)} = 1.54 + 0.026 \cdot (\alpha - 1)$$

$$\beta_{p(H2)} = 0.43 + 0.003 \cdot (\alpha - 1)$$

$$U_{0L(H2)} = 2.98 - (\alpha - 1.70)^2 + 0.32 \cdot (\alpha - 1.70)^3$$

where α – excess air coefficient.

Combustion in the 1st phase is performed to a greater extent in small-scale turbulence, and to develop models of combustion, it is expedient to apply the K.I. Schelkin's model, where the average velocity of the piston is chosen as the criterion of turbulence, and the influence of the composition of the gas fuel is reflected by the excess air ratio and the proportion of hydrogen in natural gas.

Based on this mathematical analysis, which revealed the contribution of each component in the model, and approximation of the effects of the excess air ratio on changing the flame propagation velocity, with addition of hydrogen, the empirical model was obtained, with the purpose to determine the average propagation velocity of flame front in the 1st phase when operating only on compressed natural gas (CNG) (4) and when operating on CNG with addition of hydrogen up to 15% by weight (5), when the air excess factor α varies from 0.8 to 1.4:

$$U_1 = U_{L(CH4)} \cdot \alpha^2 + U_{L(CH4)} \cdot \sqrt{\alpha} \cdot \sqrt{\left(1 + \frac{U_{mps}}{U_{L(CH4)}}\right)} \qquad (4)$$

$$U_1 = \left[U_{L(CH4)} \cdot \alpha^2 + U_{L(CH4)} \cdot \sqrt{\alpha} \cdot \sqrt{\left(1 + \frac{U_{mps}}{U_{L(CH4)}}\right)}\right] \cdot \left(1 - \frac{\Delta H}{100}\right) -$$

$$-(4.3 \cdot \alpha^2 - 10.285 \cdot \alpha + 4.6) \cdot \sqrt{\left(U_{\text{mps}} \cdot U_{L(H2)} \cdot \frac{\Delta\text{H}}{100}\right)} \quad (5)$$

where $U_{\text{mps}}$ – the average velocity of the piston; ΔH – share of the addition of hydrogen by mass in fuel, %.

The check of adequacy of the obtained models for calculation of the propagation velocity of flame in the 1st phase of combustion for UIT-85 when operating on CNG and CNG with the share of hydrogen of 5, 10 and 15%, (4 and 5) is shown in Fig. 1, when the crankshaft rotation velocity n = 900 $\text{min}^{-1}$, compression ratio ε = 7 and the ignition advance angle = 13° of the crankshaft position sensor. In average, divergence of the calculated values with the experimental ones is less than 2.5%, for α close to stoichiometric, 0.1 m/s, and the maximum discrepancy does not exceed 8%, amounting to 0.35 m/s for α = 0.8. From this it follows that by using the calculation formula accurately you can determine the average velocities of flame propagation in the 1st phase, changing the parameters of engine velocity, mixture composition and additives of hydrogen in TBC. This indicates high degree of convergence of results of calculation performed by empirical models with experimental results for the studied operating modes.

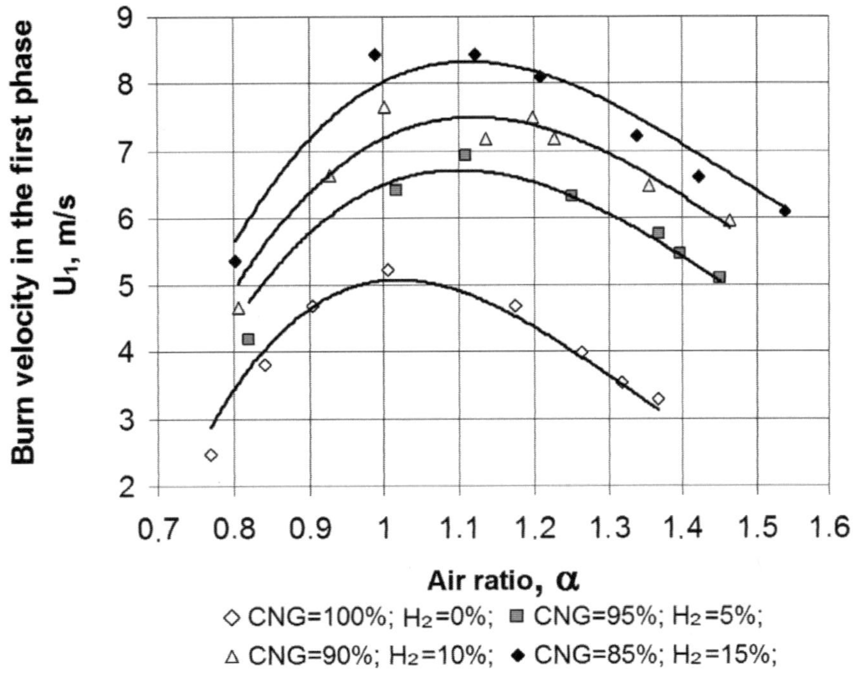


**Figure 1** – Comparison of experimental results with the model for average propagation velocity of flame front in the 1st phase of combustion in UIT-85, operating on CNG and CNG with the share of hydrogen of 5, 10 and 15%.

The main phase of combustion ($U_{1+2}$) involves the 1st and 2nd phase of combustion [6,7,8] in which combustion mainly takes place under the influence of large-scale turbulence, therefore, to develop empirical models of the propagation of the flame front in the main phase, the Damkohler – Karlovic model has been chosen [2,6,9]. In this model, the average propagation velocity of the flame front in the 1st phase of combustion was adopted as the laminar velocity, and the activating effect of hydrogen on the combustion process is considered. Such parameters of the engine as the filling ratio, which determines the throttling conditions, compression ratio and the ignition timing, calculated using the ratio of the volume at the time of supplying sparks to the operating volume, were also considered in the model. In the result the empirical relationship for calculating the average propagation velocity of the flame front in the main phase of combustion for CNG and CNG with addition of hydrogen up to 15% by weight of the fuel (6) was obtained:

$$U_{1+2} = 1.3 \cdot \sqrt{U_1} \cdot \sqrt{U_{\text{mps}}} \cdot \sqrt{\eta_V} \cdot e^{\left(\varepsilon \cdot \frac{V_\theta}{V_h}\right)} + \sqrt{2 \cdot U_{\text{mps}} \cdot U_1 - \frac{U_1^{\,2}}{1.3} \left(1 - e^{-\left(\frac{U_{\text{mps}}}{U_1}\right)}\right)} \qquad (6)$$

where $\eta_V$ – the filling ratio; $\varepsilon$ – compression ratio; $\dfrac{V_\theta}{V_h}$ – the ratio of the volume at the time of supplying sparks to the operating volume.

The convergence of the obtained model with the experimental data is presented in Fig. 2. for UIT-85 with operating mode n = 900 min$^{-1}$, ignition advance angle = 13° of the crankshaft position sensor, at operating with CNG and CNG with addition of hydrogen of 5, 10 and 15% of the fuel weight.

The average discrepancy of the calculated values with the experimental ones is about 1%, for α close to stoichiometric 0.15 m/s, the maximum discrepancy does not exceed 5%, amounting to 0.5 m/s for α = 0.8. From this it follows that by using the calculation formula accurately you can determine the average velocities of flame propagation in the main phase by changing the engine velocity, mixture composition and fraction of hydrogen in the gas composite fuel.

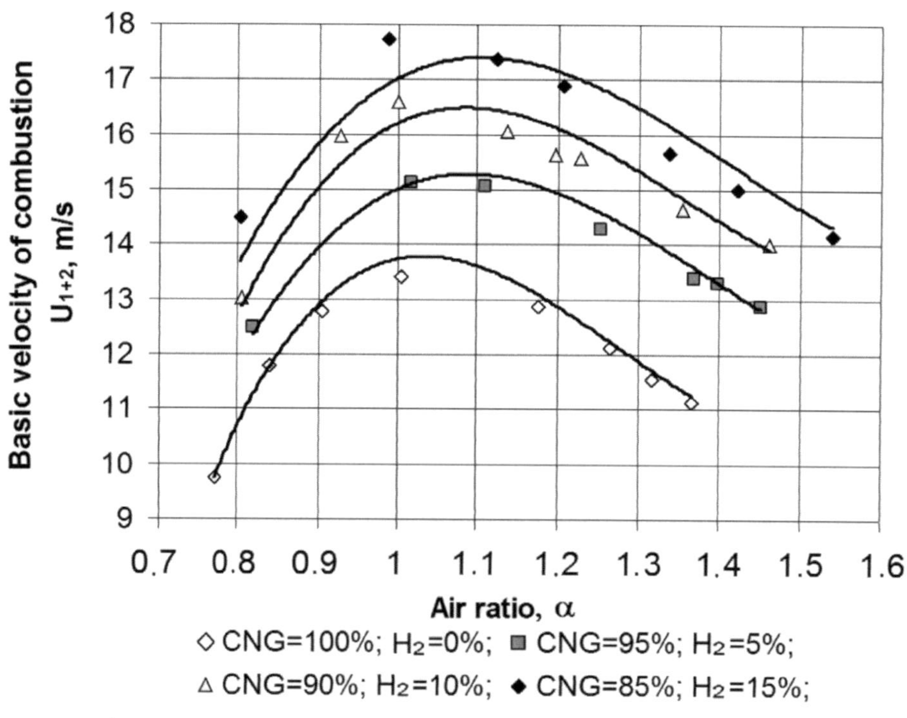


**Figure 2** – Comparison of experimental results with the model ones for average propagation velocity of the flame front in the main combustion phase in UIT-85, operating on CNG and CNG with the share of hydrogen of 5, 10 and 15%.

To determine the energy parameters of the engine it is necessary to know the characteristics of heat dissipation. In Russia, the most common is the heat dissipation characteristics, proposed by I. I. Wiebe [10,11]:

$$\chi = 1 - e^{\ln(1-\chi_Z)\left(\frac{\varphi}{\varphi_Z}\right)^{m+1}} \tag{7}$$

where $\chi_Z$ – the share of active burned mixture; $\varphi_Z$ – duration of the combustion process; $\varphi$ – the current angle from beginning of the combustion process; $m$ – the indicator of the combustion behavior.

One of the drawbacks of the model of I.I. Wiebe in relation to calculation of the heat release characteristics, is the lack of relation between parameters $m$ and $\varphi_z$ with the characteristics of flame front propagation in the conditions of ICE, especially when operating on alternative gaseous fuels consisting of natural gas and hydrogen up to 15%.

The mathematical analysis provided us the possibility to determine the indicator of the combustion process behavior $m$ and the duration of combustion $\varphi_z$, included in the dissipation model (7), depending on the average propagation velocity of flame in the 1st and main phases of combustion, which are presented in models (5) and (6).

The obtained empirical models are represented by the formulas (8) and (9):

– the indicator of the combustion process behavior:

$$m = \frac{-3.0844}{\ln\left(\frac{S_1}{S}\cdot\frac{180\cdot U_{\text{mps}}}{U_1}\right) - \ln\left[\frac{180}{S}\cdot\left(\frac{S_1\cdot U_{\text{mps}}}{U_1} + \frac{S_{1+2}\cdot U_{\text{mps}}}{U_{1+2}}\right)\right]} + 1 \tag{8}$$

– duration of the combustion process:

$$\varphi_Z = \frac{\pi}{S}\cdot\left(\frac{S_{1+2}\cdot U_{\text{mps}}}{U_{1+2}} + \frac{2\cdot S_1\cdot U_{\text{mps}}}{U_1}\right) \tag{9}$$

where $S$ – the piston stroke length, mm; $S_1$ and $S_{1+2}$ – distance of the flame moving in the 1st and main phases of combustion, respectively, mm.

The convergence of the obtained models is shown in the following ranges of operation:

– velocity range from 600 to 900 $\text{min}^{-1}$;
– ignition advance angle from 8 to 30° of the crankshaft position sensor;
– actual air excess factor from 0.8 to 1.3.

The obtained models in relation to determination of the parameters of the heat release characteristics and velocities of flame propagation have a high convergence with experimental data in the specified range.

## 4. Conclusion

1. In the result of modelling of the flame propagation velocity in the UIT-85 the empirical models of average velocities of the flame front propagation in the 1st and the main phases of combustion for natural gas with the ratio of hydrogen up to 15% by weight were obtained.

2. The dependence of the heat release characteristics from the velocities of flame front propagation in the 1st and the main phases of combustion, which is presented in the form of empirical models for determination of $m$ and $\varphi_z$ was identified.

3. The calculation method for determination of the dissipation characteristics in a low frequency range of rotation of the crankshaft, including the modes of throttling, will allow at the design stage to evaluate the influence of the composition of alternative gaseous fuels on parameters of economic and environmental performance of the engine.

**References**

[1] Smolenskaya N.M., Smolenskii V.V, Bobrovskij I., Research of polytropic exponent changing for influence evaluation of actual mixture composition on hydrocarbons concentration decreasing on deep throttling operation, IOP Conf. Series: Earth and Environmental Science 50 (2017) 012016 doi:10.1088/1755-1315/50/1/012016.

[2] Warnatz J., Maas U., Dibble R.W., Combustion. Physical and Chemical Fundamentals, Modeling and Simulation, Experiments, Pollutant Formation, 4th Edition, ISBN-13 978-3-540-25992-3 4th ed. Springer Berlin Heidelberg New York, 2006 - 378 p.

[3] Heywood J.B., Internal Combustion Engine Fundamentals, McGraw-Hill, – New York : McGraw-Hill, 1988. – 931 p.

[4] Law C.K. Combustion physics, Cambridge: Cambridge university press, 2006. – 722 p.

[5] Verhelst S. A study of the combustion in hydrogen-fuelled internal combustion engines, PhD thesis, Gent: Gent University, 2005. – 278 p.

[6] Smolenskaya N.M. Improving the efficiency of spark ignition engines through the use of composite gas fuels, dissertation of the candidate of technical sciences, Moscow: MAMI University, 2015. – 165 p.

[7] Shaikin A.P., Galiev I.R., On the effect of temperature and the width of the turbulent combustion zone on the ionization detector readings, Technical Physics, 61(8) (2016) 1206-1208. Information on https://www.scopus.com/inward/record.uri?eid=2-s2.0-84981489611&partnerID=40&md5=2a171a70052f225423d5f1082d29a746 DOI: 10.1134/S1063784216080247.

[8] Shaikin A.P., Galiev I.R., Relationship of flame propagation speed for methane-hydrogen fuel of a Internal Combustion Engine with parameters of ion current and hydrogen concentration, Russian Aeronautics (iz VUZ). 2016. V.59(2). P.249-253 DOI: 10.3103/S106879981602015X

[9] Hitrin L.N., Physics of Combustion and Explosion, MGU, Moscow, 1957,p.337 – 411

[10] Vibe I. I. The new operating cycle of the engine. Combustion rate and duty cycle of the motion engines, Sverdlovsk, Mashgiz, 1962. - 271 p.

[11] Kavtaradze, R.Z. Theory of piston engines. Special chapters, Moscow, MSTU. Bauman. 712 p.

# Impact of the Chemical Elements Upon the Convective Flows in the Molten Metal of the Weld Pool

**V D Sarychev[2], D A Chinakhov[1, a], S A Solodsky[1], A Yu Granovskii [2], S A Nevskii [2]**

[1] Yurga Institute of Technology of Tomsk polytechnic university, Russia, Yurga, Leningradskaya str., 26

[2] Siberian State Industrial University. Russia, Novokuznetsk, Kirova str.,42

E-mail: [a]chinakhov@tpu.ru

**Abstract.** World production of surfacing materials requires a large amount of alloying elements extracted from the Earth crust. Efficient use of scarce alloying elements and development of resource efficient surfacing techniques which reduce consumption of alloying elements together with improvement of the functional properties of the surfaced layer is a relevant problem of material science and preservation of the environment. In the paper the authors consider the problem of modeling the convective flows emerging under the plasma jets in the process of surfacing. The authors formulate the basic requirements to the thermal cycle of the surfacing ensuring obtaining identical functional properties of the surface while the amount of scarce alloying elements reduces.

## Introduction

At the present time the method of arc surfacing with flux cored electrodes is widely applied for producing wear resistant coatings [1]. It allows producing bimetallic products which combine high strength and low cost with increased service life although the main part of the component may be produced from the low-alloy steels. Application of the method for repairing the worn-out parts results in reduction of the amount of the spare parts of the operated equipment, reduction of delay technician time, reduction of servicing costs. On the other hand, the arc surfacing method implemented in spite of the simple design of the applied equipment and its high productivity leads to formation of coarsely dispersed heterogeneous structures with low impact strength and the quality of the surface requires after-treatment processing [2]. That is why it is necessary to find such surfacing conditions which allow forming the structure with high physical and mechanical properties. To solve this problem we need the data on the hydrodynamic processes taking place in the weld pool. The geometry of the surfaced layer, its structure, phase composition and mechanical properties depend on the characteristics of these processes [3]. Many researchers studied the hydrodynamic processes in the weld pool under the arc welding and among their papers works [4-9] are especially worth mentioning as the authors completed the numerical study of the heat exchange in the liquid and its flow under direct and pulse current welding. In [4] it has been established that the surface-tension gradient has the most significant impact over the geometric pattern of the weld pool while other factors, such as magnetic inductive capacity, heat-expansion coefficient of the material make a smaller contribution. In paper [5] the authors obtain the results demonstrating that application of pulse current allows forming the surfaced layer of higher quality.

Thermocapillar convection in the weld pool was studied in [6-8]. In [6] the authors complete the numerical study of the impact produced by Marangoni effect upon the parameters of various metals under arc welding in the enclosed volume. As a result the ranges of current intensity were established under which the thermocapillar convection plays the most important role. Further increase of the intensity of current does not lead to the increase of the depth of penetration, thus, the linear dependence of the surface tension coefficient of pure liquid metals upon the temperature deteriorates the penetrating ability of the electric arc. In [7] the authors show that for the pure metals the thermocapillar convection plays the significant role only at the initial stages of materials fusion and then the deformation of the liquid impacted by the gravity force becomes the source of convective flows. On the other hand, study of the concentrated energy flows impact upon the metals shows that under large values of power density the heat-gravitational convection is insignificant [8].

It is also necessary to take into consideration the influence of the alloying elements [8]. In [9] the influence of these elements upon the surface tension of liquid iron was studied. It was shown that addition of sulfur results first in the growth of surface tension as the temperature grows and then decreases, thus, there is non-linear dependence upon the temperature. If sulfur is introduced directly into the weld pool it results first in the growth of surface tension and the pattern of vortex flotation undergoes a significant change. The matter involved into the vortex flotation will move to the centre and then downward to the bottom of the weld pool. The downward flow of hot metal in the weld pool operates as a heat drill. The penetration depth will increase accordingly and the geometric pattern of the pool will change.

This way, when developing the model of the convective flows in the weld pool it is necessary to take the influence of the alloying elements upon the hydrodynamic flotation into consideration. It is relevant for studying the convective processes in the weld pool when surfacing with flux-cored electrodes.

The alloying elements, components of the flux-cored wire, change the character of surface tension coefficient dependence upon the temperature. This results in alteration of the pattern of the convective flotation and effects the structure-phase composition of the surfaced metal. That is why the goal of our work is to develop the mathematical model of convective flows in the weld pool which takes the presence of alloying elements into consideration.

**Problem formulation**

It is appropriate to start modeling of the convective flows from determining the bulk forces acting upon the liquid and from calculating the electromagnetic fields. The volume force acting on the liquid includes the gravity and the electromagnetic forces.

$$\vec{F} = \vec{F}_g + \vec{j} \times \vec{B} = \rho_0 \vec{g} - \rho_0 \vec{g} \beta (T - T_{ref}) + \vec{j} \times \vec{B} \tag{1}$$

We shall calculate the electromagnetic fields on the basis of Maxwell-Lorentz equations

$$\begin{gathered} \nabla \cdot \left( \sigma \nabla V + \sigma \frac{\partial \vec{A}}{\partial t} \right) = 0,\ \sigma \frac{\partial \vec{A}}{\partial t} + \frac{1}{\mu_0} \nabla \times \left( \nabla \times \vec{A} \right) + \sigma \nabla V \\ \vec{E} = -\nabla V - \frac{\partial \vec{A}}{\partial t},\quad \vec{j} = \sigma \vec{E},\ \vec{B} = \nabla \times \vec{A} \end{gathered} \tag{2}$$

With consideration to (1), (2) let us put down Navier-Stokes and thermal conductivity equations

$$\rho\left(\frac{\partial \vec{u}}{\partial t}+\vec{u}\nabla\vec{u}\right)=-\nabla p+\mu\Delta\vec{u}+\vec{F}_V,\nabla\cdot\vec{u}=0,$$
$$\rho C_p\left(\frac{\partial T}{\partial t}+\vec{u}\nabla T\right)=\nabla\cdot(k\nabla T)+Q_V \tag{3}$$

where $\vec{u}$ - velocity vector, $p$ - pressure, ρ - density, μ – dynamic viscosity, $\vec{F}_V$ - bulk forces, which are prescribed by the total of Lorentz force and gravity force, *T*- temperature, $C_p$ - specific thermal capacity, *k* – heat conductivity coefficient, $Q_V$ – volume heat sources. Only Joule effect is the volume heat source for the cathode and anode $Q_V=\vec{j}\cdot\vec{E}$.

Let us consider the impact of electric arc plasma upon the plate of hardening steel 45 (Figure 1).

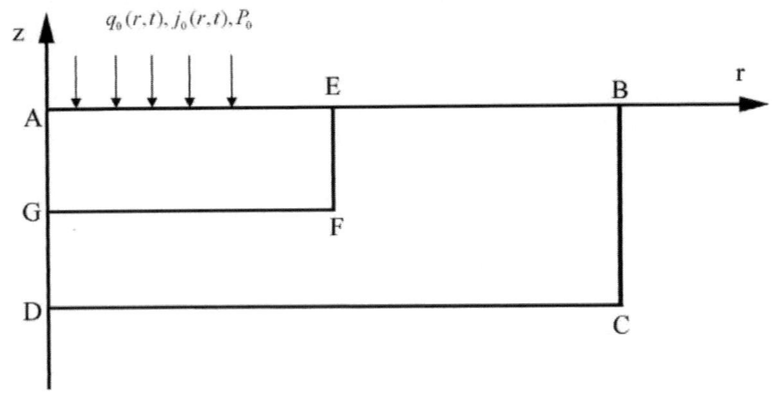


**Figure 1.** The computational region

At the top surface AB we shall set the density of current, pressure and heat flow which are the result of plasma action upon the surface of the metal, like in [3-5].

$$-\vec{n}\cdot\vec{q}=q_0(r)=\frac{dUI\eta}{\pi r_0^2}G(r,t)$$
$$-\vec{n}\cdot\vec{j}=j_0(r)=\frac{dI}{\pi r_0^2}G(r,t) \tag{3}$$
$$p=p_0(r)=\frac{\mu I}{4\pi}j_0(r)$$

where $q_0, j_0, p_0$ - heat flow, flow density, pressure accordingly; $U, I$ - voltage and strength of current; $\eta$ - performance factor; G(r,t) – the function of impact surface and time distribution:

$$G(r,t)=\exp\left(-\frac{d(r^2+(r_0-V_e t)^2)}{r_0^2}\right) \tag{4}$$

where $V_e$ - electrode feed, $d, r_0$ - distribution parameters.

On the bottom surface CD we set the condition of convection heat transfer and electromagnetic continuity;

$$-\vec{q}\cdot\vec{n}=h_0(T-T_0)$$
$$\vec{j}\cdot\vec{n}=0 \tag{5}$$

On the side boundary BC there is similar condition of convection heat transfer and equal-zero electrostatic scalar potential:

$$-\vec{q}\cdot\vec{n} = h_0(T - T_0)$$
$$V = 0 \tag{6}$$

The condition of contibuity was set for the magnetic field at all boundaries: $\vec{n}\times\vec{A} = 0$. The slipping condition was fulfilled at EF and FG boundaries for the speed.

The model of phase transition from the solid state into the liquid one was based on the concept of dynamic viscosity coefficient growth in the solid region. It looks as follows:

$$\mu = \mu_L(T)*(1 - f_s) + \mu_S f_s \tag{7}$$

where $f_s$ - the coefficient determining the solid phase, $\mu_L$ - viscosity of the liquid phase, $\mu_S$ - viscosity of the solid phase, which is large enough to stop the movement in the solid phase.

The calculations were completed according to the final element analysis in the kit Comsol Multiphysics. In Table 1 we provide characteristics of the material and the treatment conditions.

**Table 1.** Input parameters of the problem

| Designation | Description | Value |
|---|---|---|
| $U$ | Voltage | 30V |
| $I$ | Strength of current | 250 A |
| $r_0$ | The radius of efficient action of the plasma jet | 10mm |
| $T_l$ | Temperature of the liquid phase | 1723K |
| $\rho_s$ | Density of the solid phase | 7500kg m$^{-3}$ |
| $\rho_l$ | Density of the liquid phase | 6350kg m$^{-3}$ |
| $C_{pl}$ | Coefficient of heat capacity of the liquid phase | 720 J kg$^{-1}$ K$^{-1}$ |
| $C_{ps}$ | Coefficient of heat capacity of the solid phase | 602 J kg$^{-1}$ K$^{-1}$ |
| $k_l$ | Heat conduction coefficient | 20 W m$^{-1}$ K$^{-1}$ |
| $k_s$ | Heat conduction coefficient | 26 W m$^{-1}$ K$^{-1}$ |
| $\sigma$ | Electrical conductivity of metal | $7.7\times10^5\ \Omega^{-1}\,m^{-1}$ |
| $\mu$ | Dynamic viscosity coefficient | 0.05 kg m$^{-1}$ s$^{-1}$ |
| $L_f$ | Melting heat | $247\times10^3$ J kg$^{-1}$ |

In Figure 2 we provide the dependence of the surface tension upon the temperature

**Results and discussion**

It is known that the flotation in the weld pool is determined by Marangoni convection which is characterized by the temperature gradient of the surface tension $\partial\sigma/\partial T$. In Diagram 2 we can see that $\partial\sigma/\partial T > 0$ under the subcritical temperatures $T < T_c = 2300K$ and $\partial\sigma/\partial T < 0$ for $T > T_c$.

In figures 3-5 the development of the melt is shown at various time intervals. The following pattern is observed. Before the heat flow reaches its maximum $q$ the downward flow formed by two symmetric vortexes moves from the centre of the pool stirring the melt full-width the pool (Fig. 6). When the thermal influence of the plasma jet decreases the stage of cooling starts which is characterized by the motion of the downward flow towards the centre of the pool and further formation of single vortex near the symmetry point. This downward flow reaches the depth of 6 mm.

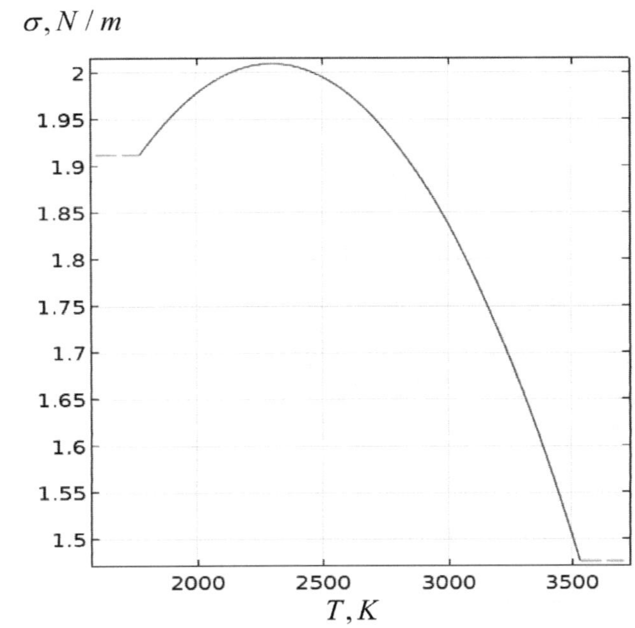


**Figure 2.** Dependence of the surface tension upon the temperature

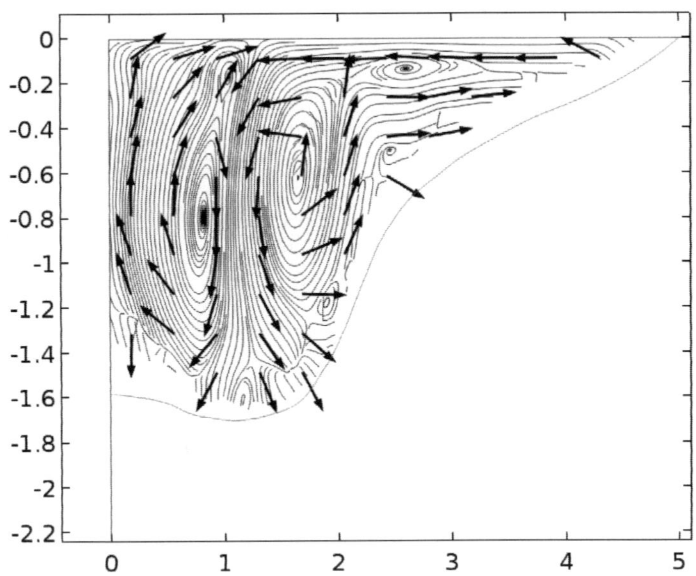


Figure 3. The lines of fluid flow in the melt and the field of velocities at time interval t=2.7s

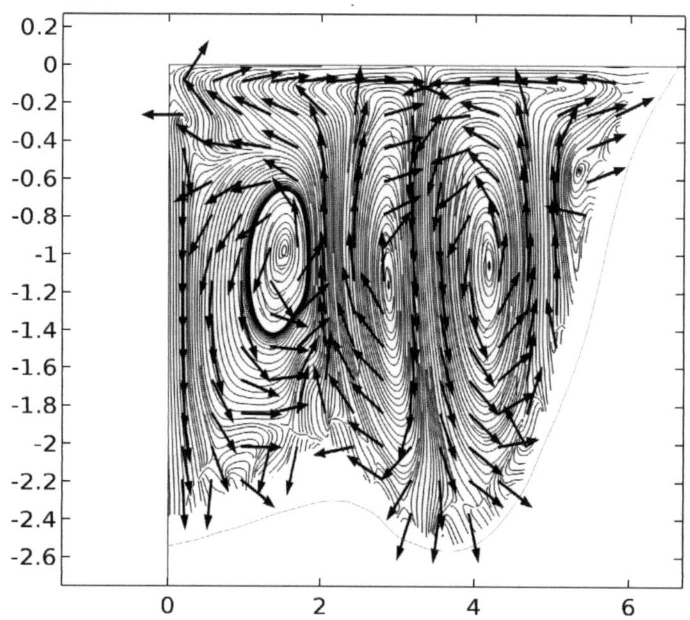


Figure 4. The lines of fluid flow in the melt and the field of velocities at time interval t=4s (the peak of heating)

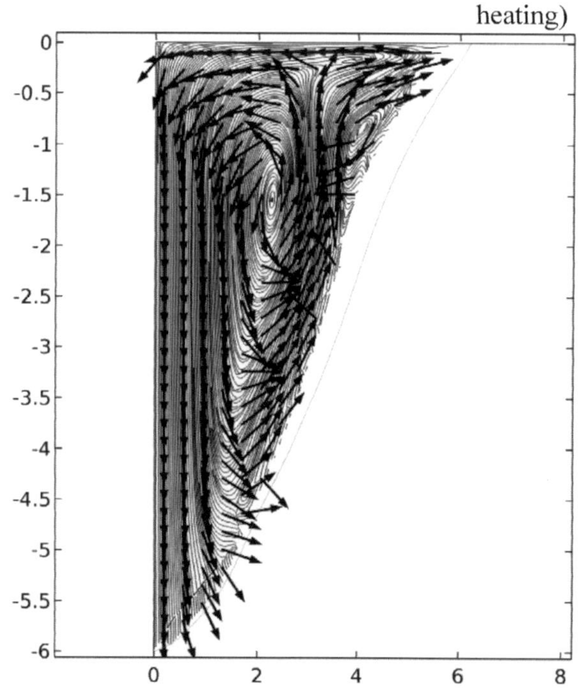


Figure 5. The lines of fluid flow in the melt and the field of velocities at time interval t=5.3s (the phase of cooling)

Formation of the given downward flow formed by two vortexes can be explained by the change of the sign of the temperature gradient of surface tension $\partial\sigma / \partial T$. When we compare Fig. 7 with the pattern of flow in Fig. 3-4 we can notice that the downward flow is where $T = T_c$. It is also proved in Fig. 8.

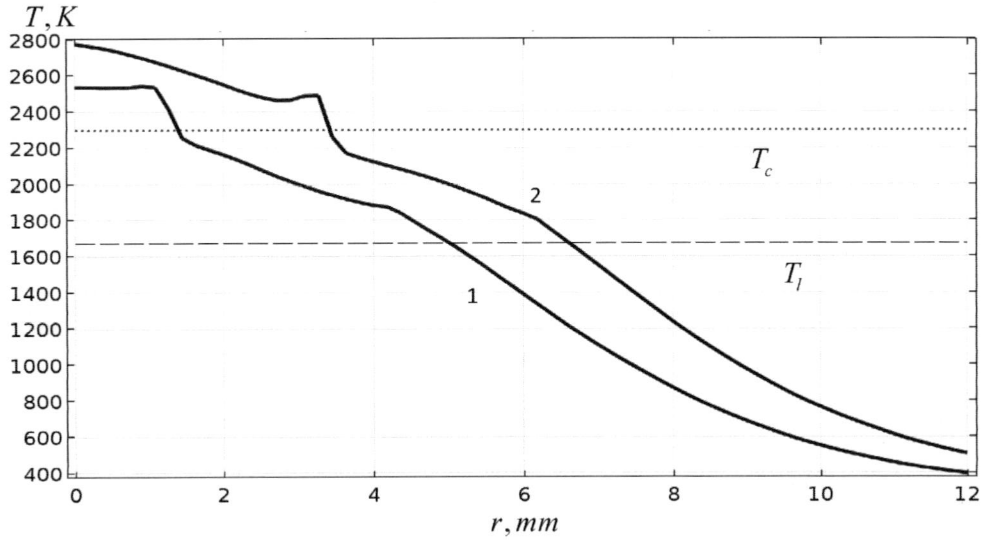


Figure 6. Dependence of surface temperature on the distance from the center. 1 – time interval t=2.7s, 2 - t=4s

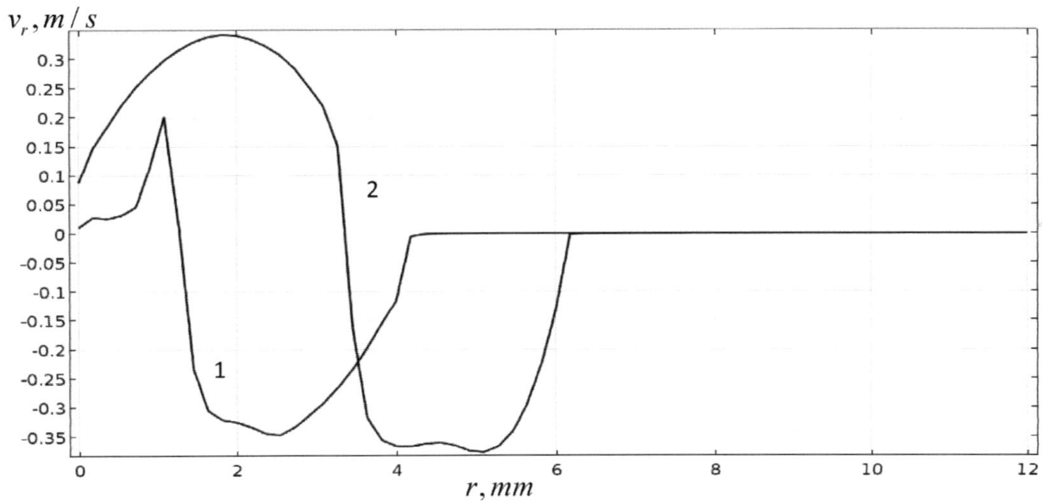


Figure 7. Dependence of radial speed on the surface of the pool upon the distance from the center: 1 – t=2.7s, 2 – t=4s

Modeling of the processes of convective flows of plasma jets [10] allowed formulating the basic requirements to the thermal cycle of surfacing for obtaining a layer surfaced with flux-cored electrodes with the smaller content of alloying elements and basic performance characteristics (hardness, thermal endurance, wear resistance) with high enough strength and viscosity.

Thus, to obtain high performance characteristics of the surfaced layer and reduce the costs on the base of modeling we can formulate the basic requirements to the thermal cycle of surfacing for efficient use of alloying elements. To do this it is necessary to develop the methods allowing combining the surfacing and the hardening. To improve the cold cracking resistance of the surfaced parts at the first stage of the thermal cycle it is necessary to prevent the overheating in the heat-affected zone [11] and formation of coarse columnar structure in the surfaced metal. This is used when welding heat-treated steels by regulating the parameters of the welding thermal cycle through reduction of the heating time and increase of the cooling rate within the temperature range of the least stability of austenite.

The specific feature of the suggested thermal cycle is in three stages of thermal heating. The first ensures the limited time of heating and the increased cooling rate in the high-temperature region, thus, preventing growth of grain and austenite transformation with formation of equilibrium low-strength structures. The first stage of the thermal cycle is achieved by application of highly concentrated heating sources and accompanying cooling of the surfaced part. The second stage of the thermal cycle ensures that the surfaced metal remains in the austenite state when adding all layers in the process of surfacing. It is achieved by application of preheating by 50-100 C over the temperature of abnormal plasticity. Cooling of the surfaced part at the third stage of the thermal cycle after the surfacing is over induces phase transformations in the chrome-tungsten surfaced metal accompanied by development of abnormal plasticity.

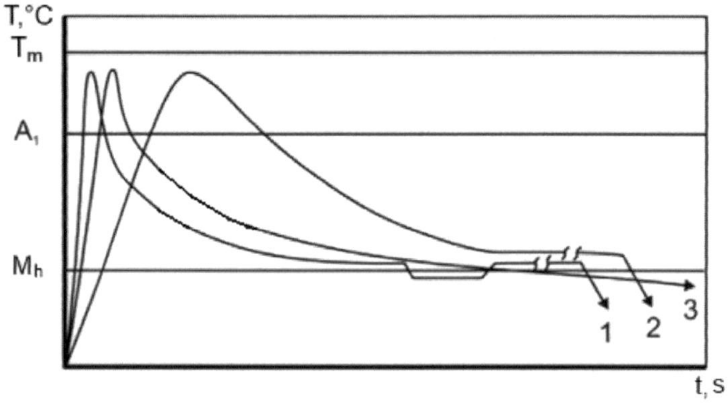


Figure 8. The welding thermal cycles: 1 – under the multiple-bead deposit of hardening steels; 2 – under surfacing thermostrengthened steels; 3 – under electron-beam surfacing

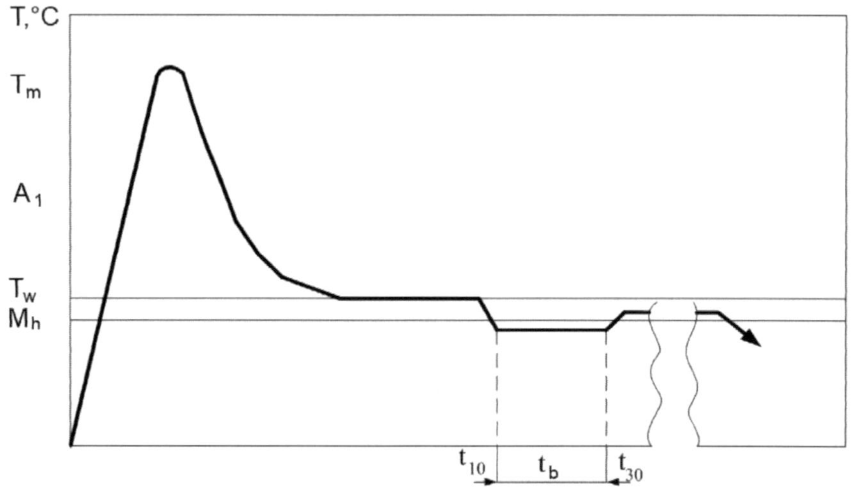


Figure 9. The diagram of the thermal cycle under plasma surfacing

Under these conditions the internal temporary stresses in the surfaced item are partly relaxed which allows obtaining the surfaced metal with the required mechanical properties under lower concentration of alloying elements in the surfacing materials with low levels of residual stresses and, thus, with low ability to cold cracking.

## Conclusion

1. Modeling of convective flows and determining the bulk forces acting on the liquid as well as calculation of electromagnetic fields allowed formulating the optimal conditions for transferring the alloying elements of the flux-cored wire onto the faced surface depending on its structure-phase state.
2. The thermal cycles of surfacing for imparting the required properties to the formed surface were determined. The specific features of the given cycles are the increased metal cooling rates in the region of minimal stability of austenite and keeping the austenite condition of metal until the crystallization process is over.

The work was completed within the grant of the Russian Science Foundation (project № 15-19-00065)

## References

[1] Influence of the structure of surfaced coatings on the abrasive wear resistance / G.G. Chernyshev, N.V. Kobernik, A.G. Orlik, T.A. Chernysheva // Physics and chemistry of treated materials. – 2011. – № 5. – P. 44–50
[2] Finkelburg W, Maecker H 1956 Elektrische bogen und thermisches plasma (Berlin: Springer).
[3] Shih T -M, Zheng Y, Arie M, Zheng J -C. 2013 Numerical Heat Transfer; Part A: Applications 64 435.
[4] Xi Chen, He-Ping Li 2001 International Journal of Heat Mass Transfer 44 2541.
[5] Granovskii A.Y., Sarychev V.D., Gromov V.E. Model of formation of inner nanolayers in shear flows of material // Technical Physics. –2013. –V.58 (10), –P. 1544-1547.
[6] Trelles J P 2014 Plasma Sources Science and Technology 23 054002.
[7] Kim Y-J, Lee J-C 2013 Journal of the Korean Physical Society 62 1252.
[8] Lelevkin V M, Semenov V F 2002 Technical Physics Letters 28 722.
[9] Sarychev V. D., Mochalov S. P., Budovskikh E. A. et al. Formation of convective structures in metals and alloys under the action of pulsed multiphase plasma jets // Steel in Translation. – June 2010. – Vol. 40. – Issue 6. – P. 531–536.
[10] Chinakhov D.A., Sarychev V.D., Granovsky A.Yu., Solodsky S.A., Nevsky S.A., Konovalov S.V. Modeling hydrodynamic flows in plasma fluxes when depositing metal layer on the surface of catalyst converters. IOP Conf. Series: Earth and Environmental Science 50 (2017) 012050.
[11] Chinakhov D.A., Chinakhova E.D., Gotovschik Y.M., Grichin S.V. Influence of welding with two-jet gas shielding on the shaping of a welding joint. IOP Conf. Series: Materials Science and Engineering 125 (2016) 012013.

# The Prospects of Application of Ashes from Combined Heat and Power Plants (Chpp) in the Primorsky Region for Creation of Protective Fibre-Reinforced Concrete with Improved Impermeability Characteristics

**R S Fedyuk[a], S V Kulichkov[b], L V Andreeva[c], N A Samko[d], P A Novikova[e]**

Far Eastern Federal University
8 Sukhanova St., Vladivostok, Russia, 690090

E-mail: [a]roman44@yandex.ru, [b]seku230@yandex.ru, [c]708738@gmail.com, natalia_samko@mail.ru,[e]sokolpolly@gmail.com

**Abstract.** The article discusses the feasibility of usage of fly ashes from the Primorsky region heat power plants in production of fibre-reinforced concrete with improved impermeability characteristics. Based on the fact that ashes created by the Primorsky region heat power stations are chemically applicable as the filler in concrete composition, those of four Primorsky region heat power plants were analyzed. It is found that ashes of Vladivostokskaya HPP-2 and Artyemovskaya HPP comply with the required indicators of specific effective activity and can be classified as the first class materials applied in all kinds of construction operations. The development of the composition binder involving acidulous ash is the most promising.

## 1. Introduction

Composition materials synthesis has come into widespread use over the past decades. It allows obtaining composition materials for construction with wide range of properties. Binding materials are nominally divided into two types: binders obtained by means of mixing several components, and binders as the product of mechanochemical activation of Portland cement or other binder with chemical modifiers containing water-reducing component, mineral admixtures and, if necessary, special admixtures for obtaining binder with special properties.

To obtain composition binder raw materials available in the Primorsky region, such as fly ashes from HPP, were used. Composition binder being developed is designed for application as part of protective fibre-reinforced concrete with improved impermeability characteristics.

For this purpose we studied in the paper physical-mechanical and chemical properties of fly ash from the largest HPP in the Primorsky region: Vladivostokskaya HPP-2 and Artyemovskaya HPP, Primorskaya GRES, and Partizanskaya GRES. Possibility of dry separate screening was an important factor in ash selection. Now such method is implemented at the mentioned heat and power plants.

Wastes of the heat power plants are mainly divided into two groups: bottom ash and fly ash, which are different in the method of removal. Fly ash is a more efficient admixture for the concrete composition than bottom ash. One of essential characteristics of fly ash is its high hidraulicity conditioned by chemical interaction of silicone oxides and aluminum oxides with calcium hydroxide that is released during hydrolysis of clinker minerals with development of hydrated silicates and hydrated calcium aluminates that reflects upon hydrated cement strength.

Composition and structure of ash depends on a complex of cofactors: type and morphologic features of the fuel burned, fineness of grinding in the process of preparation, ash content, chemical composition of the mineral matter of fuel, temperatures in the burning area, particle residence time, etc.

## 2. Equipment and devices used in studies

Thermal analysis of ashes was carried out on thermal gravimetric analyzer Shimadzu DTG-60H at the rate of temperature elevation 20 deg./min., in the interval of 20 – 1,100 °C. Quality and quantity formula as well as properties of the source materials and composition binder were studied with the help of standard methods. Mineral composition and structure were analyzed by X-ray diffraction analysis on X-ray powder diffractometer D8 Advance by Bruker AXS. Figure 2.3 - Scanning election microscope CarlZeissCrossBeam 1540XB

Specific volume activity of beta- and gamma-emitting nuclides in loads was determined by spectrometric method with the help of multifunctional spectrometric complex MSC "Gamma Plus". The study was conducted according to the requirements of the following reference documents: GOST 27451-87 "Ionizing radiation measuring means. General specifications", GOST 26864-86 "Ionizing radiation power spectrometers. Methods of basic parameters measurement", Specifications 4362-002-46554900-06 (PLUS.412131.002TU) "Multifunctional spectrometric complex MSC "Gamma Plus". Specifications".

## 3. Results and discussion

According to the microstructure analysis, fly ash is represented in the form of hetero-granular finely-dispersed spherical particles with different size of granules beginning with nano-sized ones (fig. 1).

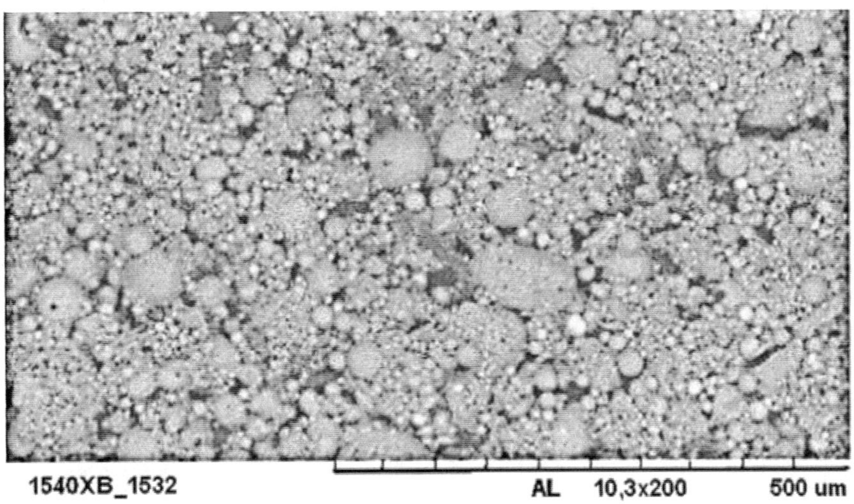


a)

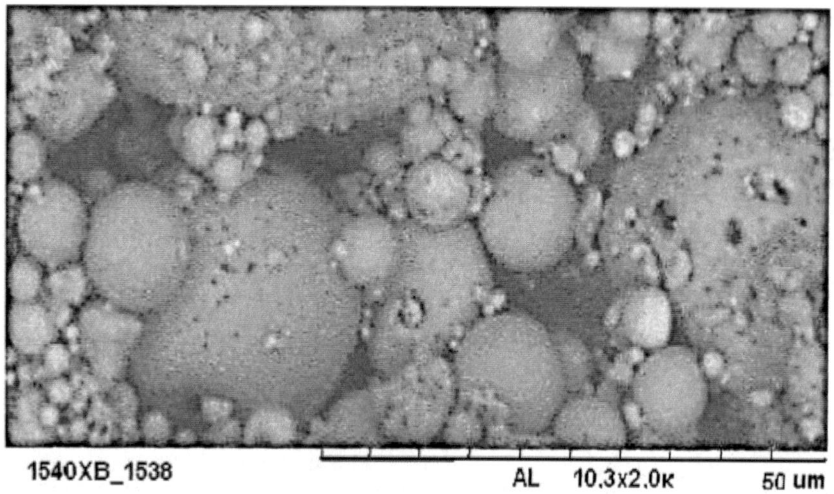


b)

**Figure 1** – Micrographs of fly ash from Vladivostokskaya HPP-2.
200X (a) and 2000X (b) magnification

High dispersion ability of ash is the forecasting factor of its high activity in respect of the components of the binder in case of hydration. Polydisperse spheres present in ash have smooth glassy surface. According to the literature data, glassy phase, proportion of which depends on the burning conditions and the fuel used, is represented by X-ray amorphous aluminosilicate compounds.

Experimental tests have allowed defining the values of the main characteristics of fly ash that determine the possibility of its usage for construction materials production. Chemistry data of ashes indicate the differences by content of certain oxides, as a result of pulverized combustion of different types of coal (table 2.12). So, quantity of $SiO_2$ varies from 47.4 % to 63 %, $Al_2O_3$ – from 12.6 % to 29.3 %, CaO – from 3.4 to 12.5 %. This fact reflects upon ash properties and determines the field of its application as part of construction materials.

**Table 1** – Chemistry of ashes at the Primorsky region HPP

| | | Heat and Power Plant | | | |
|---|---|---|---|---|---|
| | | Primorskaya GRES | Vladivostokskaya HPP-2 | Artyemovskaya HPP | Partizanskaya GRES |
| Prevailing coal type | | Luchegorsky and Bikinsky brown | Primorsky brown (Pavlovsky coal mine) | Black | Neryungrinsky black |
| Chemical composition calculated using oxides, % | $SiO_2$ | 55.3 | 63.0 | 48.1 | 47.4 |
| | $TiO_2$ | 0.5 | 0.5 | 0.0 | 0.9 |
| | $Al_2O_3$ | 12.6 | 21.4 | 29.3 | 22.3 |
| | $Fe_2O_3$ | 10.7 | 7.5 | 6.5 | 19.6 |
| | CaO | 12.5 | 3.4 | 9.7 | 4.8 |
| | MgO | 3.5 | 2.1 | 1.8 | 2.8 |
| | $K_2O$ | 1.0 | 1.3 | 1.2 | 0.1 |
| | $Na_2O$ | 0.4 | 0.3 | 0.2 | 0.4 |
| | $SO_3$ | 3.4 | 0.6 | 2.3 | 1.62 |
| | Free CaO | 1.0 | 0.4 | <0.1 | - |
| | o.i. | 2.3 | 1.4 | 0.6 | <5 |

According to the requirements of GOST 25818 – 91, by the type of coal combusted the ashes are divided into:
– acidulous – anthracitic, black coal and brown coal, containing up to 10 percent of calcium oxide (Vladivostokskaya HPP-2, Artyemovskaya HPP, Partizanskaya GRES);
– base – brown coal, containing more than 10 percent of calcium oxide by weight (Primorskaya GRES).

High content of $Al_2O_3$ (up to 29.3 %) and $SiO_2$ (up to 63 %) in ash can be the reason of crystallization of mullite-based compounds. Based on determination of losses on ignition that comprise less than 5 percent, there are traces of fuel remains in the ashes. This fact suggests positive influence on increase of hydrated cement strength in case of substitution of Portland cement by up to 50 percent of fly ash.

The results of thermal analysis are graphically represented at Fig. 2.

During heat treatment of ash in the range of 40–200 °C water loss occurs due to adsorption by finely-dispersed surface of particles. Carbonates are degraded into CaO and $CO_2$ at 712 °C. Intense endoeffect with weight loss at 500–700 °C testifies to the burn-up of residual fuel that is probably represented by coal particles as well as coke and semi-coke residues. Relatively small exothermic effect with maximum at 932 °C reflects crystallization of mullite-based compounds in aluminosilicate phase.

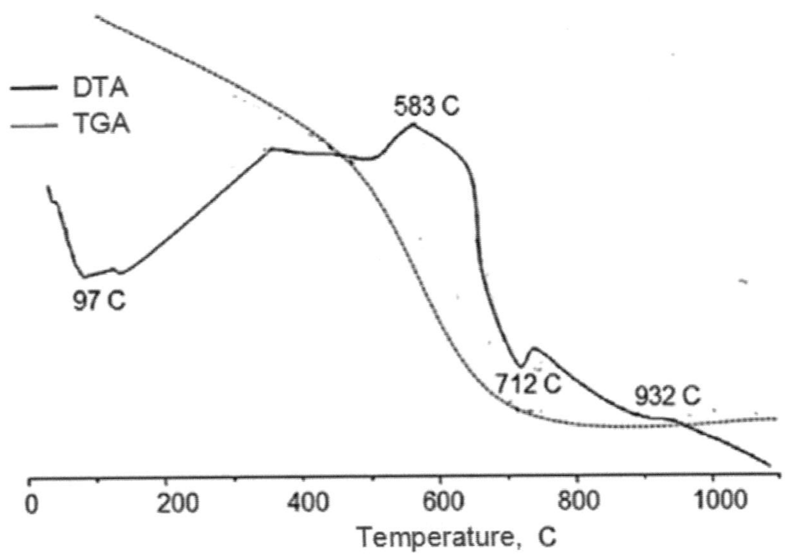


**Figure 2** – The results of DTA and TGA of fly ash from Vladivostokskaya HPP-2

According to the results of X-ray diffraction analysis, diffractive crystalline reflections of mullite phase are identified in addition to quartz crystals (fig. 3).

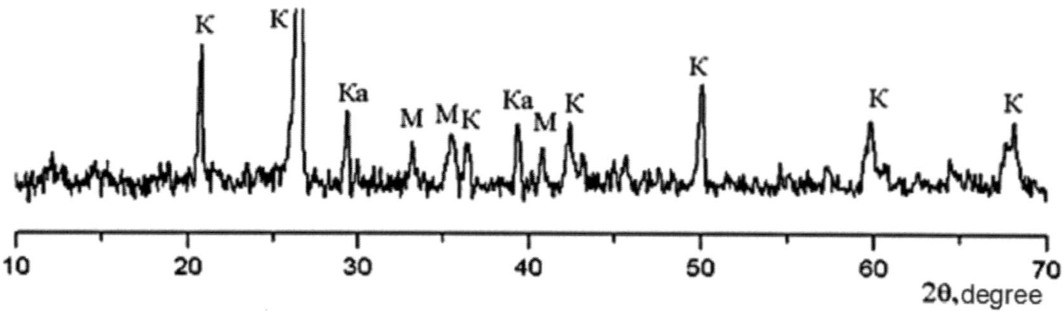


**Figure 3** – The results of X-ray fluorescence analysis of fly ash from Vladivostokskaya HPP-2. K – quartz, Ka – calcite, M – mullite

So, the ash under discussion is very close to aluminum silicates due to high content of silicone oxides and aluminum oxides up to 80-90 percent, of which silicone oxide is about two-thirds. Fly ash almost has no unburnt particles where harmful components are as a rule concentrated. Ash consists of crystalline and noncrystalline phases. Crystalline phase is represented by quartz, field spars, mullite and others, while noncrystalline phase is in the form of glass. So, one may suppose that ashes from the Primorsky region heat power stations are chemically applicable for usage as the filler in concrete composition.

According to the classification of GOST 24640-91 the ashes under discussion are low-calcium (acidulous) and may be used as active mineral admixtures possessing pozzolanic properties.

In accordance with pozzolanic properties (as per classification of RILEM association) fly ashes from Vladivostokskaya HPP-2 and Artyemovskaya HPP may be taken for further studies.

During the following studies a radioactivity background of ash was estimated (table 2).

Table 2 – Specific effective activity of fly ash at the Primorsky region HPP

| Item | Measurement (A), Bq/kg | | | |
|---|---|---|---|---|
| | Primorskaya GRES | Vladivostokskaya HPP-2 | Artyemovskaya HPP | Partizanskaya GRES |
| Activity $^{40}$K | 496.9±101 | 392±89 | 342±68 | 516.9±101 |
| Activity $^{232}$Th | 153.6±20.3 | 31.5±19.7 | 29.5±15.7 | 193.2±22.3 |
| Activity $^{226}$Ra | 163.1±9.36 | 37.63±6.32 | 27.23±5.93 | 113.1±6.37 |
| $A_{eff.} = A_{Ra} + 1.31A_{Th} + 0.085A_K$ | >398 | 80±30 | 93±20 | >410 |

## 4. Conclusions

Analyzing the results, it should be noted that specific effective activity of ashes at Vladivostokskaya HPP-2 and Artyemovskaya HPP is less than 370 Bq/kg. So, according to GOST 30108-94 "Building materials and elements. Determination of specific activity of natural radioactive nuclei" they may be classified as the first class materials applied in all kinds of construction operations.

Radioactive background of ashes at Primorskaya GRES and Partizanskaya GRES exceed the limits, so their usage in the construction industry should be restricted.

So, according to GOST 25592-91 "Mixes of fly-ash and slag of thermal plants for concretes" ashes from Vladivostokskaya HPP-2 and Artyemovskaya HPP comply with the indicators of specific effective activity.

As further studies showed, the use of HPP waste as nanodispersed component of composition binder is more efficient as compared to commercial additives. It is conditioned by the additional useful function – clearing of ash disposal areas. In addition to the above, if viewed from that angle, development of the composition binder involving acidulous ash is the most promising.

As a result, a composition binder consisting of cement (55%), acidulous ash (40%), and lime-stone (5%) was developed by means of intergrinding with superplasticizer up to specific surface of 550 kg/m$^2$ with activity of 77.3 MPa. It makes possible to develop a cement stone of a denser structure.

**References**

[1] Lesovik V.S. Use of industrial wastes in production of construction materials / V.S. Lesovik. – M.: Vyssh. shk., 1987. – 111 P.

[2] Lesovik R.V. *The use of anthropogenic sands for production of fine grained concrete* / R.V. Lesovik // Construction materials. 2013. – No. 9. –P. 78.

[3] Ovchinnikov R.V. Modification of cement concretes structure with the fillers of bottom ash wastes of Novocherkasskaya GRES: diss. … PhD Tech. 05.23.05 / Ovchinnikov Roman Valeriyevich. – Novocherkassk. South-Russian State Polytechnic University, 2014. – 180 P.

[4] Fedyuk R.S. Increase of impermeability of fibre-reinforced concretes on the basis of composition binder: diss. … PhD Tech. 05.23.05 / Fedyuk Roman Sergeyevich. – Belgorod. Belgorod State Technological University named after V.G. Shukhov, 2016. – 184 P.

[5] Falikman V.R. Nanomaterials and nanotechnologies in modern concretes / V.R. Falikman // ALITINFORM International analytical review. 2011. No. 5-6 (22). P.34-48.

[6] Bogusevich V.A. Fine grained concrete on the basis of anthropogenic sands of Kursk Magnetic Anomaly for cold-weather concreting: diss. … PhD Tech. 05.23.05 / Bogusevich Viktor Alexandrovich. – Belgorod. BSTU named after V.G. Shukhov, 2014. – 172 P.

[7] Khozin V.G. Feasibility of application of fly ash from Gusinoozerskaya GRES as part of low water demand cements / V.G. Khozin, O.V. Khokhryakov, A.V. Bituev, L.A. Urkhanova // Construction materials. 2011. – No. 7. - P. 76-78.

[8] Pukharenko Yu.V. Scientific and practical basis for formation of structure and properties of fibre-reinforced concretes: diss. … PhD Tech. 05.23.05 / Pukharenko Yuriy Vladimirovich. – St. Petersburg: SPSUACE, 2005.

[9] Tugarina A.O. Fibre-reinforced gas-ash concrete with application of plant polymer products: diss. … PhD Tech. 05.23.05 / Tugarina Anna Olegovna. – St. Petersburg: SPSUACE, 2009. – 166 P.

[10] GOST 25592-91 "Thermal plant fly-ashes for concretes." Specifications (as amended, No.1) – M.: Publishing and Printing Complex Standards Publishing House, 2003.

[11] GOST 25592-91 "Mixes of fly-ash and slag of thermal plants for concretes". Specifications (as amended, No.1) – M.: Gosstroy of Russia, SUE Centre of Construction Design Products, 2002.

# Model of Environmental Development of the Urbanized Areas: Accounting of Ecological and other Factors

**E N Abanina[a], K G Pandakov[b], D A Agapov[c], Yu V Sorokina, E H Vasiliev**

Federal State Budget Educational Institution of Higher Education "Saratov State Law Academy"
Volskaya st., 1, Saratov, Russia,410056

E-mail: [a] elena-abanina@yandex.ru, [b] k_zp@ssla.ru, [c] agapovda@mail.ru

**Abstract**. Modern cities and towns are often characterized by poor administration, which could be the reason of environmental degradation, the poverty growth, decline in economic growth and social isolation. In these circumstances it is really important to conduct fresh researches forming new ways of sustainable development of administrative districts. This development of the urban areas depends on many interdependent factors: ecological, economic, social. In this article we show some theoretical aspects of forming a model of environmental progress of the urbanized areas. We submit some model containing four levels including natural resources capacities of the territory, its social features, economic growth and human impact. The author describes the interrelations of elements of the model. In this article the program of environmental development of a city is offered and it could be used in any urban area.

## 1. Introduction

Nowadays from about 7 billion population of the planet over a half (52%) live in the urbanized space, and the growth rates of number of city dwellers (on average annually 2.2%) considerably exceed rates of increase of mankind [1]. In many cities poor administration and wrong policy lead to quality degradation of the environment, poverty growth, lower economic growth and social isolation. For the purpose of coordination of such activities as settlements development in the UN system the United Nations Human Settlements Program (UN-Habitat) is founded. One of the priority UN-Habitat directions is a permanent urban development.

So for the solution of the above-mentioned problems the following programs are founded in the Russian Federation: social development, ecological development, economic development, development of science, development of industry, development of agriculture, etc. However, these separate programs don't solve a general problem of permanent development of the region taking into account all problems. The small exception is constituted by such strategies of regions development in which along with social and economic development actions of nature protection are also stated [2]. But even such strategy in general don't answer the common goal proclaimed UN-Habitat. The single Program as system of the interconnected subprograms taking into account all peculiarities of the region is necessary for achievement of permanent urban development.

The aim of this work is to develop a theoretical model of the program of environmental development of the urban areas taking into account ecological, human, social, economic factors of the specific territory.

Work tasks: to develop theoretical content of each "layer" of model that will act as a basis for the program of environmental development; to show the connection of all elements of each layer among themselves and between "layers"; to present the structure of this program based on the developed model.

## 2. The research object and work progress

Materials for research are data from the official sources about the state of environment, social and economic constituent. In this research the basis for layers are the maps developed for creation of the general plan of Saratov city formed bases for layers in this work. The content of each layer in various regions (even in various administrative territorial units within one city) will differ, but the shown connections of elements of all layers will be the same for all regions. This model can become a copy for working out the development programs of any urbanized territory.

The object based on which the model will be created is Saratov city. Saratov is the administrative center of the Saratov region. Location – the right coast of the Volgograd reservoir.

Work progress:

1) the natural layer is developed. Natural and resource potential of the territory is determined; all natural resources are graphical displayed on the map.

2) the social group is developed. Human potential is determined; are graphical displayed on the map of a zone of density of population, incidence, and employment.

3) the economic layer is developed. The managing type, types and the level of development of all industries is determined. Because the creation of layers is subordinated to one purpose – creation of the program of environmental development, on this layer are displayed graphically area industry rankings, depending on level of impact on the environment.

4) the layer "an environment condition" is developed. Environmental problems are defined; are graphical displayed on the map of a zone with a formation on the level of pollution of all natural resources: free air, water objects, soil, and snow cover.

5) all layers are compared; the model of the program of environmental development is formed.

## 3. Results and discussion

The first layer is natural. Its constituent elements is a natural variety: presence of all or parts of natural components – surface and underground water, the woods, mountains, ravines; climatic features (temperature condition, humidity of air, etc.). By drawing up model of development of the whole region, on the map-layer the habitats of wildlife are noted, objects that are under the threat of disappearance (loss of a biodiversity) are noted. In region borders the biodiversity purchases great importance, especially recently when the steady tendency to reducing a biodiversity remains: in case of present methods of production and consumption by 2050 in comparison with the beginning of the 21st century the world can lose two thirds (from 61 to 72%) of flora and fauna [3].

The reasons of it are generally connected with human activities (pollution of the environment, destruction of separate animal species, climate change) [4]. Considering the fact climatic conditions in Saratov belong to moderate and continental and are characterized by minor deviations therefore aren't displayed on the map.

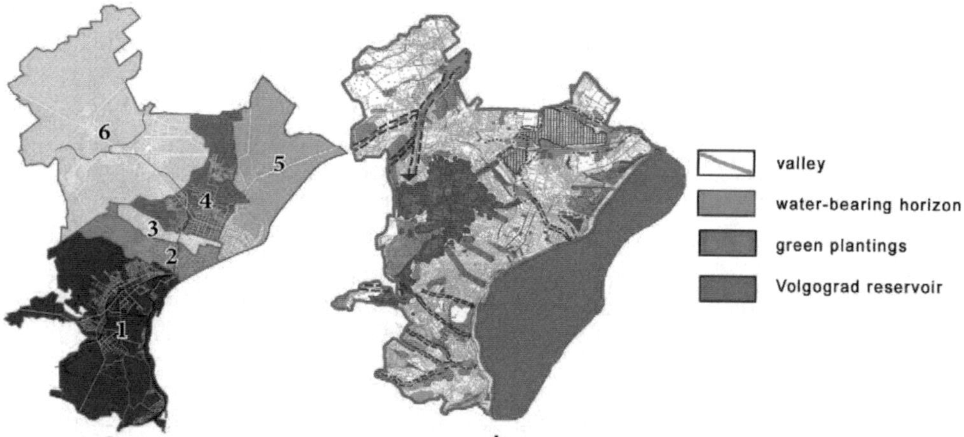


a b

**Figure 1** – An object is Saratov city. a – An administrative-territorial division of the city: 1 – Zavodskoy district, 2 – Oktyabrsky district, 3 – Frunzensky district, 4 – Kirovsky district, 5 – the Volga district, 6 – Leninsky district. b – Model of environmental development. 1 layer – natural

On this layer, it is possible to see not only the security with natural resources of all city but the common problems that are typical for whole territory. For example, by comparison of a layer of "b" and the map "a", we can draw a conclusion on small security with green plantings of the city. The woods settle down in areas 1, 2, 6. Therefore, it is necessary to provide gardening of areas 3,4 in the program of environmental development.

The second layer is social. Treats a social group: structure of labor resource potential; demographic situation; employment rate of the population. A very important indicator is an incidence rate. Types of the incidence are marked on the map: the general incidence, an infectious incidence, incidence by the major not epidemic diseases, and incidence with temporary disability. Serious threats for development of the region are mass spread of HIV infection, tuberculosis, drug addiction. These diseases are the indicator of social trouble, depression of the standard of living, deterioration of sanitary and epidemiologic control in regions [5]. At identification of environmental problems and creation of the program of environmental development paramount value gains population density. The limits of population density characterize the degree of familiarity of the territory, its industrial transformation. The second layer – the map of population density is given below.

The third layer – economic. It includes the type of the specification of the region; managing type. As our task is not the creation of complete model with all these factors, and an illustration of a possibility of creation of such model, during creation of an economic layer we will consider only one factor – production. Maintenance of ecological balance must be based on assessment of activities of all production enterprises occupied in all industries of economy [6]. In structure of the industry in Saratov region the greatest specific weight belongs to fuel and energy complex (45,5%), mechanical engineering (19,1%), chemical and petrochemical (15,6%), food (9,2%) industries.

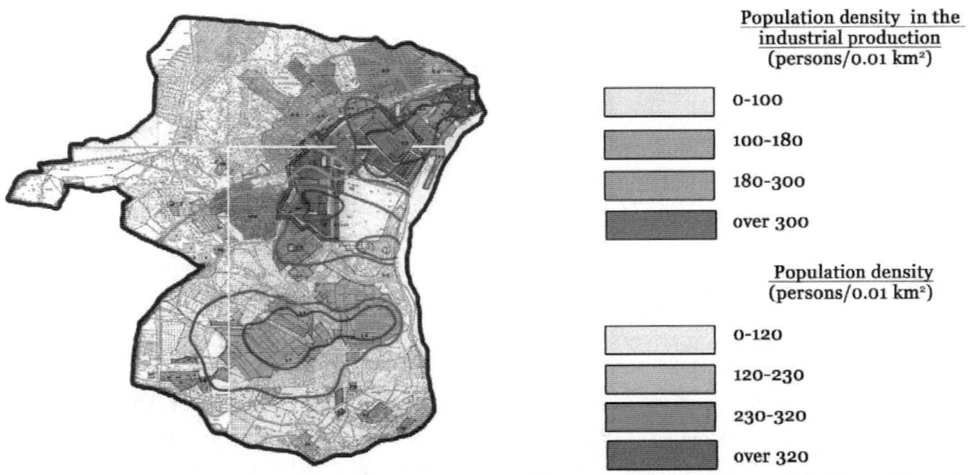


**Figure 2** – Model of environmental development. The second layer is social. The object is the administrative and territorial unit – Zavodskoy district of Saratov city

The modern Russian ecological legislation orders to economic activity subjects which makes considerable negative impact on the environment, to apply the best available technologies to prevent and reduce negative impact on the environment. However not all entities have technical ability to use such technologies. Besides, as the results of researches show that specifics of a regional natural background influence the size of reduction of negative impact on the environment [7]. Therefore, despite all legislative and economic conditions, you can see areas with the objects making considerable negative impact on the environment from the maps of economic development. In model, we note locations of objects of production on a rank depending on the level of negative impact on the environment. Such objects are subdivided into four categories: the objects making significant negative impact on the environment – category objects I; moderate negative impact, – category objects II; minor and minimal negative impact, – category objects III and IV.

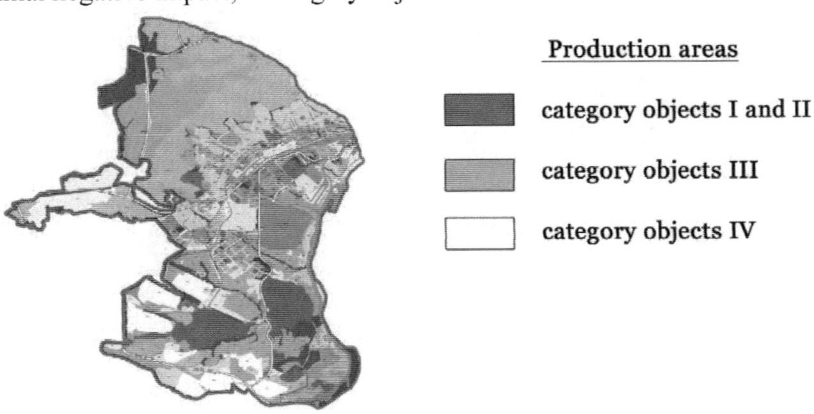


**Figure 3** – Model of environmental development. The 3rd layer is economic. The object is the administrative and territorial unit – Zavodskoy district of Saratov city

The fourth layer – «an environment condition». It includes results of economic activity impact on natural resources: surface and underground water objects, soil, free air. It is one of the most important layers. Based on data of monitoring it is necessary to establish precisely the polluted city zones therefore on the map zones of environmental problems of the urbanized area "appear". These zones influence the amount of financing environmental measures. Lack of attention to environmental problems leads to worsening of influence of pollution on the environment and human health. It is especially important to set precisely zones and limits of air pollution and water objects. As for free air, data of daily monitoring taking into account atmospheric conditions are of great importance. In the presence in the urban areas of dangerous productions, as a rule, there are excessive emissions of pollutants. Need for more complete information demonstrates the fact that about 6% of city fatal cases can be connected with emissions of firm particles [8] to air pollution. The most dangerous elements to the person are carbon oxide and sulfur dioxide. Certain compounds of these substances are toxic: cause changes in blood circulation, strike nervous system, lead to allergic damages of skin [9]. However it is difficult to make sure generalization concerning quality of air in general as the scope of monitoring stations is poor, researches data are limited, and we can rely on government reports (at which there is a restriction of information) [8].

On Figure 1 (b map), you can noted the first water-bearing horizon where the question of safety and security becomes a priority task. As we see in Figure 3, in some places, this water-bearing horizon settles down in zones with intensive negative impact on the environment, and water is, as a rule, polluted there. For example, in 2014 in the Saratov region 1819 underground sources of water supply were operated. On chemical indicators the percent of non-standard tests has made 28,8%. Water pollution became a planetary problem. In science provisions strategies of water security are developed. Threats to water security differ geographically [10]. The feature of the explored city is its favor on the bank of a reservoir; both all-underground and surface waterways get to a reservoir. In addition, it must be taken into account when developing the program of the region.

After identification of environmental problems, it is necessary to perform ranging of these problems on the importance of their consequences and extent of manifestation. Taking into account cumulative impact of economic activity on the environment, depending on pollution level, the territory is divided into zones: I – catastrophic level of pollution; II – crisis, III – critical, IV – rather satisfactory. Zones of warehousing of production wastes and consumption influence the level of pollution. We intentionally do not include this indicator that there was no layer heap, but actually to create the program of environmental development, on this layer we add a zone of placement of waste.

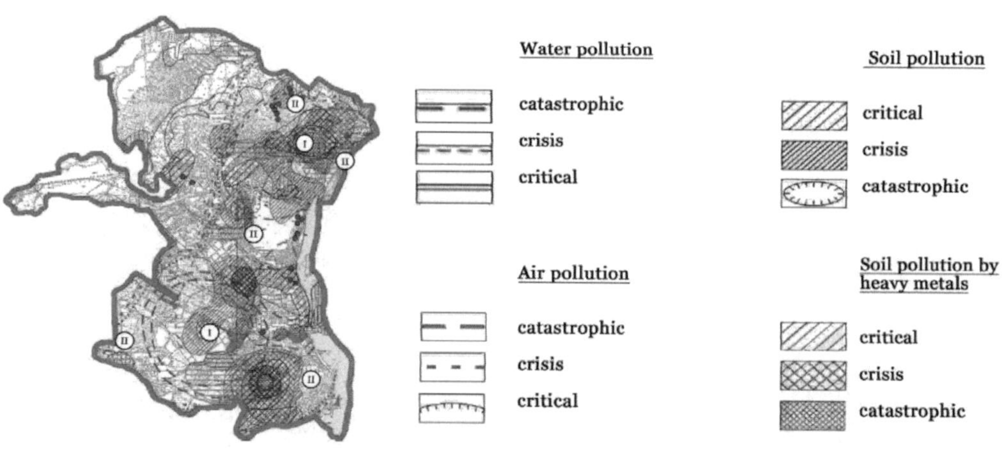


**Figure 4** – Model of environmental development. The 4th layer is an environment condition. The object – Zavodskoy district of Saratov city.

It is necessary to pay attention that availability of information on concentration of pollution, without any subsequent actions in itself won't lead to anything. It is necessary to understand who is a pollution source to have an opportunity to take measures to the entity-pollutant. Therefore, this layer shall be analyzed in a sheaf with a layer of 3 (Figure 3).

Having received four layers (map) with various content, we can make the Model program of environmental development. The main work of this stage is careful comparison of layers of model.

IOP Conf. Series: Earth and Environmental Science **66** (2017) 012019 doi:10.1088/1755-1315/66/1/012019

It is known that than more small scale, especially problems are generally formulated and the most significant problems come out on top. At increase in scale up to regional level, the problems appear inherent in all this territory. Both of these rules are applicable concerning creation of model. Working in small scale, by comparison of all points of maps layers we try not only to define environmental problems, but also we establish connection between these problems and existence of technological and social influence, and we predict possible responses of the environment to negative impact. All this is necessary to define the system of the program of environmental protection and calculation of financial support of such activities.

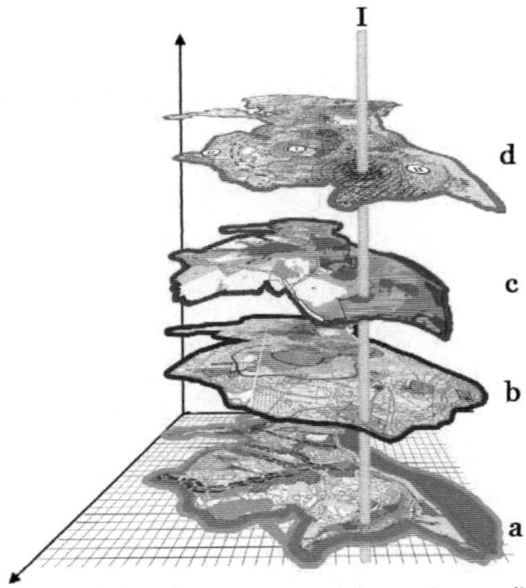


**Figure 5** – Model of environmental development. a – 1 layer (natural); b – 2nd layer (social); c – 3rd layer (economic); d – 4th layer (environment condition). The object – Zavodskoy district of Saratov city

On Figure 5, we determine the characteristic of the certain place.

Point of intersection of the line I and layer "a". In this place there are no forest plantings, the area is crossed by ravines, underground there passes the underground river, level of ground waters – less than 1 meter.

Point of intersection of the line I and layer of "b". In this place small density of population – to 120 people on 1 hectare (foot $^2$). The increased incidence of adult population: diseases of respiratory organs; diseases of digestive organs; blood circulatory system diseases; diseases of endocrine system, frustration of food and metabolic disorder.

Point of intersection of the line I and layer "with". In this place, the production objects making considerable and moderate negative impact on the environment are located. Economy industry types – chemical and petrochemical.

Point of intersection of the line I and layer of "d". The place is characterized by dangerous pollution of the soil, at the same time dangerous and very dangerous pollution of the soil by heavy metals, and pollution of snow cover heavy metals (Zn, Pb, Cu, Ni) – dangerous and very dangerous is noted. Pollution of water objects is catastrophic. Free air pollution – crisis.

All these layers have mutual impact at each other. Dangerous production of environmental pollution increase in the incidence of possible responses of the environment – pollution of the adjacent areas (pollution of a water storage basin surface and underground water) reducing population recession of region's economic development. Thus, the plan of ecological development of this specific place should include the landing of forest plantings, strengthening of ravines, the creation of systems of a water disposal for a water flooding exception, the maximum arrangement of industrial facilities treatment by sewage disposal plant for reducing environmental pollution.

In the same way there is a scrupulous combination of all points of layers of model. As a result we receive the detailed ecological analysis (taking into account economic, social, human factors) of any urbanized area in small scale. In large scale, we can already create the directions of ecological development. Moreover, based on this model to create a basis of the program of environmental development, considering all researched factors (not just ecological). The program includes main units:

1. State of environment of the urbanized area. The basic principle of filling of the block – accuracy of the information.
2. Ecological strategic objectives and tasks of development of the urbanized area (proceeding from permanent development goals). The environmental goals and tasks for the first period (it is created of the directions of development because of work in large scale). Concerning Saratov city, it is reducing of free air pollution, soil, and water objects.
3. Subprograms with indication of the specific nature protection and other actions directed to goal achievement and tasks. Concerning Saratov city: Protection of water resources; Protection and reproduction of forest resources; Free air protection; Development system of the treatment of production wastes and consumption; Ecological education.
4. Indicators of environmental development: amount of emissions of pollutants from stationary sources; the population living in adverse conditions; population disease rate; a share of the utilized and processed production wastes; the number of the held events for ecological education.
5. Financing of ecological actions. The main difference from all existing programs of environmental protection where the direction without binding to the area is financed, has to be stated financing of problem ecological situations based on comparison in this program and the analysis of all layers of model.

## 4. Conclusions

Environmental degradation, poverty growth, decrease in rates of economic growth in many respects depend on incorrectly chosen strategy and settlement development program. Incorrect arrangement of priorities, lack of environmental problems accounting in case of the solution of social and economic problems leads to deterioration in a situation. Based on the assumption of interrelation of ecological, economic, social factors we developed a theoretical model of environmental development of the urbanized area. It is established that all layers of the model exert mutual impact at each other and have a tough binding to the area. Therefore financing of ecological actions should not be the subject (to all areas the identical or conditional amount), and item-by-item: for a specific area within each administrative unit strictly calculated amount taking into account the revealed environmental, social and economic problems. Based on the obtained data the structure of the program of environmental development of the city is provided, can become the framework for Development programs of any urban area.

## References

[1] Bobylev S N, Kudryavtseva O V, Solovyova S V. Sustainable development indicators for cities. Economy of Region. 2014. No 3. pp. 101-110.
[2] Burmatova O.P. Nature conservation strategy for regional socioeconomic development. Regional Research of Russia. 2015. Vol 5. No 3. Pp. 286-297.
[3] Potravny I M, Novoselov A L, Gengut I B. Formalization of the General Model of the Green Economy at the Regional Level. Economy of Region. 2016. No 2. pp. 438-450.
[4] Pereira M, Navarro L M, Martins I S. Global biodiversity change: the bad, the good, and the unknown Henrique. Annual Review of Environment and Resources. 2012. No 37. pp. 25–50.
[5] Chereshnev V A, Verzilin D N, Maksimova T G, Verzilin S D. Environmental and socio-economic development of regions: evaluation of regional differentiation. Economy of Region. 2013. No 1. pp. 33-46.
[6] Karbetova Z, Karbetova S, Otyzbayeva K, Daribayeva A, Dulatbekova Z, Tastanbekova K.

Strategic priorities and challenges of environmental management in Kazakhstan. Journal of Advanced Research in Law and Economics. 2016. Vol 7. No 5. pp. 1058–1065.

[7] Zhuravel' N M. Systematic evaluation of environmental and economic effectiveness of the best available technologies: regional dimension. Regional Research of Russia. 2016. Vol 6. No 2. pp. 193–202.

[8] Henry L A, Douhovnikoff V. Environmental Issues in Russia. Annual Review of Environment and Resources. 2008. No 33. pp. 437–460.

[9] Glyzina T S, Matugina E G, Bagamaev B M, Tokhov Yu M, Kolbysheva Yu V, Gorchakov E V, Sotnikova T V, Shilova A S. Environmental monitoring of natural waters in Krasnodar and Stavropol Territories. IOP Conf. Series: Earth and Environmental Science. 2016. Vol 33. No 1. pp. 12021-12025.

[10] Garrick D, Hall J W. Water Security and Society: Risks, Metrics, and Pathways. Annual Review of Environment and Resources. 2014. No 39. pp. 611–639.

# Optical Method For Monitoring Tool Control For Green Burnishing With Using Of Algorithms With Adaptive Settings

**A A Lukyanov[1,a], S N Grigoriev[2,b], I N Bobrovskij[3,c], P A Melnikov[3,d], N M Bobrovskij[3,e]**

[1]JSC AEROMAT,
Leninskij prospect, 18, Togliatti, Russian Federation
[2]Moscow State Technological University Stankin,
Vadkovskij per.1, Moscow, Russian Federation
[3]Togliatti State University,
Belorusskaya st. 14, Togliatti, Russian Federation

*E-mail:* [a]a.lukyanov92@tehnomasch.ru, [b]Rector@stankin.ru, [c]bobri@yandex.ru, [d]topavel@mail.ru, [e]bobrnm@yandex.ru

**Abstract.** With regard to the complexity of the new technology and increase its reliability requirements laboriousness of control operations in industrial quality control systems increases significantly. The importance of quality management control due to the fact that its promotes the correct use of production conditions, the relevant requirements are required. Digital image processing allows to reach a new technological level of production (new technological way). The most complicated automated interpretation of information is the basis for decision-making in the management of production processes. In the case of surface analysis of tools used for processing with the using of metalworking fluids (MWF) it is more complicated. The authors suggest new algorithm for optical inspection of the wear of the cylinder tool for burnishing, which used in surface plastic deformation without using of MWF. The main advantage of proposed algorithm is the possibility of automatic recognition of images of burnisher tool with the subsequent allocation of its boundaries, finding a working surface and automatically allocating the defects and wear area. Software that implements the algorithm was developed by the authors in Matlab programming environment, but can be implemented using other programming languages.

## 1. Optical method of control as part of green techs

The Human Development Index (HDI) is a summary measure of average achievement in key dimensions of human development: a long and healthy life, being knowledgeable and have a decent standard of living [1]. One of the highest HDI is in United States, Russian Federation is on 50 place. The advantage of the HDI leader's is achieved by concentrating resources in the areas with the highest added value: research, development and design. Production (material stage) has the lowest added value and does not improve

IOP Conf. Series: Earth and Environmental Science **66** (2017) 012020 doi:10.1088/1755-1315/66/1/012020

HDI (for example, the people's Republic of China – 90 place). Thus, improving the wealth of citizens is impossible without the development of the country's science complex.

Creating advanced manufacturing technologies (AMT) is the main directions of research and development in postindustrial countries. AMT represents complex design and manufacturing processes in the modern technological level of individualized material objects (goods) of varying complexity, the cost of which is comparable with the value of the goods of mass production, including in countries with cheap labour [2]. AMT allows to achieve an advantage over countries with cheap labour that is needed to create not only advantageous external conditions to attract investment, but also the competitiveness of technologies in global markets.

In public the analytical report on the development of new production technologies, presented in October 2014 by Skoltech there are four areas that form the shape of a promising production in the world [3]:

1) computer technologies for modeling and production;
2) industrial and service robotics;
3) additive manufacturing;
4) new materials.

One of the main mechanism for increasing of HDI is by creating AMT related to computer technologies for modeling and production. The way for achieving new technologies now realized with National Technology Initiative (NTI) [4]. Roadmap of NTI contains new markets which will use AMT, one of them is TechNet [5]. Platforms for the implementation of new technologies will be created on large industrial enterprises so called "Factories of the future".

The authors have developed two technologies of surface plastic deformation (SPD). One of them is wide burnishing (WB), the main advantage: hypercapability (processing is carried out for several revolutions of workpiece). Other technology associated with the implementation of modern technologies in the sphere of industrial vector – «Green» technologies: burnishing (or dry burnishing) without using of metalworking fluids (MWF) [6]. This technology aims to eliminate negative effects on environment and health of workers, which many oils based on MWF in Russia Federation (based on results of ecological and toxicological characteristics). All parts of new techs must meet the requirements of AMT, including operation of tools control. The mechanical assembly productions applies organoleptic method ("by-sight") condition monitoring of working surface of machining tool that have low confidence. The result completely depends on the qualifications of inspector - a "human factor". The constant routine work degrades the human sensory organs work, contributing to increased measurement accuracy and increasing the probability of defect products. The purpose of the authors was to improve old [7] semiautomatic algorithm. The lack of a formalized description of the key factors contributes to the uncertainty in the process and does not allows to create a universal algorithm for detecting objects on the image. Only for graphics and text segmentation tasks on digital images of documents invented hundreds of fundamentally different algorithms.

## 2. The concept of the algorithm

The reason for the difficulty of establishing methods and algorithms of image recognition is the lack of information that have semantic content images causal or dynamic model formation, the content of the image is not created under the influence of physically laws, which can be described in a mathematical equation, a manifests itself in the form of endless brightness-geometric structures diversity, spawn model of which are missing [8]. In the coming decades occurrence of system "understanding" any visible scene is not expected. Different groups of images of model objects are separate areas of research, such as analysis of objects on aerial photographs [9]. Currently, the field of study is not formed and includes a number of outstanding theoretical tasks and a large number of ideas and approaches.

However, the important task of machine vision can be reduced to a simple issue of detection and recognition or measurement across multiple image objects that match the model description [10]. There are different types of models from simple indicative descriptions to highly specialized and sophisticated structural models, however, there is no general principle of such functioning models in each case, therefore, it is necessary to carry out research. The development and use of models, suitable for effectively addressing the challenges of detecting relevant objects remains on the verge of science and art and represents a «know-how», or knowledge of the subject area, reflecting years of research experience on solving particular tasks.

The algorithm is applied to the field of machine vision – range of engineering technologies, methods and algorithms associated with the task of interpreting the scenes observations on its two-dimensional projections, as well as the practical application of the results of interpretation.

Algorithm aim is to «understand» images of a specific subject area – burnishing tool.

In accordance with the adopted modular D. Marr paradigm [11] image processing is performed on several consecutive levels, from the iconic representation of the object (bitmap, not structured) to a symbolic representation (vector and attributive data, relational structure) and includes the following steps (which can be divided into the processing of lower, middle and high level):

- Image preprocessing;
- Segmentation;
- Selection of geometric structures;
- Determination of the relative structure and semantics.

The algorithm is not related to computer vision systems (discipline, studying theory and basic algorithms to analyze images and scenes), but to computer vision (covers the technological scope of systems development issues: choice of lighting scenes, investigated choice characteristics of sensors, their quantity and geometry of the image sizing issues and orientation, selecting hardware for digitizing and processing, development of algorithms and their implementation) and includes the top-level algorithms (for "understanding" images) using existing low-level algorithms (for "image processing"). The algorithm can also be attributed to the more narrow field of "robots vision" on the required characteristics of the system: the need to build backward linkages by results of understanding images taking into account requirements in time (in the specific case - production time or cycle time).

Factors influencing the understanding of images cannot be formalized due to: interference and noise, occlusion effect, distorting the optical effects, abrupt change of lighting variability in objects, effects of surroundings changes between sensors and objects etc [12]. In this regard, to address similar challenges at the present time there are dozens of algorithms, each of which checks on samples of actual data covering all the negative effects. In the future it is planned to ensure sustainability of the algorithm i.e. robust which is necessary for its commercialization.

Despite the development of computer technology and the creation of specialized processing base for image simple processing, basic computation-intensive algorithms occupying seconds of computer time, that is unsuitable for mass production (important for WB). The practical tasks of image processing, as set forth in this work include many additional degrees of freedom, such as the search bright-geometric structure in the image that can be not only have an arbitrary location, orientation, and scale, but is subject to change, not only is affine or projective, but warping.

The algorithm is a development of previously application software for estimation of wear of the working part of the tool, that algorithm is presented in [7]. It allows to calculate the area of the worn working parts of any instrument by applying two primitives: rectangular and ellipse. The main disadvantage of this algorithm is that the recognition and selection of tools wear is impossible in automatic mode without operator intervention.

The presented algorithm in this paper implements automatic mode to define the boundaries of the tool and its working surface. Calculation of fields wear is made using the methods of the bounds selection on image, such as operators of Prewitt, Canny, Sobel, etc. [13, 14].

At the initial stage of analysis for performance of image processing is a gray gradation - color mode in which images are displayed in shades of gray. Grayscale mode is used to determine the brightness of each pixels color of image on the intensity of the color components in the formula (1) [11]:

$$\mathrm{Lum} = \mathrm{Red}*0.299+\mathrm{Green}*0.587+\mathrm{Blue}*0.114, \tag{1}$$

where $Lum$ – light intensity operator; Red, Green, Blue – components of red, green, and blue, respectively; 0.299, 0.587, 0.114 - the coefficients used for converting colors.

Formula (1) can be represented as integer (2):

$$\mathrm{Lum} = \frac{\mathrm{Red}*77+\mathrm{Green}*150+\mathrm{Blue}*29}{256}. \tag{2}$$

Next, the resulting brightness value Lum is assumed for all color components by the formula (3):

$$\mathrm{Red} = \mathrm{Green} = \mathrm{Blue} = \mathrm{Lum} \tag{3}$$

The basement of the mathematical model for determining the area of tool wear is Prewitt operator [13].

Prewitt operator – the method of boundaries selection in image processing which calculates the maximum response on multiple kernels convolutions to find local targeting boundary at each pixel.

The operator uses two cores 3×3, compressing the original image to calculate approximate values of derivatives: one horizontal and one vertical.

Set: A - original image, and $G_X$, $G_Y$ - two images, where each pixel contains horizontal and vertical derivative approximation which is calculated as follows:

$$G_X = \begin{bmatrix} -1 & 0 & +1 \\ -1 & 0 & +1 \\ -1 & 0 & +1 \end{bmatrix} * A \quad u \quad G_Y = \begin{bmatrix} -1 & -1 & -1 \\ 0 & 0 & 0 \\ +1 & +1 & +1 \end{bmatrix} * A \tag{4}$$

Each point of the image is an approximate value of magnitude of the gradient. It can be calculated, using the approximation derived:

$$G = \sqrt{G_X^2 + G_Y^2}. \tag{5}$$

Gradient direction:

$$Q = arctg\left(\frac{G_Y}{G_X}\right). \tag{6}$$

Under converting a vector graphic to a bitmap occurs error associated with an approximation of smooth borders area, complained of a particular function, discrete boundaries of the bitmap. The transition from a vector graphic to a bitmap is carried out according to the Bresenham's line algorithm [15].

General algorithm of the program is presented in the following block schematic diagram (Fig. 1).

This algorithm can be functionally divided into 2 blocks: 1) determination of the boundaries of the tool and its working surface; 2) analysis of the working surface, determination of defects and wear. Taking into account that on the analyzed instrument image there are 2 circles (the tool's border and the boundary of the working surface), the search for the circles is performed using the Hough transform for circles searching. Considering the uneven lighting and the universality of the algorithm, cyclic search of circles is performed with automatic change of sensitivity settings of Hough transform function.

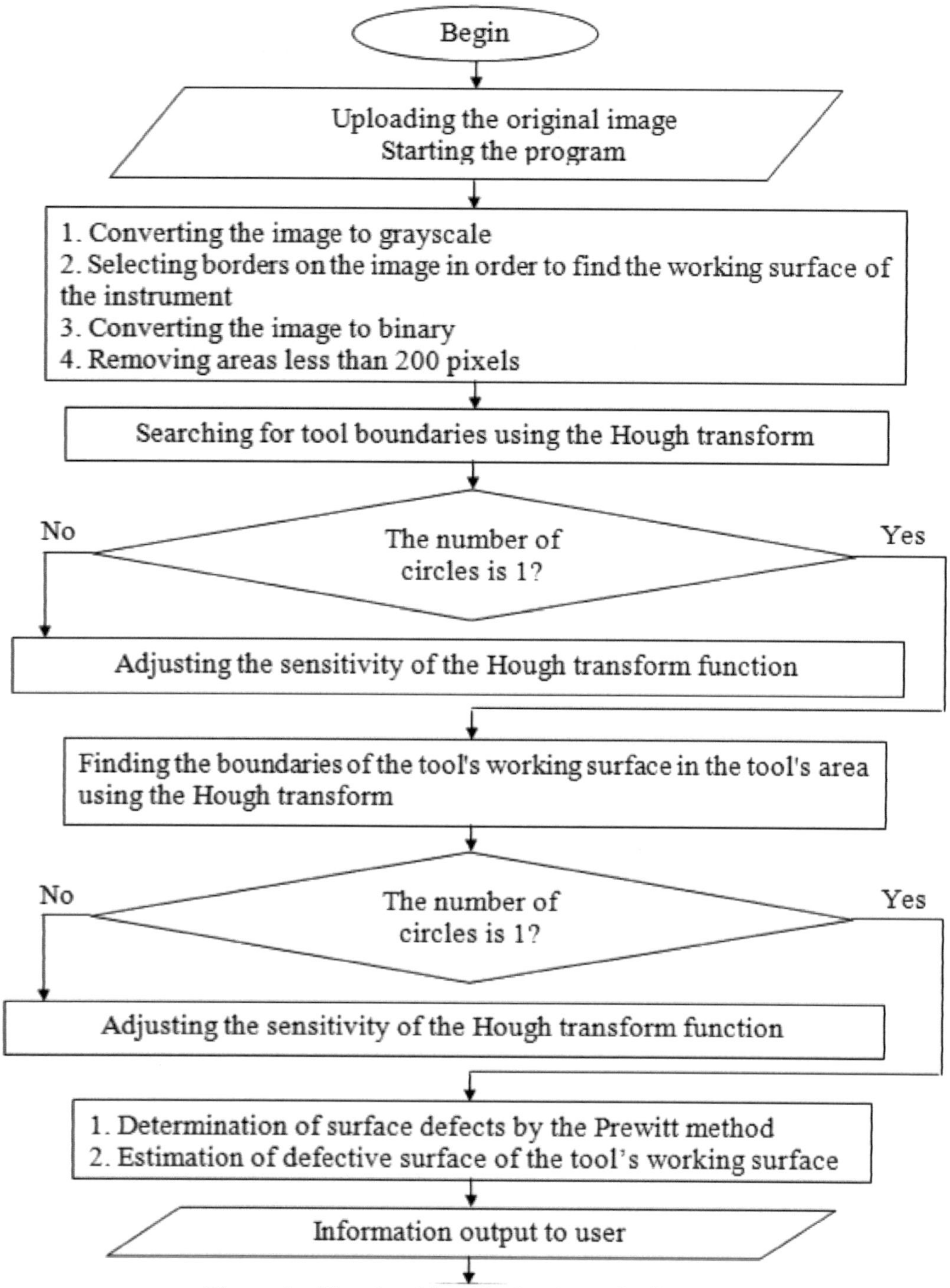


**Figure 1** – The algorithm block schematic diagram

To visualize the work of algorithm the basic steps presented in the following figures.

**Figure 2** – The original image of burnisher

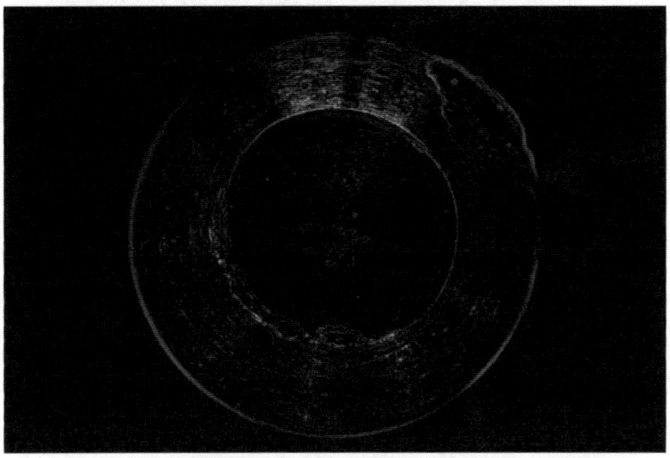

**Figure 3** – Edge detection

The definition of the boundaries of the tool and its working surface is carried out using the Hough Transform (Fig. 4). Method is used to retrieve items from the image. It is designed to search for objects belonging to a certain class of shapes, using the voting procedure. The voting procedure is applied to the parameters from which objects obtain a certain class of shapes on a local maximum in the so-called accumulator space, built during calculating the Hough transformation [16].

The simplest case of Hough transform is detecting straight lines. In general, the straight line $y = mx + b$ can be represented as a point (b, m) in the parameter space. However, vertical lines pose a problem. They would give rise to unbounded values of the slope parameter m. Thus, for computational reasons, Duda and Hart proposed the use of the Hesse normal form:

$$r = x\cos\Theta + y\sin\Theta \quad (7)$$

where r is the distance from the origin to the closest point on the straight line, and $\theta$ is the angle between the x axis and the line connecting the origin with that closest point.

It is therefore possible to associate with each line of the image a pair (r, θ). The plane (r, θ) is sometimes referred to as Hough space for the set of straight lines in two dimensions. This representation makes the Hough transform conceptually very close to the two-dimensional Radon transform [16].

**Figure 4** – Tool's border detection using the Hough Transform

**Figure 5** – Working surface area of wear defined automatically

The above steps are hidden from the user, the program displays only the end result image analysis (Fig. 5). The program window displays information about the percentage of the wear area to the entire working surface of the tool, the program adopted a decision on the possibility of further operation of the instrument based on the previously conducted life tests of different types of a burnishing tool [17].

## 3. Results and Discussion

Software has a high precision on the burnishers cylindrical surface defects. It is excludes influence of human factor on making decision of tool's further exploitation. Adjustable threshold in finding the field of wear allows to optimize the cost of processing a square surface area details depending on customer requirements. Hardware requirements are not high (was used: processor Intel Core i7-2630QM CPU 2.00 GHz, RAM 4.00 Gb), However, the performance of the hardware determines the program execution speed (performance comparison was not conducted due to the variety of configurations), and depends on the customer's needs (based on output).

## 4. Conclusions

A range of methods of surface analysis is appeared due to the increasing growth of computing power, the average level of which is constantly growing. Currently represented algorithm is at the stage of registration of property rights, in consequence this article just the basic components of algorithm are presented in the form of flowcharts. It can be argued, that at the current stage the algorithm is "equipment reliable" if there is no negative factors polluting the surface of instrument or environment (MWF couples), that meets the conditions of work within the green technologies, and can be applied in green burnishing technology processing.

## 5. Acknowledgments

This work is supported by Foundation for assistance to small innovative enterprises in the scientific and technical sphere (The Foundation for promotion of innovation), program «START», project No17680 [18].

## References

[1] I.S. Maiorov, S.Y. Golikov, E.A. Tikhomirova Control of natural management on marine coasts by simulation modeling, International Journal of Circuits, Systems and Signal Processing, 10 (2016) 101-107.

[2] O.V. Egorova, M.Yu. Egorov Development trends in light microscopy, Journal of Optical Technology (A Translation of Opticheskii Zhurnal), 78(1) (2011) 7-18.

[3] A. Babenko, A. Slesarev, A. Chigorin, V. Lempitsky Neural codes for image retrieval, Lecture Notes in Computer Science (including subseries Lecture Notes in Artificial Intelligence and Lecture Notes in Bioinformatics), 8689 LNCS (PART 1) (2014) 584-599.

[4] A. Klimova, E. Rondeau, K. Andersson, J. Porras, A. Rybin, A. Zaslavsky An international Master's program in green ICT as a contribution to sustainable development, Journal of Cleaner Production, 135 (2016) 223-239.

[5] A. Nemov, V. Modestov, I. Buslakov, I.N. Loginov, I.V. Ivashov, A. Lukin, A.I. Borovkov, M.M. Kochergin, E.E. Mukhin, A.E. Litvinov, A.N. Koval, S.Y. Tolstyakov, P. Andrew Development of ITER divertor Thomson scattering support structure design on the basis of engineering analyses, Fusion Engineering and Design, 89(7-8) (2014) 1241-1245.

[6] N.M. Bobrovskij, P.A. Melnikov, S.N. Grigoriev, I.N. Bobrovskij Aspects of thermal field by wide burnishing, IOP Conference Series: Materials Science and Engineering. 91(1) 2015 012035. ISSN: 1757-899X. URL: http://stacks.iop.org/1757-899X/91/i=1/a=012035 DOI: 10.1088/1757-899X/91/1/012035

[7] P.A. Melnikov, I.N. Bobrovskij, S.N. Grigoriev, N.M. Bobrovskij Burnishing Tool Face Optical Control Method, Applied Mechanics and Materials. 770 (2015) 248-252. URL: http://www.scientific.net/AMM.770.248. DOI: 10.4028/www.scientific.net/AMM.770.248

[8] H. Kauppinen, H. Rautio, O. Silvén Non-segmenting defect detection and SOM based classification for surface inspection using color vision, Conference on Polarization and Color Techniques in Industrial Inspection (SPIE 3826), June 17-18, Munich, Germany, 1999, pp. 270-280.

[9] S.Yu. Zheltov, Yu.B. Blochinov, A.A. Stepanov, A.V. Sibiryakov Computer 3D Site Model Generation Based On Aerial Images, SPIE Proceedings. 3084 (1997).

[10] U. Seeger, R. Seeger Fast corner detection in grey level images, Pattern Recogn. Lett. 15(7) (1994) 669–675.

[11] D. Marr.Vision: a computational investigation into the human representation and processing of visual information. Freeman, San Francisco, 1982.

[12] V.A. Soifer, L.L. Doskolovich, D.L. Golovashkin, N.L. Kazanskiy, S.I. Kharitonov, S.N. Khonina, V.V. Kotlyar, V.S. Pavelyev, R.V. Skidanov, V.S. Solovyev, G.V. Uspleniev, A.V. Volkov Methods for computer design of diffractive optical elements. New York, John Wiley & Sons, Inc., Edited by Victor A. Soifer, 2002, 765 p.

[13] S. Mondal, I. Ali An efficient algorithm for boundary detection, Proceedings - 1st International Conference on Computational Intelligence and Networks, CINE 2015, art. no. 7053825 (2015) 184-187.

[14] D.K. Chaudhary, R. Lal, N. Kashyap, T. Choudhury Hybrid edge detection technique for digital images, Proceeding - IEEE International Conference on Computing, Communication and Automation, ICCCA 2016, art. no. 7813883 (2016) 1116-1121.

[15] X.-W. Liu, K. Cheng Three-dimensional extension of Bresenham's algorithm and its application in straight-line interpolation, Proceedings of the Institution of Mechanical Engineers, Part B: Journal of Engineering Manufacture, 216 (3) (2002) 459-463.

[16] L. Calatroni, van Y. Gennip, C.-B. Schönlieb, H.M. Rowland, A. Flenner Graph Clustering, Variational Image Segmentation Methods and Hough Transform Scale Detection for Object Measurement in Images, Journal of Mathematical Imaging and Vision, 57(2) (2017) 269-291.

[17] I.N. Bobrovskij, S.N. Grigoriev, P.A. Melnikov, N.M. Bobrovskij Research Of Hardalloyed Burnishing Tool Durability With Coatings By Ion-Plasmous Sputtering Method, Applied Mechanics and Materials. 770 (2015) 274-278. DOI: 10.4028/www.scientific.net/AMM.770.274

[18] Information on: http://www.fasie.ru/fund/

# Feed Additives Production Out of Dairy Industry Waste

**EV Ulrikh**

Kemerovo State Agricultural Institute
Markovtseva Street, 5, Kemerovo, Russia, 650056

E-mail: elen.ulrich@mail.ru

**Abstract.** Application of macro- and microelements in animal feed is the most effective in the case of their industrial brining in mixed feeds, feed mixes, and protein-vitamin supplements in the form of various complex salts. Application of the product contributes to the body's needs of broiler chickens in vitamins and minerals, normalization of metabolism, and ensures a high rate of growth and development. The composition of the premix can be adjusted depending on the actual proportion of biologically active substances in the feed used by a consumer. It is possible to include in the premix other biologically active substances. Assessing the slaughter qualities of experimental pigs, it was found (Table. 2) that the pigs of group II has a tendency toward greater weight of hot carcass (4.5 kg), of slaughter yelts (by 3.83%) and toward a smaller thickness of fat over the spinous processes of the 6-7th thoracic vertebrae (1.67 mm). The performed investigations have established that there is no significant difference between groups I and II in the content of certain amino acids, however, group I shows poorer results in the content of valine, isoleucine, leucine and lysine by 0.16 g / 100 g of protein ($P > 0.999$) 0.2 ($P > 0.90$), 0.46 ($P > 0.999$) and 0.39 ($P > 0.999$) g / 100 g protein respectively.

## 1. Introduction

Toxicity of feed and feed additives is determined by their ability to have an adverse effect on an animal organism [1].

Unfortunately, feed poisoning still cause significant economic damage to the national economy (death of animals, loss of their productivity). This damage is aggravated by poisoning people who consume milk, meat and eggs produced by these animals. The more significant the damage is, the fewer breeders know about these issues, so they mix toxicosis with other non-communicable or contagious animal diseases. As a result of such incorrect actions the diseases are not eliminated but, on the contrary, are promoted to be spread. [2]

Feed poisoning can be caused by some poisonous plants, poisonous seeds, feed means that clearly obtain toxic properties if they are used incorrectly, as well as animal feed contaminated with certan agrochemicals. In addition, poisoning can occur in the case of using food, affected by fungi and other macro- and micro-organisms [3].

At the first glance the major number of poisoning is harmless and caused by feed additives used incorrectly. This is poisoning with sodium chloride, salts of copper, molybdenum, selenium and fluoride, salts of heavy metals, which often contaminate feed additives of mineral origin, and many feeds of industrial origin.

The following premixes are distinguished: vitamin, mineral, vitamin-therapeutic, vitamin and mineral premixes and others [5].

One of the conditions for obtaining low-cost high-quality products is to use minerals and biologically active substances in animal feeding rations balanced on a large number of nutrients. At the same time a special attention is paid to premixes, mineral and vitamin mixtures [6].

Every year practice of animal feed significantly expands the scope of various feed additives and premixes in particular, mineral and vitamin mixtures. Vitamins and minerals perform a variety of functions, participate in the biosynthesis and maintain animal vital activity. Highly productive animals often experience lack of calcium, phosphorus, magnesium, sodium, sulfur, iron, copper, zinc, manganese, cobalt, iodine, selenium, and vitamin A, D, E, K and group B [7].

Premixes are used for the production of protein-vitamin-mineral supplements (PVMS) and for guaranteed enrichment of mixed fodders, feed mixtures. Following special recipes premixes are produced for all kinds of animals and then marked with alphabetic letters in accordance with the prescription code. The premix recipe is a set of biologically active substances (BAS) and filler, expressed in mass proportion of the individual components in the final mixture.

Premixes contribute to the prevention of diseases associated with a lack of vitamins and minerals. They increase the digestibility of feed nutrients; improve the nutritional and technological value of milk, meat, eggs, fell quality of fur-bearing animals; harden strength of bones and eggs shell [8].

The Russian Federation mainly produces 1% vitamin and mineral premixes with stabilized (protected) forms of vitamins and enzymes. Premixes with increased concentration (0.2-0.5%) should be produced as vitamin concentrates without bringing micronutrient salts. Contact between vitamins does not lead to negative consequences (except for the contact of choline-chloride with vitamins A, $B_1$, $B_6$, K and C, because it can lead to a decrease in vitamins activity). [9]

Application of macro- and microelements in animal feed is the most effective in the case of their industrial brining in mixed feeds, feed mixes, and protein-vitamin supplements in the form of various complex salts [10].

Biological properties of the premix are specified with the presence of vitamins and minerals in its composition in the optimum ratio. Application of the product contributes to the body's needs of broiler chickens in vitamins and minerals, normalization of metabolism, and ensures a high rate of growth and development. The approximate composition of the premix for broiler chickens is shown in Table 1.

The composition of the premix can be adjusted depending on the actual proportion of biologically active substances in the feed used by a consumer. It is possible to include in the premix other biologically active substances (enzymes, amino acids, flavorings, digestion intensifier, chelate compounds of microelements, coccidiostatics, etc.).

The intensity of swine breeding depends largely on full animal feed. To make up for nutritional deficiencies in pigs diet it is used different supplements, including premixes.

There were experimental groups to study ethological reactions of pigs: quantitative in fighting, its duration, duration of free time, feed intake by the procedure of V.I. Velikzhanina (1979) with calculations of the index of functional activity (IFA).

When yelt reached 105 kg in weight, it was slaughtered for assessment of slaughter (meat) qualities. 24 hours later after a slaughter (post-mortem ageing) the tests were taken out of 10 carcasses of pigs in group I and 10 – group II of muscle (rib eye) and fat (subcutaneous fat) tissues at a level between the 9th and 12th thoracic vertebrae to study pork quality.

**Table 1** – The content of micro- and macro-elements per 1 k1 of premix

| | broiler chickens S (1-4 weeks) | broiler chickens G (≥ 5 weeks) |
|---|---|---|
| vitamin A | 2 400 000 IU | 2 000 000 IU |
| vitamin D3 | 700 000 IU | 650 000 IU |
| vitamin E | 25 g | 15 g |
| vitamin B1 | 600 mg | 600 mg |
| vitamin B2 | 1,5 g | 1 g |
| vitamin B6 | 500 mg | 500 mg |
| vitamin B12 | 5 mg | 5 mg |
| vitamin K3 | 600 mg | 600 mg |
| vitamin H (biotin) | 20 mg | 20 mg |
| nicotinic acid (B5) | 6 g | 6 g |
| Pantothenic acid (B3) | 3 g | 2 g |
| Folic acid (Bc) | 150 mg | 140 mg |
| choline cloride | 60 g | 60 g |
| vitamin C | 10 g | 10 g |
| manganese | 20 g | 20 g |
| cobalt | 200 mg | 200 mg |
| copper | 2 g | 0.6 g |
| zinc | 12 g | 12 g |
| iodine | 400 mg | 300 mg |
| iron | 6 g | 6 g |
| selenium | 60 mg | 60 mg |
| Natuphos | 20 g | 20 g |
| enzyme (on request) | + | + |
| antioxidant | + | + |

## 2. Equipment and devices used in studies

The content of amino acids was determined with an amino acid analyzer LC 5700 produced by "Eppendorf-Biotonis" (Germany). For the separation of amino acids it used a standard buffer system. Solutions before introduction into the analyzer were filtered through a Teflon filter with a pore size of 0.22 microns. The work of the analyzer was performed on the standard hydrolyzate program, which includes a stepwise change of the buffer mode and of the chromatographic column temperature in each stage. Standard analysis time from sample introduction to the moment of peak output of arginine took 50 minutes. While the flow rate of eluent was 0.22 ml / min.

The results were biometrically processed by according to standard procedures on a PC using the Excel program.

Considering the dynamics of the linear measurements, it was found that a 2-months. age piglets of group II compared to group I had an advantage in high withers - 19.3%; width - 12.45% (P> 0.99) and chest girth behind the shoulders -16.56%, metacarpus - 16.68% (P> 0.95).

## 3. Results and discussion

Feeding qualities of studied yelts are shown in in Table 2.

**Table 2** – Feeding qualities of studied yelts

| group | age of getting 105 kg in weight, days | average daily gain with fattening, g. | Expense for 1 kg of fattening, fodder unit |
|---|---|---|---|
| I | 256.3 ±3.50 | 455.8 ±17.8 | 4.95 ±0.23 |
| II | 201.4 ±2.70 | 609.1 ±12.9 | 4.32 ±0.19 |

Assessing the slaughter qualities of experimental pigs, it was found (Table. 2) that the pigs of group II has a tendency toward greater weight of hot carcass (4.5 kg), of slaughter yelts (by 3.83%) and toward a smaller thickness of fat over the spinous processes of the 6-7th thoracic vertebrae (1.67 mm).

The slaughter quality of studied yelts are presented in Table 3.

**Table 3** – Slaughter quality of studied yelts

| | Quantity of yelts | yelts body weight before the sluaghter, kg | weight of hot carcass, kg | slaughter yelts, % | thickness of fat over the spinous processes of the 6-7th thoracic vertebrae, mm | weight of back third of semi-carcass, kg |
|---|---|---|---|---|---|---|
| I | 10 | 105.46±3.0 | 68.10±1.05 | 64.49 | 33.48 ±1.10 | 10.43±0.54 |
| II | 10 | 106.24±3.2 | 72.58±2.56 | 68.32 | 31.81 ±0.94 | 11.35 ±0.69 |

Based on the analysis of international experience and existing realities it is believed that the development of new and improvement of existing content and fattening animals technologies it is necessary to must take into account productivity and economic efficiency indexes of meat production, as well as quality indicators.

An important component of meat productivity is an assessment of pork quality. After carrying out evaluation of the physical and chemical properties of muscle tissue of the studied yelts (Table 3) it was determined that the animals of group I show higher pH level (by 0.152 acid. units), water binding capacity (by 3.76%), color intensity (by 2.98 units of extinction), fat proportion (2.1%), and less proportion of protein (0.86%) and lactic acid (at 50.1 mg%) comparing to group II. Physical and chemical properties of yelts muscle tissue are presented in Table 4.

**Table 4** – Physical and chemical properties of yelts muscle tissue

| № group | moisture ,% | fat, % | protein, % | pH acid. unit. | nitrogen, mg/% | lactic acid , mg/% | water binding capacity, % | color intensity, units of extinction x 1000 |
|---|---|---|---|---|---|---|---|---|
| I | 45.62 ±0.26 | 40.78 ±0.52 | 13.6± 0.29 | 5.87± 0.06 | 39.97± 0.19 | 409.9±12.25 | 65.35±0.3 | 57.7±0.63 |
| II | 46.87 ±0.25 | 38.68 ±0.32 | 14.5± 0.14 | 5.72 ±0.02 | 41.46± 0.32 | 460±14 | 61.6±1.1 | 54.7±0.84 |

In general, the physical and chemical indicators of muscle tissue of all carcasses of studied yelts are characterized by good quality. It should be noted that amino acid composition of muscle tissue is a very important indicator of meat quality. Essential amino acids are not synthesized in animal cells, but get into a body as part of food proteins. The deficit of the diet on the essential amino acids leads to a stunted growth, a drop in body weight, and various metabolic disorders and even to death in acute deficiency.

The performed investigations have established that there is no significant difference between groups I and II in the content of certain amino acids, however, group I shows poorer results in the content of valine, isoleucine, leucine and lysine by 0.16 g / 100 g of protein (P> 0.999) 0.2 (P> 0.90), 0.46 (P> 0.999) and 0.39 (P> 0.999) g / 100 g protein respectively.

In the proposed premixes are found some macro- and microelements: calcium, phosphorus, sodium, chlorine, iron, copper, zinc, manganese, cobalt, iodine, etc.

These enrichment mixtures - premixes - provide high economic efficiency: they reduce the feed consumption to 12%. This fact is proved by the data of Table 5.

**Table 5** – Costs of fodder per day for calves and poultry for different diets

| | Kind of feeding | Feed consumption, kg |
|---|---|---|
| 1 | Unbalanced food:<br>For calves<br>For poultry | <br>7-8<br>3 |
| 2 | Full balanced feed for all nutritional factors:<br>For calves<br>For poultry | <br>6-7<br>2.6 |

Premixes are available in the form of powder products. The size of the powder particles of milk-protein premixes is $10^{-4}$ cm. It is proposed to produce protein-fat-vitamin-mineral and carbohydrate premixes. When calculating the prime cost of prime, the cost of raw materials, depreciation of equipment, energy costs, commodity profit are taken into account. Based on the studies performed in this section, it was found that after a 7-month storage, the losses of methionine, calcium pantothenate, folic acid and vitamin A were small, therefore, the storage times of protein and carbohydrate premixes with a moisture content of 10-13% were determined; They are 5-9 months. Premixes are stored in dry, clean, cool, well-ventilated rooms. Can be stored under a canopy.

The results of using premixes for feeding farm animals and poultry are presented in the table 6. Types of farming animal's and birds

**Table 6** – Results of use of premixes for feeding cattle and poultry

| Types of farming animals and birds | Consumption Premix per day per 1 animal, g | Weight gain,% | Advantages of using premixes |
|---|---|---|---|
| Calves | 70-350 | 22.2 | Increasing body weight, reducing the incidence of young animals, increasing biological value and organoleptic indicators of meat, reducing fattening costs, reducing the age of the first calving, increasing the thermal stability of milk. |
| Laying hens | 20-100 | 7-10 | Increase the egg-laying rate of chickens from 213 to 320 eggs per year, reduce morbidity, mortality, embryo freeze, improve the quality of egg protein, the strength of the shell, improve protein-fat metabolism in the body. |
| Broiler chickens | 30-150 | 14 | Increasing muscle mass, reducing the cost of feeding by 10-12%, reducing the incidence and mortality of chickens, improving the quality of meat, improving the immunity and digestibility of nutrients. |

All the studied indicators in the experimental batches of calves, chickens and chickens significantly exceed these parameters in control batches of animals and birds.

## 4.Conclusions

The problem of proper feeding of cattle is particularly relevant today, as the cost of production, the share of expenditure on food occupies larger proportion and according to various sources animal productivity in our plants depends on the feeding factor by 70-90%. Today unbalanced feeding inhibits expression of high genetic potential of animals. in recent decades scientific advances in animal nutrition is far ahead of the practical success of livestock breeding, and for most practitioners there is a lack of current knowledge about normalized animal nutrition.

Premixes do not include antibiotics, tranquilizers, flavors, flavoring agents, stabilizers, detergents, antioxidants, preservatives, vegetable protein and fat, meal, cake, meat-bone and fish meal, and other components usually included in premixes. However, this does not reduce their nutritional and biological value and the economic effectiveness of the resulting protein and lactose premixes.

It has been additionally established that carbohydrate premixes stimulate the development of bifidogenic microflora of the intestines of animals, reduce diarrheal phenomena, which positively affects the vitality of chickens and the growth of productivity of cattle.

As a result of the studies, a technical specification was developed and a sanitary and hygienic certificate for the use of modified flocculants in industry.

The concept of creating a non-waste technology for the processing of whey and dairy flushing water has been developed and the possibility of creating a water rotation cycle at dairy enterprises has been proved by purposeful and complete isolation of fats and proteins from whey and dairy flushing water, followed by processing of the solid phase for vitamin-protein premixes , And the liquid fraction for milk sugar – lactose.

## References

[1] Zhou Y.X. Determination of Kanamycin A in Animal Feeds by Solid Phase Extraction and High Performance Liquid Chromatography with Pre-Column Derivatization and Fluorescence Detection / Y.X. Zhou, W.J. Yang, L.Y. Zhang & Z.Y. Wang // Journal of Liquid Chromatography & Related Technologies. 2007. V. 30. P 1603-1615

[2] Ajila C.M. Bio-processing of agro-byproducts to animal feed / C.M. Ajila, S.K. Brar, M. Verma, R.D. Tyagi, S. Godbout & J.R. Valéro // Crit Rev Biotechnol. 2012. V. 32. P. 382-400

[3] Cowieson A.J. Prediction of ingredient quality and the effect of a combination of xylanase, amylase, protease and phytase in the diets of broiler chicks. 1. Growth performance and digestible nutrient intake / A.J. Cowieson, D.N. Singh & O. Adeola // British Poultry Science. 2006. V. 47. P. 477-489

[4] Jönsson L. Mussel meal as a replacement for fish meal in feeds for organic poultry – a pilot short-term study / L. Jönsson & K. Elwinger // Acta Agriculturae Scandinavica, Section A – Animal Science. 2009. V. 59. P. 22-27

[5] Mendoza Germán D. Considerations on the Use of Exogenous Fibrolytic Enzymes to Improve Forage Utilization / Germán D. Mendoza, Octavio Loera-Corral, Fernando X. Plata-Pérez, Pedro A. Hernández-García, and Mónica Ramírez-Mella // The Scientific World Journal. 2014. Article ID 247437, 9 pages. http://dx.doi.org/10.1155/2014/247437

[6] Nyannor E.K.D. Corn Expressing an Escherichia Coli-Derived Phytase Gene: Comparative Evaluation Study in Broiler Chicks1 / E. K. D. Nyannor, O. Adeola // Poult Sci. 2008. V. 87 (10). P 2015-2022

[7] Gómez S. Combination of an Enzymatically Hydrolyzed Yeast and Yeast Culture with a Direct-fed Microbial in the Feeds of Broiler Chickens / S. Gómez, M. L. Angeles, M. C. Mojica, S.

Jalukar // Asian-Australasian Journal of Animal Sciences (AJAS). 2012. V. 25(5). P. 665-673

[8] Rhodes Melanie A.Use of Dried Fermented Biomass as a Fish Meal Replacement in Practical Diets of Florida Pompano, *Trachinotus carolinus* / Melanie A. Rhodes, Yangen Zhou & D. Allen Davis // Journal of Applied Aquaculture. 2015. V. 27. P. 29-39

[9] Al-Harthi M.A. Impacts of dried whole eggs on productive performance, quality of fresh and stored eggs, reproductive organs and lipid metabolism of laying hens / M.A. Al-Harthi, A.A. El-Deek & Dr. Y.A. Attia // British Poultry Science. 2011. V. 52. P. 333-344

[10] Botsoglou E. Oxidative stability and microbial growth of turkey breast fillets during refrigerated storage as influenced by feed supplementation with olive leaves, oregano and/or α -tocopheryl acetate / E. Botsoglou , A. Govaris, A. Moulas & Dr N. Botsoglou // British Poultry Science. 2010. V. 51. P 760-768

# Developing the System of Monitoring and Diagnostics to Increase the Availability of Equipment

**M S Kovalchuk[a] and A N Skamyin**

Saint-Petersburg Mining University,
2, 21 LineAve., Saint-Petersburg, Russia, 199106

*E-mail:* [a] cheremushkina-ms@yandex.ru

**Abstract**. The purpose of undertaken work, part of which was presented in this paper is to develop a system of collecting and processing information for monitoring, diagnosing, and predicting the future condition of the equipment that is needed to increase the coefficient of readiness and, ultimately, the reliability of restored equipment.

## 1. Introduction

With increasing operating time of the conveyor line, its reliability is reduced in comparison with the project due to depreciation of conveyor parts and handling stations. This is not taken into account in the design of conveyor lines. With the increase of operational time of conveyor lines increases the number of failures, cost of recovery and downtime of conveyor lines and associated loss of coal mining. Therefore, the research and development of methods and technical tools to ensure the required level of reliability of mining equipment and, in particular, the belt conveyor, is an actual task.

Currently undertaken researches and experiments to expand the implementation of the principle of technical operation of components, units, assemblies and subassemblies "on-condition maintenance".

The duration of technical operation for the elements is determined by the limit value of the controlled parameters of the uptime and the amount and frequency of work depends on the value of the controlled parameter. Scheduled maintenance is the only work to control technical condition. Non-routine maintenance and repair, be determined by the results of the control, i.e. the main purpose of the transition to the principles of "technical condition" service:

• improving reliability;
• reduction in maintenance operation.

Efficient method for improving reliability is the repair of failed devices. One means of increasing reliability is to reduce recovery time.

The transition to the maintenance on the "technical condition" principle requires significant improvement in the quality of works performance on technical inspection. Here it is necessary to solve problems related to failure detection and the search of failed elements. The efficiency of diagnosis increases with the use of automated control systems.

Creating monitoring systems is a logical consequence of technological development on the modern stage. Today, the system of automatic control is considered not only as a system of passive supervision and warning, but as an intelligent diagnostic system of monitoring and control.

IOP Conf. Series: Earth and Environmental Science **66** (2017) 012022 doi:10.1088/1755-1315/66/1/012022

## 2. Results and discussion

Availability factor Kg is the probability that the product will be operational at a randomly selected point in time except for planned periods during which the targeted application of the system is not assumed.

Statistically, the availability factor is defined as the ratio of time of correct operation to the total time of correct operation and downtime of the system taken for the same calendar period:

$$K_a = \frac{t_u}{t_u + t_r} \tag{1}$$

where $t_u$ - is the uptime of the system; $t_r$ - the recovery time, i.e. the time spent on preventive maintenance and repair of system.

The efficiency of the conveyor lines operation is greatly influenced by their reliability: in case of failure of conveyors or handling stations decreases the throughput of the transport system due to the reduction of conveyors operational time. When designing conveyor lines the availability factor of the individual conveyor take an equal to , but during operation, it reduced due to physical deterioration of conveyors. The average daily operational time of the conveyor line is reduced in some cases up to 12 hours against 16-18 h, embedded in the design.

Currently, network technologies are becoming more widespread in digital drive control systems of industrial plants. This approach is proposed to use for the diagnosis, monitoring and control of belt conveyors.

Automatic control system must perform the following functions:

- measurement of physical values characterizing the operation of the conveyor;
- a comparison of the values of the measured variables with their defined boundary values (admission control).
- generation with the results of admission control the control actions to the actuators to ensure the maintenance of values of these variables are within specified admissions;

Control system of coal mine includes the monitoring of technological processes and the environment; ensure the collection, transmission, processing, display, recording and printout of data on the main production sectors, and outputs control signals to the underground and overground transport systems of the mine. At the same time monitoring of the mine atmosphere, ventilation systems, temperature, etc.

High-speed CAN bus includes two signal lines CAN_H and CAN_L line and potential-GRD (Fig. 1). The transmission of information is carried out by changing the potential difference between the signal lines. This structure allows to reduce the influence of electromagnetic interference, since the physical bus is done in the form of twisted pairs and common mode interference does not change the potential difference between the lines. To reduce the influence of noise on data transmitted at the ends of the bus terminator resistors are installed. CANbus can be in two States: dominant – corresponding to logical zero and the recessive is a logical one. In case of simultaneous placing on the bus the recessive and dominant status of different nodes on the bus will be dominated by a dominant level. Thus there is the concept of non-destructive bus arbitration when simultaneous exposure of packages.

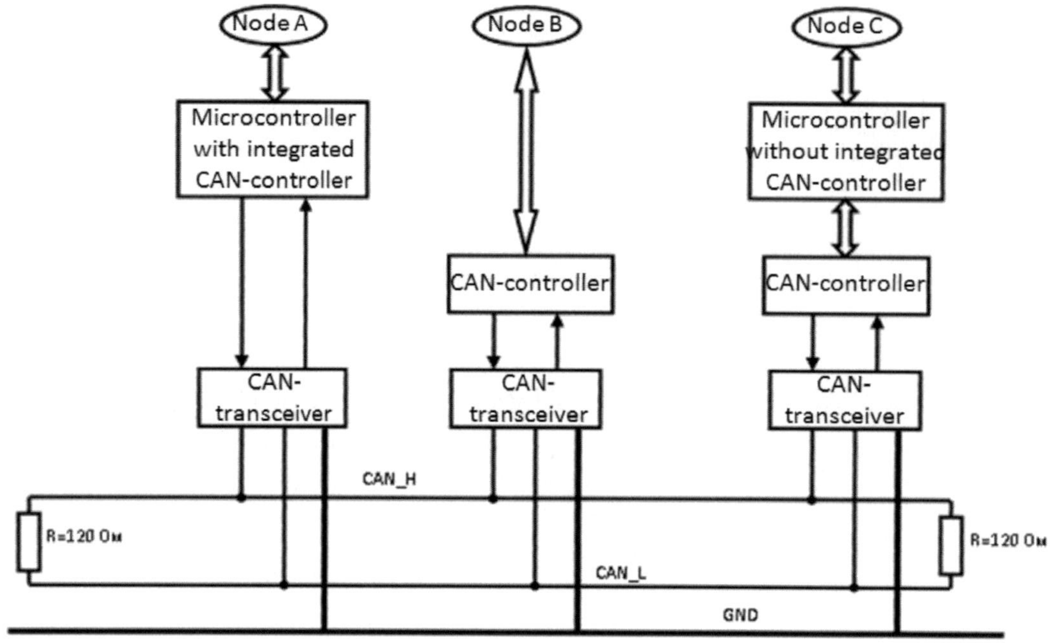


**Figure 1** – Networking based on the CAN interface

Node A contains the microcontroller (with integrated CAN–controller) and the CAN transceiver.

The node B CAN–controller and the CAN transceiver.

A node C - microcontroller (without integrated CAN–controller), external CAN controller and CAN transceiver.

Characteristics of the CAN Protocol show that the application of this Protocol in ACS conveyor transport will provide [6]:

• high reliability of the communication channel;
• built-in self-test;
• registering and recording messages;
• ability to remotely modify (upgrade) the software.

The electric drive system of the conveyor is multi-engine. The system comprises two or more of the motor converters of electrical energy. For example, consider the work of the conveyor shown in Fig. 2.

The flexibility and high noise immunity of the communication system CAN allows to combine the whole electrical and control systems and control units of the conveyors in a common control system. The core of the control system will be an industrial computer that performs the following functions: processing of information (download of elementary information, the task mode, report generation, data exchange with the subsystems, etc.), visualization of status and parameters of the object nodes for staff, communication with external devices (ACS top level of the hierarchy), the formation of local task control systems. The role of slave devices in industrial network will carry the controllers of the individual systems – control systems of drives, control systems of tape tension, monitoring load, etc.

The final goal of diagnostic testing as the last phase of the diagnostic control – determine the type and amount of repairs to increase a residual resource or to restore an operable state [7].

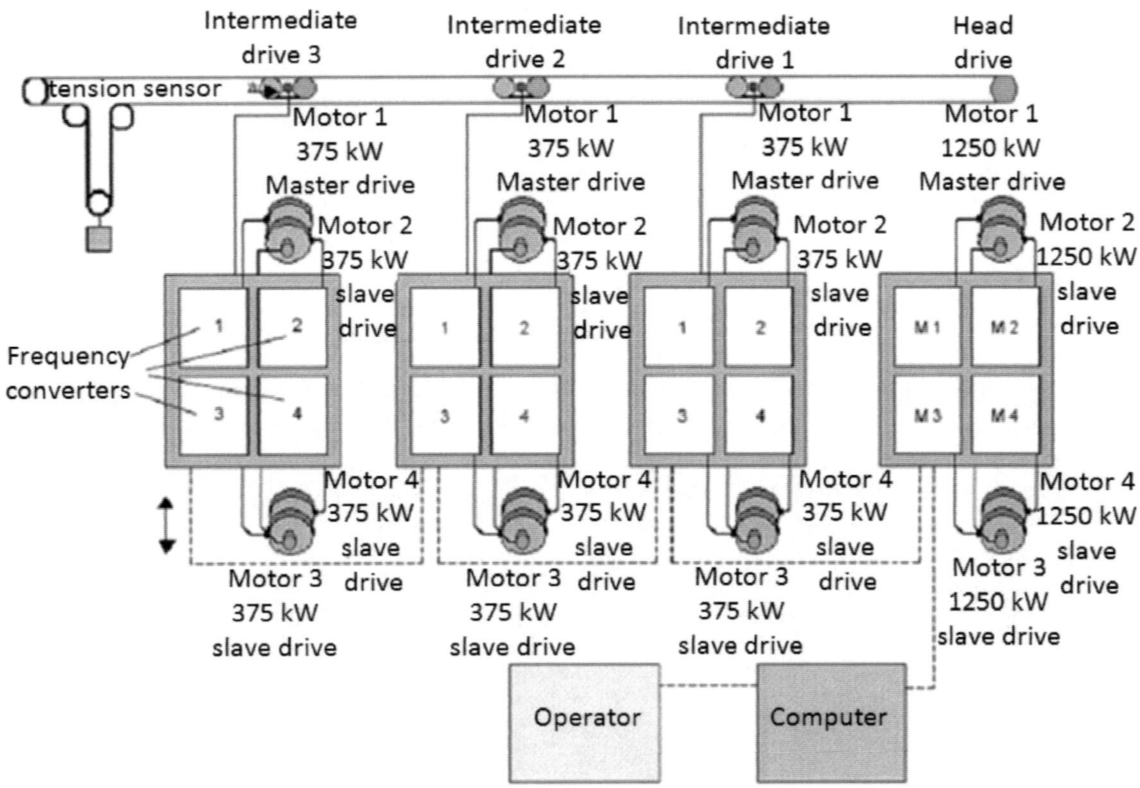


**Figure 2** – The control system of conveyor.

Tasks of the diagnostic condition monitoring of the equipment of coal mine belt conveyor are solved in two ways [8] (Fig.3):

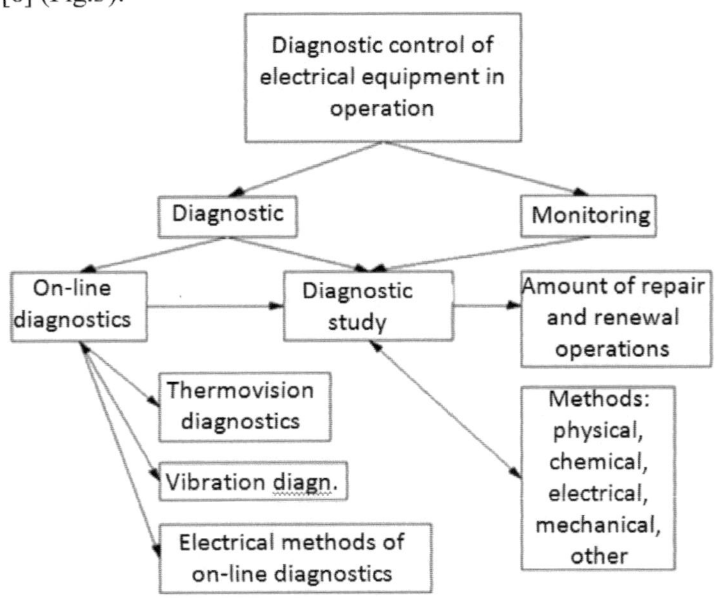


**Figure 3** – Diagnostic control of electrical equipment in operation

- through occasional determination of the most important settings through the diagnostic.

through continuous supervision of the most informative parameters by means of monitoring.

Consider the diagnosis of one of the types of belt conveyors equipment - induction motors. Analysis of factors affecting the reliability of induction motors and the well-known and promising methods of diagnosis, led to the conclusion about expediency of use their main electrical parameters as an informative diagnostic parameters of engines (voltages and currents of phases), the ambient temperature, as well as vibration displacement of the machine casing and, in addition, requires additional parameters which are calculated using existing mathematical models [9].

A number of studies of the damage modes of AC motors has the following statistics /1-2/:

— Damaged bearing elements - 40%
— Damaging the stator - 38%
— Damage to the components of the rotor - 10%
— Other damages - 12%

To study various faults of induction motors, collection of statistical information on failures and future forecasting has been created a laboratory stand consisting of electrical machines, DAQ board and a personal computer.

The scheme of the stand (Fig. 4):

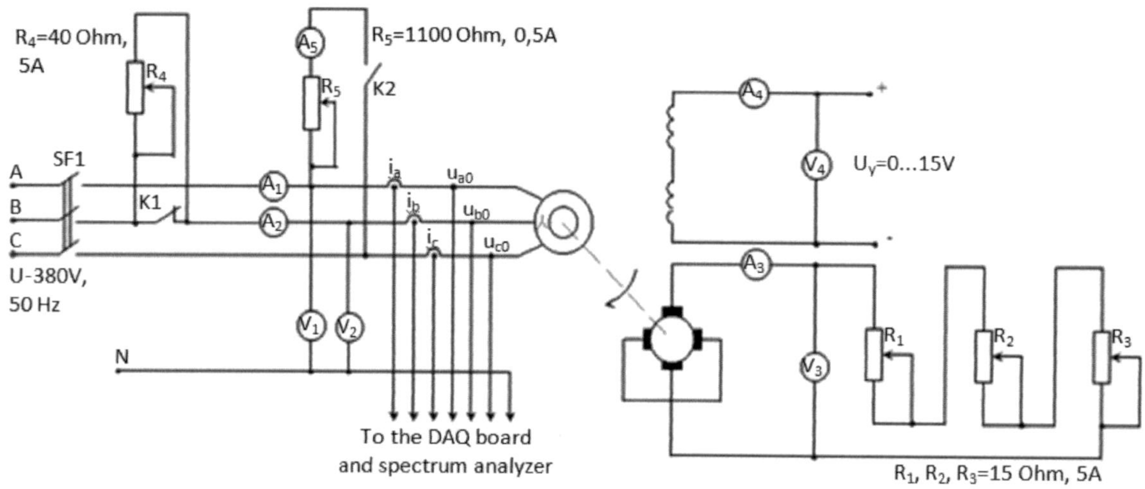


**Figure 4** – The scheme of the stand

The physical principle underlying the operation of a diagnostic system lies in the fact that any disturbance in operation of electrical and/or mechanical parts of the motor and the associated device result in changes in the magnetic flux in the gap of the electric machine and therefore for the weak modulation of consumed electric current.

Thus, the presence in the spectrum of the motor current characteristic (not identical) frequencies of specific size indicates the presence of damage to electrical and/or mechanical parts of the motor and associated mechanical devices.

For example, the breakage of the rods of the rotor winding reduces the starting torque and increases the slip, while the frequency of rotation at no load does not reach the nominal and under load start may not take place.

Breakage of the rods is accompanied by vibration and noise changes periodically at a frequency equal to twice the slip frequency, which depends on the load and the number of damaged rods. At the breakage of two, four and six rods the slip will increase by 8, 17 and 41%, respectively [10].

Fault diagnosis of AD according to the analysis of harmonic spectral components in the voltages and currents of the stator in the electromagnetic and vibro-acoustic processes of functioning of asynchronous drive in artificially-defined faults (various types of short circuits in the stator winding,

IOP Conf. Series: Earth and Environmental Science **66** (2017) 012022 doi:10.1088/1755-1315/66/1/012022

broken rotor bars, eccentricity of the rotor) dedicated a series of publications [2, 4-6, etc.], which presents evidence that the most informative for these purposes, is a 3rd harmonic component.

As an example, below are the results of the spectral analysis of currents.

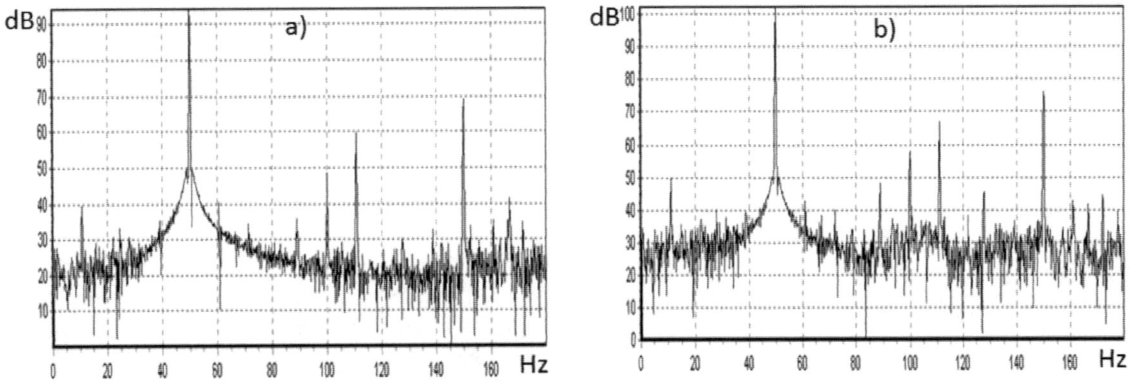


**Figure 5** – Range of voltage in logarithmic scale (a - new motor, b - long operated motor)

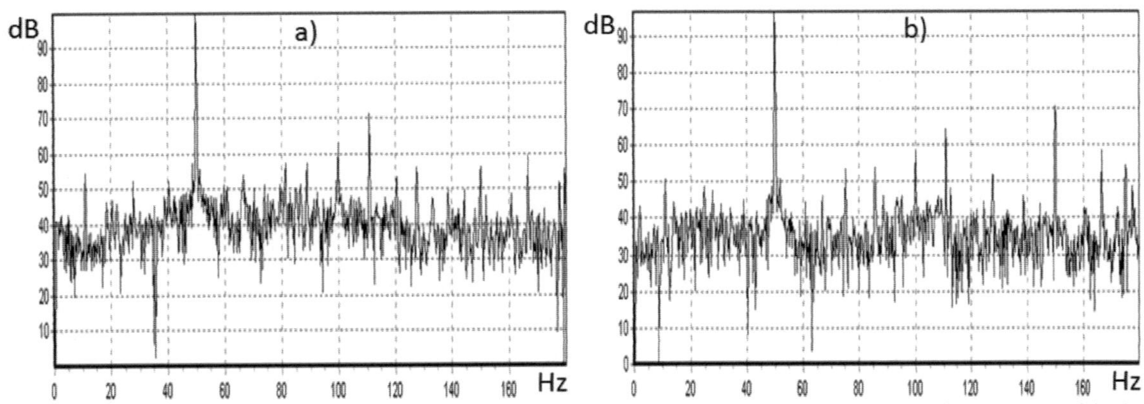


**Figure 6** – Range of currents in logarithmic scale (a - properly functioning motor, b - motor with the broken rod)

In the above drawings it is seen as an increase of the number of frequency bands corresponding to different types of damage and the growth of their value (with meaningful level of signals in the frequency range from -80 dB) for the electric motor, a long time in work (Fig. 5).

Damage to the rotor of the motor (broken rods, loose rods to the slip rings, the hidden defect the casting). This breakdown is detected by the presence of 2 symmetric peaks in relation to the mains frequency in the current spectrum (Fig. 6).

## 3. Conclusions

In a workable system it is necessary to know its behavior in the future. The answer to this question is predictive control. It is the result of the measurement. The received control data is used for planning the future operation of the system. The task of forecasting is extremely important, particularly, for organization of technical maintenance of objects in terms of condition (instead of maintenance by time or by resource).

Well-organized diagnostic support of object with the storage of all previous diagnostic results can provide useful and objective information, representing the background (dynamics) of the process of changing the technical characteristics of the object in the past, which can be used for the systematic correction of the forecast and improve its accuracy.

## References


[1] Baburin S V, Ustinov D A 2016 *International Journal of Applied Engineering Research***11(1)**749-755

[2] Baburin S V, Ustinov D A 2016*International Journal of Applied Engineering Research***11(9)** 6402-6406

[3] Branishtov S A, Tumchenok D A, Shirvanyan A M and Vershinin Y A 2016*International Conference on Mechanical Engineering, Automation and Control Systems (MEACS)***1** 3-8

[4] Abramovich B N, Sychev Yu A, Ustinov D A and Shkljarskiy A Ya 2014 *Neftyanoe khozyaystvo - Oil Industry***8**, 110-112

[5] Liu Q, Peng L, Kang Y, Tang S, Wu D and Qi Y 2014 *IEEE Transactions on Industrial Electronics***61(8)** 4000-4010

[6] Bardanov A I and Skamin A N 2015 *International Conference on Mechanical Engineering, Automation and Control Systems (MEACS)***1** 1-3

[7] Araujo-Vargas I, Salas-Duarte S, Ramirez-Hernandez J, Del-Muro-Cuellar B and Rivera M 2015 *IEEE International Symposium on Predictive Control of Electrical Drives and Power Electronics (PRECEDE)***1** 79-84

[8] Chen Z, Wang Z and Li M 2014 *The40th Annual Conference of the IEEE Industrial Electronics Society (IECON)***1** 1041 - 1047

[9] Jimenez F R, Salamanca J M and Cardenas P F 2014 *IEEEThe5th Colombian Workshop Circuits and Systems (CWCAS)* **1** 1-10

[10] Kopteva A V 2015 Nondestructive measurement of the oil density by using radioisotope radiation*Exposition Oil & Gas. Russia***3(42)**58 - 61

IOP Conf. Series: Earth and Environmental Science **66** (2017) 012023 doi:10.1088/1755-1315/66/1/012023

# Risk Assessment at the Cosmetic Product Manufacturer by Expert Judgment Method

**A N Vtorushina, E V Larionova[a], I L Mezenceva, E D Nikonova**

Ecology and Human Safety Department, Tomsk Polytechnic University
Lenin av. 34, Tomsk, Russia, 634050

E-mail: [a] evl@tpu.ru

**Abstract.** A case study was performed in a cosmetic product manufacturer. We have identified the main risk factors of occupational accidents and their causes. Risk of accidents is assessed by the expert judgment method. Event tree for the most probable accident is built and recommendations on improvement of occupational health and safety protection system at the cosmetic product manufacturer are developed. The results of this paper can be used to develop actions to improve the occupational safety and health system in the chemical industry.

## 1. Introduction

Professional risk management is a part of occupational health and safety management system and refers to organization and sustention of preventive actions focused on optimization of hazards and risks such as prevention of accidents, injuries and occupational diseases. Professional risk management includes complex of actions to identify, assess and reduce the level of professional risk. Implementation of professional risk management is essential to the chemical industry because of increasing hazards of modern process and facilities and more over harmful substances can emit to air or discharge to water or dispose with waste [1,2]. Therefore professional risk management must be a core of preventive health and safety protection system in the chemical industry. There is a variety of methods that have been developed to analyse risk originating from hazardous facilities and substances. Good review of qualitative, semi-quantitative, quantitative and hybrid methods of risk assessment can be found in [3]. Papers [1,2,4] are devoted to analysis of risk assessment methods applied in the chemical industry.

In case of time series accident data over several years are not available the expert judgement method can be recommended for risk assessment [5]. The expert judgement method makes valuable use of prior knowledge and experience of experts and can be made easily and speedily. The disadvantage, however, is that it relies on subjective opinion, hunch and intuition of experts. Limitation of the method is the uncertainties associated with the results [6,7].

The aim of this work is to assess risk of accidents by the expert judgement method and to develop recommendation on improvement of occupational health and safety protection system in a cosmetic product manufacturer. The company is located in Russia, Altai region. The main activity of the company is the manufacture of perfumes, cosmetics and household chemicals.

Legislation in Russia establishes the requirements to risk assessment of accidents and their threats of occurrence, to analysis of the causes of accidents, their factors and risks associated with them, as well as the selection of risk analysis methods for subsequent evaluation. It is known that the reasons for the implementation of accident and injury was a fault of the process equipment, as well as the human factor such as the wrong organization of work, carelessness, neglect of use of personal protection equipment [8]. However, in the Russian legislation in the field of risk analysis the human risk factors are not considered. Thus, the relevant issue is the harmonization of Russian legislation with international standards and practice in the field of risk assessment, such as OHSAS 18001, ILO-OSH 2001, AFS 2001 (AFS 2008:15), ISO 31010:2011.

This paper consider the risks arising both at the fault of process equipment due to its failure or malfunction; and, the human factor, risks associated with the neglect of safety regulations, ignoring the use of personal protective equipment, violation of fire and electrical safety rules.

## 2. Results and discussion

2.1. Risk assessment of accidents by expert judgment method

To evaluate the risk of accidents in the cosmetic product manufacturer the expert judgment method is used. We used the individual opinions of the expert group members based on the preliminary gathering of information from experts that were interviewed independently of each other with the following treatment of the data. In this work a questionnaire method is used. For this purpose the questionnaire is developed. Experts was asked to rate the probability of the events and the severity of accidents on a scale from 1 to 5. According to the process features there are following possible accidents:

1. Electric shock due to the failure of process equipment or insulation.
2. Thermal injure received from the hot components of process equipment.
3. Falling from a height in warehouse.
4. Injuries received from the rotating parts of process equipment.
5. Chemical injury, caused by accident in production process or in analytical laboratory.
6. Injuries caused by the explosion of the equipment which is under pressure.

There are no objective criteria for proclaiming a person an expert [9]. Experts have been selected as objectively as possible among professionals who really stand out for their knowledge and experience.

Risk is evaluated as the product of the probability by the severity of accident. The assessment of accident probability and severity is exposed on a 5-point scale, which is presented in Table 1.

**Table 1** – Evaluation scale

| Probability of the event | The accident severity | Scale |
|---|---|---|
| Very low, the event is unlikely to occur (the probability is from 1 to 20%); | Minor injuries, no need to go to the medical center | 1 |
| Low, the event is unlikely to occur (the probability is from 21 to 40%) | Moderate injuries without disability, appealing to the medical center | 2 |
| Middle, the event is likely to occur (the probability is from 41 to 60%); | Accident at workplace with a temporary disability | 3 |
| High, the event is very likely to occur (the probability is from 61 to 80%) | Accident at workplace with a heavy injuries, with full disability (amputation, etc.) | 4 |
| Very high, the event is likely to happen earlier than it is supposed (the probability is more than 80%); | Fatal accident | 5 |

The results of the questionnaire of experts are presented in Table 2. Then scores of risk are ranked provided that the rank of one is less probable event and the rank of six is the most likely event. If scores of risk received the same ranks, then they were replaced by "standardized ranks" calculated as the average amount of ranks divided among the ranks with the same incidents. The results of the ranking of risk scores are presented in Table 3.

**Table 2** – Results of scoping of the probability accident (P), the accident severity (S) and the risk (R)

| Expert | Accident | | | | | | | | | | | | | | | | | |
|---|---|---|---|---|---|---|---|---|---|---|---|---|---|---|---|---|---|---|
| | 1 | | | 2 | | | 3 | | | 4 | | | 5 | | | 6 | | |
| | P | S | R | P | S | R | P | S | R | P | S | R | P | S | R | P | S | R |
| 1 | 3 | 1 | **3** | 1 | 1 | **1** | 3 | 3 | **9** | 2 | 3 | **6** | 2 | 4 | **8** | 1 | 2 | **2** |
| 2 | 1 | 1 | **1** | 3 | 2 | **6** | 2 | 4 | **8** | 3 | 4 | **12** | 3 | 3 | **9** | 2 | 5 | **10** |
| 3 | 2 | 2 | **4** | 1 | 1 | **1** | 3 | 3 | **9** | 2 | 5 | **10** | 4 | 3 | **12** | 2 | 4 | **6** |
| 4 | 3 | 2 | **6** | 1 | 1 | **1** | 2 | 2 | **4** | 3 | 3 | **9** | 3 | 2 | **6** | 1 | 1 | **1** |
| 5 | 1 | 3 | **3** | 3 | 3 | **9** | 3 | 4 | **12** | 4 | 2 | **8** | 4 | 5 | **20** | 3 | 3 | **9** |
| 6 | 1 | 2 | **2** | 1 | 2 | **6** | 3 | 3 | **9** | 2 | 3 | **6** | 4 | 3 | **12** | 2 | 2 | **4** |
| 7 | 2 | 1 | **2** | 3 | 2 | **6** | 2 | 1 | **2** | 4 | 3 | **12** | 3 | 3 | **9** | 2 | 4 | **8** |

**Table 3** – Results of risk assessments ranking

| Expert | Accident | | | | | |
|---|---|---|---|---|---|---|
| | 1 | 2 | 3 | 4 | 5 | 6 |
| 1 | 3 | 1 | 6 | 4 | 5 | 2 |
| 2 | 1 | 2 | 4 | 6 | 5 | 3 |
| 3 | 2 | 1 | 4 | 5 | 6 | 3 |
| 4 | 4.5 | 1.5 | 3 | 6 | 4.5 | 1.5 |
| 5 | 1 | 3.5 | 5 | 2 | 6 | 3.5 |
| 6 | 1 | 2 | 5 | 4 | 6 | 3 |
| 7 | 1 | 2 | 1 | 5 | 4 | 3 |

Kendall's coefficient of concordance is used to assess consistency of experts' judgments. Kendall coefficient of concordance essentially denotes the average rank order correlation between the cases. If the coefficient is greater than 0.4, it is considered that the judgments of experts have the consistency [10].

In order to estimate significant differences in risk assessments of accidents Friedman ANOVA by ranks is used. In the test the null hypothesis is formulated as an equality of medians in experts' assessments on each accident. The conclusion on rejection or not rejection of the null hypothesis is done by comparing of the achieved significance level $p$ with the fixed one $\alpha = 0.05$. Thus, if the null hypothesis is not rejected then there are only random differences between the risks of accidents. If the null hypothesis is rejected then there are significant differences between the risks of accidents.

The data processing was performed by the Statistica program. The obtained results are presented in Table 4. In addition for the visualization of the data a box plot is constructed (figure 1).

The table 4 shows that the concordance coefficient is equal to 0.59, which indicates a high consistency of experts' judgments. Friedman ANOVA results show that $p < \alpha$, therefore the null hypothesis is rejected. Thus, according to experts' judgments the risk of accidents is significantly different. As it can be seen from the figure 1 and table 4 the most likely accident is the chemical injury.

To identify the main causes and factors of the most likely events realization (that were identified earlier) it is advisable to use an event tree analysis. The algorithm of the event tree construction is the coherent definition of events following from the main event.

**Table 4** – Results of Friedman ANOVA and Kendall's coefficient of concordance

| ANOVA Chi Sqr. ($N = 7$, df $= 5$) $= 20.66390$ $p = 0.00094$ Coeff. of Concordance $= 0.59040$ Aver. rank $r = 0.52213$ | | | | |
|---|---|---|---|---|
| | Average Rank | Sum of Ranks | Mean | Std.Dev. |
| 1 | 2.000000 | 14.00000 | 1.928571 | 1.367131 |
| 2 | 2.000000 | 14.00000 | 1.857143 | 0.852168 |
| 3 | 4.071429 | 28.50000 | 4.000000 | 1.632993 |
| 4 | 4.714286 | 33.00000 | 4.571429 | 1.397276 |
| 5 | 5.357143 | 37.50000 | 5.214286 | 0.809174 |
| 6 | 2.857143 | 20.00000 | 2.714286 | 0.698638 |

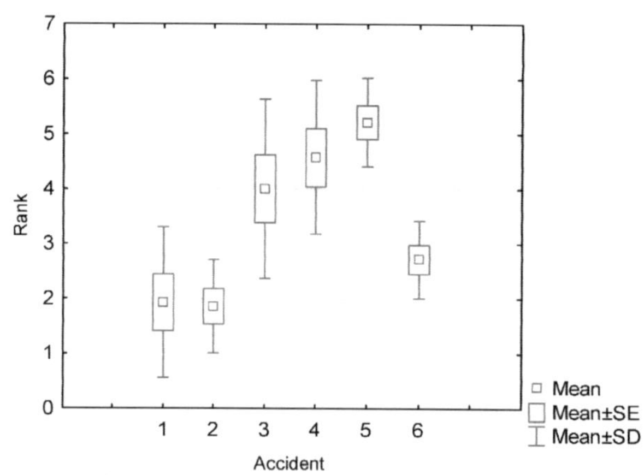


**Figure 1** – Box and whisker plot of risk assessment at the cosmetic product manufacturer

Event tree provides an opportunity to strictly record the sequence of events and identify the relationship between the initiation and subsequent events, the combination of which results in the chemical injury. Thus the event tree is developed in order to model the scenario of the most likely event at the cosmetic product manufacturer (Figure 2).

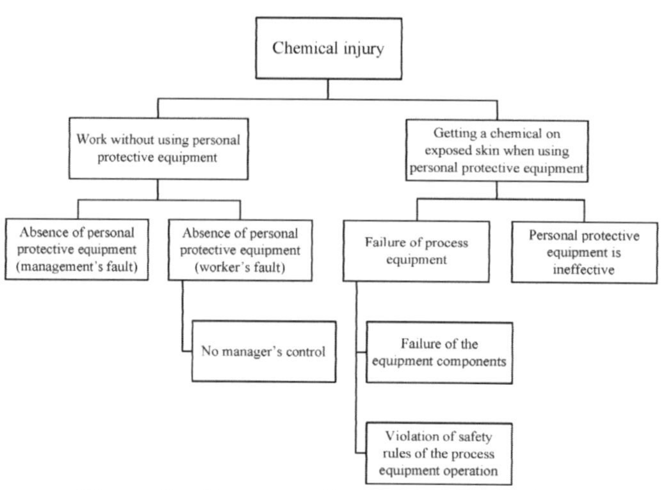


**Figure 2** – Event tree for the chemical injury.

IOP Conf. Series: Earth and Environmental Science **66** (2017) 012023 doi:10.1088/1755-1315/66/1/012023

According to the event tree it can be concluded that the events leading to the realization of the accident depends on technical and organizational factors.

2.2. Recommendations on reducing risk of accidents in the cosmetic product manufacturer

To increase the effectiveness of occupational safety and health system and provide safe working conditions for staff, as well as reduce the risk of occupational injuries, the following organizational and technical actions should be performed:

- organization of periodic checks of expiration date and effectiveness of personal protective equipment;
- organization of staff occupational safety training in accordance with the decision of Russian legislation;
- organization of access control of people and motor transport to the company territory;
- providing conditions for the effective and continuous monitoring of the company area security.

In addition in order to avoid industrial injuries, occupational diseases and to minimize the impact of hazardous and harmful factors at the cosmetic product manufacture it is proposed introduce the following recommendations in the field of occupational safety and health management systems:

- to eliminate sources of hazards and reduce risk to an acceptable level;
- to limit the negative impact of hazards by using of collective protective equipment. For example: to use soundproof covers on the machines to reduce noise;
- to minimize the risk by designing process systems with a limited time of contact with harmful and dangerous process factors or to organize shift work for staff;
- to pay more attention to training of staff in accordance with applicable legislation and local regulations;
- not to neglect the use of personal protective equipment, including protective clothing, despite the fact that PPE do not eliminate the hazards however using it helps to reduce the negative impact of harmful factors down to acceptable.

To improve the overall level of safety should be also taken into account following organizational and technical actions:

1. Industrial safety:

- carrying out a timely technical inspection of process equipment;
- organization of effective control of the implementation of rules, regulations and guidelines;
- organization of staff training in the field of industrial safety;
- exclusion of the possibility of any unregulated work.

2. Fire safety:

- organization of staff training in the field of fire safety;
- fire prevention discipline at operation and maintenance of process equipment;
- completion of the Russian legislation in the field of fire safety.

3. Ecological safety (according to Russian legislation): organization of development and coordination of environmental plan at the cosmetic product manufacturer.

4. Civil and territory protection (according to Russian legislation):

- development a comprehensive plan of the company security improvement;
- providing staff with personal protective equipment and its storage areas;
- organization of training activities for staff to prepare their actions in the case of emergency;
- organization and conducting of the trainings and alarms for the staff to provide possible actions in the case of accident;
- audit of staff knowledge of the rules of action in the circumstances of the accident in the company.

## 3. Conclusion

In this paper we have realized the risk assessment at the cosmetic product manufacturer by expert judgement method. We have preliminary determined the possible accidents at the manufacturer. The data of experts' assessments is processed by statistical methods. Opinion consistency of experts is assessed by Kendall's coefficient of concordance. Significance of differences in accident risk assessments is estimated by Friedman ANOVA by ranks. The experts demonstrated appropriate consistency in their evaluation of risk. It is concluded that the risk of accidents is significantly different and the accident with the highest risk according to experts is the chemical injury. We have built and analyzed the event tree for the chemical injury. Based on the risk and event tree analysis we have offered organizational and technical actions to reduce the risk of implementation of occupational injuries.

## References

[1] Villa V, Paltrinieri N, Khan F, and Cozzani V 2016 Towards dynamic risk analysis: A review of the risk assessment approach and its limitations in the chemical process industry *Safety Science* **89** 77
[2] Sengupta A, Bandyopadhyay D, Van Westen C J and Van Der Veen A 2016 An evaluation of risk assessment framework for industrial accidents in India *Journal of Loss Prevention in the Process Industries* **41** 295
[3] Khan F, Rathnayaka S and Ahmed S 2015 Methods and models in process safety and risk management: Past, present and future *Process Safety and Environmental Protection* **98** 116
[4] Garrick B J 1988 The approach to risk analysis in three industries: Nuclear power, space systems, and chemical process *Reliability Engineering and System Safety* **23**(3) 195
[5] Rosqvist T 2003 On the use of expert judgement in the qualification of risk assessment. *VTT Publ.* **507** 3
[6] Goossens L H J and Cooke R M 1997. Applications of some risk assessment techniques: formal expert judgment and accident sequnce precursors. *Saf. Sci.* **26** 35
[7] Paltrinieri N and Khan F 2015 Dynamic risk analysis in the chemical and petroleum industry (Elsevier)
[8] Reniers G L L 2010 A novel methodology for evaluating the change of existing operational staffing levels (MCSL) in the chemical industry *Safety Science* **48**(2) 179
[9] Goossens L H J, Cooke R M, Hale A R and Rodić-Wiersma L 2008 Fifteen years of expert judgement at TUDelft *Safety Science* **46**(2) 234
[10] Mannan S 2012 Lees' Loss Prevention in the Process Industries (Fourth Edition) vol 1 (Elsevier)

IOP Conf. Series: Earth and Environmental Science **66** (2017) 012024 doi:10.1088/1755-1315/66/1/012024

# The Environment Friendly Power Source for Power Supply of Mobile Communication Base Stations

**N V Rudenko[a], V V Ershov[b], V V Evstafiev[c]**

Don State Technical University
Gagarin Square, 1, Rostov-on-Don, Russia, 344000

*E-mail:* [a]rnv.2017@mail.ru, [b]ervv46@yandex.ru, [c]ebor63@mail.ru

**Abstract**. The article describes the technical proposals to improve environmental and resource characteristics of the autonomous power supply systems of mobile communication base stations based on renewable energy sources, while ensuring the required reliability and security of power supply. These include: the replacement of diesel-generator with clean energy source - an electrochemical generator based on hydrogen fuel cells; the use of wind turbines with a vertical axis; use of specialized batteries. Based on the analysis of the know technical solutions, the structural circuit diagram of the hybrid solar-wind-hydrogen power plant and the basic principles of the algorithm of its work were proposed. The implementation of these proposals will improve the environmental and resource characteristics.

## 1. Introduction

The condition of the normal and smooth functioning of any communication object is to ensure high quality of its electricity supply. Therefore, when designing and constructing the special attention is paid to the power supply elements such as a cellular network base stations (BS), base station controllers (BSC), etc. According to regulatory documents [1], BSC refers to the first, and BS to the second categories of technological power consumers on power supply reliability. This includes the absence of interruptions in electricity supply, the presence of two independent mutually reserving power supplies, as well as presence of two power supplies from the electrical networks of the power system [1].

In practice, there are many situations when different subscriber groups are geographically distant from areas with established structure of the State power system. To date, over 65% of the territory of our country is a zone of decentralized power supply. More than 15 million people are residing in this zone. Of the 225 million kW of installed capacity of power plants in Russia - 17 million kilowatt is accounted to power plants working in the area of decentralized power supply. In the northern parts of our country there are more than 6,000 diesel power plants (DPP), which have a capacity of more than 3 million kW. About half of these DPPs do not work reliably due to interruptions in the supply of fuel and high prices on imported fuel. To the remote areas of the Far North and Far East, the fuel is delivered by rail or by road, and sometimes by helicopter, as well as by river and sea transport with limited navigation period. Such deliveries are unreliable and expensive. The considerable damage is caused to the ecology through inefficient burning of fossil fuels in boilers and diesel power plants

IOP Conf. Series: Earth and Environmental Science **66** (2017) 012024 doi:10.1088/1755-1315/66/1/012024

(DPPs) of low technical level. In the far north it is aggravated by the removal of the empty fuel barrels [2].

In addition, connection to the existing power networks is very costly. Thus, according to the existing rates, according to the indicative calculations, more than 1.5 million roubles will be required for the construction of overhead power lines in the Rostov region on the first level of voltage in the length of 5 km, for organizational and technical measures, as well as for the issue of the relevant project documentation.

Therefore, the construction of the national cellular communication networks with power supply from base stations based on renewable energy sources (RES), taking into account environmental requirements in the regions of Russia, where there is no centralized power supply, is an actual scientific and technical task.

## 2. Results and discussion

The problem is formulated as follows. Develop technical proposals to improve the environmental and resource characteristics of autonomous power supply systems of base stations (APSS BS) of mobile communication based on renewable energy sources, while ensuring the required reliability and security of power supply.

To solve this problem, the preliminary study of the following issues is required:

- the choice of renewable energy sources for the construction of APSS BS;
- development of proposals to improve the environmental and resource characteristics;
- analysis of the known technical solutions;
- development of technical recommendations to the designed installation.

*A. Selection of the renewable energy sources*

This selection is expedient to carry out in view of the particular geographic region. Analysis of the renewable energy resources of Russia leads to the following conclusions:

- areas of effective application of wind turbines (WPP) are regions on the subjects of the Russian Federation: Arkhangelsk, Astrakhan, Volgograd, Kaliningrad, Kamchatka, Leningrad, Magadan, Murmansk, Novosibirsk, Rostov, Tyumen; krais: Krasnodar, Perm, Primorye, Khabarov, republics: Dagestan, Kalmykia, Khakassia, Sakha (Yakutia); autonomous regions: Nenets, Chukotka, Yamalo-Nenets [3,4];
- the most promising regions in the use of solar batteries (SB): Kalmykia, Stavropol region, Rostov region, Krasnodar krai, Volgograd region, Astrakhan region and other regions in the south-west, Altai, Primorsky region, Chita region, Buryatia and other regions of the West, East Siberia, and the Far East [3, 4];
- today, hybrid solar-wind turbines are the most popular, they are a combination of solar panels and wind generators, and often supplemented by a diesel generator. They successfully replace the gas turbines of small power, oil boilers and diesel generators, especially located in the area of decentralized energy. By 2020, the world market of such installations could reach 65 billion dollars. Their use will allow increasing the share of renewable sources in electricity production from 5% to 15% by 2035 [5].

Thus, to ensure the required reliability and security of electricity supply it is advisable to consider the possibility of designing APSS BS on the basis of a hybrid solar-wind-diesel power plant (HSWDPP), because it uses all the advantages, and disadvantages of individual sources are compensated.

*B. Development of proposals to improve the environmental and resource characteristics.*

For the formation of these proposals it is advisable to perform the analysis of the environmental properties of the elements of HSWDPP: diesel power plant (DPP), wind power plant (WPP), solar power plant (SPP).

DPP has the following disadvantages [2]:

- the risk of fire;
- the presence of exhaust gases;

IOP Conf. Series: Earth and Environmental Science **66** (2017) 012024 doi:10.1088/1755-1315/66/1/012024

- the generation of noise and vibration;
- accumulation of empty fuel barrels.

WPP has the following disadvantages [2]:

-presence of noise and vibration;
- generation of infrasonic vibrations by wind turbines with rotor diameter more than 10 m These vibrations have adverse effects on humans and animals;
- the occurrence of electromagnetic interferences in the reception of TV and radio transmitting stations.

Using SPP necessarily implies the presence of sufficient storage capacity of electricity. As a rule, these are ordinary batteries, which pollute the environment.

In order to develop proposals for improving the environmental characteristics, consider the available possibilities. Disadvantages of DPP can be compensated in the following ways:

- fire hazard is reduced by correct installation of diesel generator and the use of fire-prevention automatics;
- exhaust emissions can be reduced by using high-quality motor oil and fuel, regular replacement of fuel and oil filters, the use of the ventilation system, as well as the use of filters in the exhaust system;
- noise and vibration can be reduced by placing a diesel generator in the block container, or in a room with special sound and vibration dampening walls;
- removal of empty fuel barrels can be arranged using the transport that delivers fuel.

All these measures lead to a significant increase in capital and operating costs, which ultimately can not solve the environmental problems associated with the use of DPPs. The improvement of the environmental characteristics of the power plant can be radically achieved only as a result of replacing the diesel generator with the ecological clean energy source - hydrogen fuel cells (FC), in which only hydrogen and oxygen contained in the air are involved in the reaction, and the reaction product is water vapor [6].

Ecological characteristics of WPP can be improved by the use of wind turbines with a vertical axis. It has the following advantages [7, 8]:

- reduced initial speed (1-3 m/s);
- it does not create a vibration on the ground;
- it does not require the installation on high masts;
- harmless (noise level does not exceed 35 dB);
- it does not require orientation to the wind;
- it uses energy not only of horizontal wing, but also updraft wind.

Environmental issues associated with the use of SPPs are currently successfully being addressed as follows:

- the environmentally friendly technologies of production and utilization of solar panels containing cadmium telluride were developed [9];
- the market offers a wide range of specialized batteries for operation in the composition of SPP. These are gel and AGM batteries having the following properties [9]:

a) use in any position (there is no leakage of electrolyte);
b) does not require topping up of water and electrolyte replacement;
c) sufficiently resistant to deep discharges;
d) doubled the cycle life: over 600 cycles of discharge in 100% depth;
e) do not emit hazardous noxious gases into the atmosphere;
f) resistant to strong frosts;
g) the quick commissioning: 105-107% capacity in the first cycle;
h) the service life - 16 years.

Thus, summarizing all the above steps, on can identify the main proposals for improving the environmental and resource characteristics of hybrid power systems for designing and operating organizations. These include the following:

IOP Conf. Series: Earth and Environmental Science **66** (2017) 012024 doi:10.1088/1755-1315/66/1/012024

- replacement of diesel generators with the clean energy source - electrochemical generator (ECG) based on hydrogen fuel cells;
- the use of wind turbines with a vertical axis;
- use of specialized batteries for operation in the composition of SPP.

Taking into account the recommendations made to improve environmental and resource characteristics, provision of the required reliability and security of electricity supply, APSS BS is expedient to design based on hybrid solar-wind-hydrogen power plants (HSWHPPs).

*C. Analysis of the known technical solutions*

One of the examples of the use of hydrogen fuel cells is the base station of the St. Petersburg branch of OJSC "VympelCom" [6]. Tests of a power plant (PP) on hydrogen fuel cells, natural cooling system and special software for remote monitoring of the equipment complex were conducted. For the conduction of tests, the 48 V power system was installed based on the Dantherm Power hydrogen power fuel cells. Two hydrogen fuel cells with a capacity of 1.6 kW, and rectifier unit with the supercapacitors block are placed in a 19" rack. The rack also houses the control and safety unit of the power plant. An important feature of FC Dantherm Power is the air cooling system of the fuel cell. In general, we received a very compact and reliable solution that is not afraid of lower temperatures, as the cooling is provided only by the fan. The hydrogen PP can operate simultaneously as an uninterruptible power supply and backup power supply for all the BS consumers. Supercapacitors (electric double layer capacitors) are intended to compensate for the short-term voltage dips. In the course of tests, they ensure reliable operation of the base station mode of frequent switching on and off of power supply from the city network for 10 minutes at intervals of 30 seconds.

During the tests, the equipment RBS900 / 1800, 3G, PPC transport hub, obstruction lights and air conditioner were connected to the plant. When the air conditioning compressor was turned on, the overload protection of the plant Dantherm Power was tripped. It was decided to disable split-system and connection of free cooling plant Nordic Blue. At repeated starts, the system of alternative uninterruptible power supply and monitoring system equipment were operated normally. During the tests, the compliance with the standards on assurance the smooth operation of the telecommunication equipment RBS 900/1800 - 4 hours, PPC - 24 hours, protecting fires - 48 hours was conducted.

Another example of the use of hydrogen fuel cells is the base station of PJSC "Mobile TeleSystems" in the Noginsk district of Moscow region [10]. This is Russia's first base station in the LTE network, which runs on hydrogen-air fuel. In the hybrid power supply system: an ordinary electrochemical battery and the backup power supply system is used with the use of the hydrogen-air fuel cells, which power is 10 kW. In case of a power failure or voltage drop, the hydrogen unit is started up in a few seconds automatically. It converts the energy released by the chemical reaction of hydrogen with oxygen compound into electricity. The equipment, components and fuel for the hydrogen supply system are made in Russia. The costs of installation of the hydrogen supply system are by 10-30% higher than the conventional power supply systems, but such systems are generally advantageous. Their lifetime is 15 years, which is five times longer than conventional batteries, and maintenance costs are less. Platform "Astra", in contrast to the "traditional" emergency and backup power systems based on batteries and diesel generators, combines high environmental friendliness due to the lack of exhaust, odour, vibration and noise, safety and reliability, simplicity and modularity, and allows easy scalability and low operating costs. Unlike traditional systems, an important feature of the platform "Astra" is the lack of moving-rubbing parts, the ability to provide "long reserve" and exclude fuel logistics and remote monitoring. The obvious advantage is the high energy efficiency - efficiency of power generation is 60%, as well as the lowest operating costs of all types of backup power systems. The ASTRA platform is almost not serviced (replacement of air filters - every 2 years).

*D. Development of technical advice to the projected installation*

In development of the technical advice the given above proposals should be considered to improve the environmental and resource specifications, as well as the following requirements and assumptions [2, 5, 7, 8].

1. Selection of the installed capacity of wind WPP, SPP and electrochemical generator (ECG) based on hydrogen fuel cells should be made taking into account the fact that ECG should generate all the energy flow in the absence of wind and sun, and the installed capacity of WPP and SPP can share nominal one according to specific climatic conditions. This arises from the need to ensure the required reliability and security of electricity supply to BS power consumers. Given that the work of ECG is supposed to be separated from joint work of WPP and SPP, the total nominal installed capacity is assumed equal, i.e.

$$P_{N.ECG} = P_{N.WPP} + P_{N.SPP}. \quad (1)$$

Thus, the nominal capacity of WPP and SPP is understood as power developed by WPP at the estimated wind speed and power of SPP with the estimated flow of solar energy.

2. HSWHPP consisting of WPP, SPP and ECG, should provide uninterrupted electricity supply of cellular base station during the warranty life of at least 5 years with the required quality parameters. The generated total capacity must be at least 5 kW, voltage should be a single-phase/three-phase 220/380 V of nominal value and have a frequency of 50 Hz. Converting AC into current with the characteristics and parameters required for operation of the telecommunication equipment (usually a direct current voltage of 48 V). The power supply not only converts, but also stabilizes parameters within a rather rigid levels - up to 1-2%.

3. There are also consumers of AC and DC. The telecommunication equipment system are provided with DC, consumers of own needs with AC.

4. HSWHPP control should be based on a balance between consumed and generated energy.

On the basis of these technical requirements, as well as the known technical solutiontion [11, 12] a block diagram of ASPP BS is developed and based on HSWHPP shown in Figure 1.

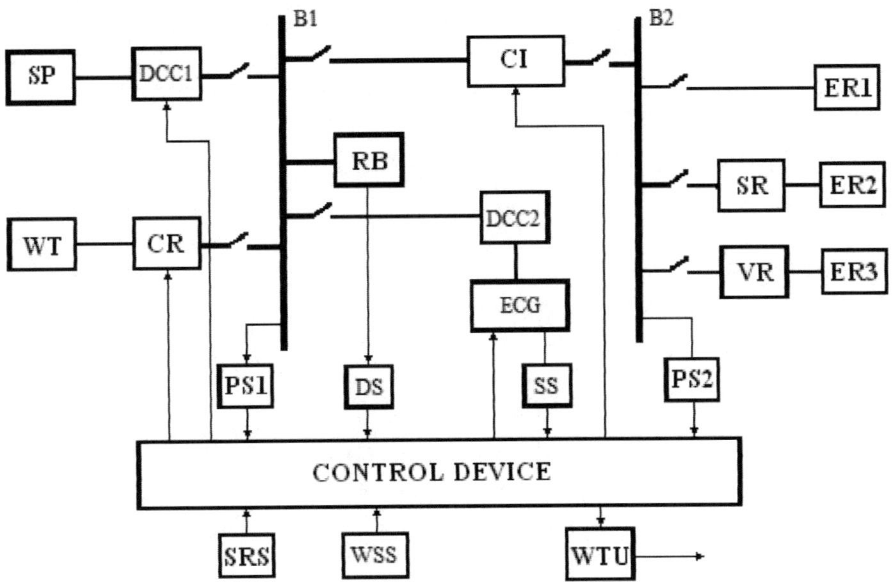


**Figure 1** - Block electric diagram of a hybrid solar-wind-hydrogen power plant: SP - solar panels; WT - wind turbine; DCC1 and DCC2 - DC converters; CR - controllable rectifier; PS1 and PS2 - power sensors; SRS- solar radiation sensor; RB- rechargeable battery; DS - discharge sensor; WSS - wind speed sensor; CI- controllable inverter; ECG - electrochemical generator; SS - ECG status sensor; WTU - wireless transmission unit; SR- stabilizing rectifier; VR - voltage regulator; B1 - DC bus; B2 - AC bus; ER1, ER2, ER3 - electricity receivers

The algorithm of control over electric power sources work of HSWHPP should involve implementation of the principle of adaptability, according to which power supply of receivers with variable load schedule is provided by renewable energy sources, current values of the output parameters of which correspond with the required values, taking into account current environmental conditions.

Wind energy is expedient to estimate by the current value of electrical power generated by WPP ($P_{WPP}$). Energy of the sun by the current value of electrical power generated by SPP ($P_{SPP}$). The total consumed one by load of electric power in each time can be estimated by required current value ($P_{\Sigma.R}$).

Proceeding from the above, the HSWHPP control algorithm can be generally described as follows. If the current value of electrical power of WPP ($P_{WPP}$), greater than the current required value $P_{\Sigma.R}$, AC and DC receivers can be provided by WPP. Similarly - if the current value of electric power of SPP ($P_{SPP}$), greater than the current required value $P_{\Sigma.R}$, all the receivers can be provided by SP.

In general, if the total current value $P_{WPP}$ and $P_{SPP}$ exceed the current value of power $P_{\Sigma.R}$, the level of power generated by each of these sources is provided by respectively regulators to ensure the balance of generated and consumed powers, that is

$$P_{WPP} + P_{SPP} = P_{\Sigma R}. \tag{2}$$

With the low total current value $P_{WPP}$ and $P_{SPP}$ puts into operation and provides ECG station receivers. Thus, ECG as a backup source is activated only in case of impossibility to cover the required current power $P_{\Sigma.R}$ from WPP and SP. It also runs the balance of generated and consumed powers, i.e.

$$P_{WPP} + P_{SPP} + P_{ECG} = P_{\Sigma R}. \tag{3}$$

The installation in contrast to known circuits is introduced with additional sensors of current capacity of renewable power sources PS1 and current load power sensor PS2. To calculate the balance of power conditions and control energy conversion channels, the control device is used.

To ensure the required quality of electric power on DC buses (B1) and AC (B2) in the circuit, there are DC converters (DCC1 and DCC2), controllable rectifier (CR) and controllable inverter (CI).

Receivers of electric power ER2 and ER3 that require high quality electricity (electronic communication equipment) are connected to the power supply line B2 through stabilizing rectifier (SR) and through voltage regulator (VR).

In order to transmit signals to the dispatcher about launch of ECG, end of stock of fuel for ECG, discharge of battery RB etc., wireless transmission unit WTU is introduced.

## 3. Conclusions

1. The following recommendations for improving the environmental and resource specifications of hybrid power plants are proposed: replacement of diesel generator for clean energy source - electrochemical generator based on hydrogen fuel cells; use of wind turbines with a vertical axis; use of specialized batteries.

2. The cost of installation of hydrogen supply system is higher by 10-30% than normal ones, but in general such systems are profitable. Their lifetime is 15 years, which is five times longer than conventional batteries and maintenance costs are lower.

3. The proposed block electric diagram of environment friendly power sources of energy and basic principles of the algorithm of its work allow realizing adaptive control over operating modes with a view to ensuring the balance of power between sources and receivers at the current weather conditions. This control makes it possible to increase the energy conversion efficiency, power reliability, and efficient use of resources throughout the installation.

## References

[1] Vorobyov A.U. Power supply of computer and telecommunications systems. Moscow: EkoTrendz, 2002. 60 p.

[2] Tolmachev V.N., Orlov A.V., Bulat V.A. Efficient Use of Wind Power in the Autonomous Power Supply Systems. Military Engineering-Technical University. Saint-Petersburg. 2002. 203 p.

[3] Bezrukikh P.P. Wind Power. (Reference and Handbook). M.: PH ENERGIIA. 2010, 320 p

[4] Resources and Efficiency of the Use of Renewable Energy Sources in Russia / Group of Authors. Saint-Petersburg: Nauka, 2002. 314 p.

[5] Rudenko N.V., Sukiyazov A.G., Shokova Yu.A. Adaptive control efficiency enhance-ment for hybrid solar-wind energy farms // 7th International Scientific Conference Science and Society. Held by SCIEURO in London. 25-26 November 2014. P.43-50.

[6] Vishnevskii V.P. [et al.] Fuel cells in the backup power system of base stations. "X-Media." magazine. 2010. No.9. [Electronic source]: URL: // http://www.iksmedia.ru/articles/3434337-Toplivnye-elementy-v-sisteme-rezerv.html (date of request 25.02.2017)

[7] Rudenko N.V. [et al.] On the feasibility of using alternative energy sources for uninterrupted power supply of base stations of cellular communication // Scientific almanah. 2015. No.7 (9). P. 790-801. DOI: 10.17117/na.2015.07.790 [Electronic source]: URL: // http://ucom.ru/doc/na.2015.07.790.pdf (date of request 25.02.2017)

[8] Rudenko N.V. The Use of Alternative Energy Sources for Autonomous Power Supply of Base Stations of Cellular Communication // Application of Innovation in the Development of Radio Systems: Collective Monograph / under the general editorship of Doctor of Physical and Mathematical Science M. Zvezdina. M .: Publishing House of the Academy of Natural Sciences, 2015. P. 25-43. ISBN 978-5-91327-349-9. DOI: dx.doi.org/10.17513/np.126

[9] Shen W.X. Optimally sizing of solar array and battery in a standalone photovoltaic system in Malaysia. Renewable Energy. 2009; 34(1):348–52.

[10] News Telecom. Portal "Russian producer of telecommunications equipment." [Electronic source]: URL: // https://www.zelax.ru/news/telecom/2015/12/08/mts-poximichila-nad-oborudovaniem/(date of request 26.12.2016)

[11] Baranov N.N. Non-Conventional Sources and Energy Conversion Methods: Manual for Higher Educational Establishments. M .: Publishing House MEI, 2012. 384 p.

[12] Pandey K., Banerjee P., Mathur D. Study and Modelling of Green Energy based Micro-Grid for Rural Area. Indian Journal of Science and Technology. 2016 June, 9(21), DOI: 10.17485/ijst/2016/v9i21/91568

IOP Conf. Series: Earth and Environmental Science **66** (2017) 012025 doi:10.1088/1755-1315/66/1/012025

# Electromagnetic Processing as a Way of Increasing Microbiological Safety of Animal Waste

**O M Soboleva[1,a], M M Kolosova[1], L A Filipovich[1], V A Aksenov[2,b]**

[1]Kemerovo State Agricultural Institute,
Markovtseva St., 5, Kemerovo, Russia, 650056

[2] EkoMashServis, Avtozavodskaya St., 29, Novokuznetsk, Russia, 654038

*E-mail:* [a]meer@yandex.ru, [b]va.aksenov@mail.ru

**Abstract.** The article shows the possibility of using the electromagnetic field of ultrahigh frequency (EMF UHF) for drying and disinfecting of such animal waste as pig manure and poultry droppings. The studied modes included the following options: processing exposure of 60, 90, 120 sec, the capacity of 60 kW, the frequency of 915 MHz. The method of UHF processing of manure and poultry droppings is environmentally safe and effective in neutralizing the pathogenic microflora, as well as larvae and eggs of worms. The following processing mode of animal waste in the electromagnetic field of ultrahigh frequency was recognized as optimal: exposure of 90 seconds, the capacity of 60 kW, the frequency of 915 MHz. This option leads to the complete destruction of pathogenic and conditionally pathogenic microorganisms, as well as the eggs and larvae of worms. As a result of this processing, a high level of microbiological safety of pig manure and poultry droppings is achieved that allows using them as organic fertilizers. The peculiarities of some species of pathogenic fungi developing on the surface of the wheat grain are shown. Pre-processed animal waste (pig manure and and poultry droppings) were applied in experimental variants. Used organic fertilizers underwent electromagnetic processing of ultra-high frequency. The qualitative composition of the microflora on the surface of the grain depends on the type of animal waste (manure or droppings) and used dose. The safest part of the microflora of grain was marked with the application of the UHF-processed pig manure and poultry droppings in doses of 10 t/ha.

## 1. Introduction

Intensification of livestock production causes a greater unfavourable impact on the environment [1, 2]. Problems of storage, processing, disposal and use of the livestock industry waste of agricultural production are among most urgent nowadays [3-5].

One of the most common ways of their solution is to use livestock manure and poultry droppings as organic fertilizers for restoring and / or improving the fertility of agricultural land. However, this significant biological resource is used by no more than 25-30% because of the lack of efficient and economical technologies of preparation liquid and semi-liquid waste of animals and poultry to be used as organic fertilizer.

In agriculture, livestock waste as fertilizer are used in various forms beginning from fresh manure up to materials that undergone a profound transformation such as biological [6] (eg, after the anaerobic digestion for obtaining biogas) [7], the chemical (by formalin, ammonia, ozone, chlorine, etc.) or physical (by ionizing radiation) [8]. One tries most often to add materials that have passed at least the minimal processing, for example, partially decomposed materials in clamps instead of fresh ones [9].

It is proved that the use of organic fertilizers such as pig manure is not inferior in their effectiveness to application of fertilizers [10]. Today, there are a lot of environmentally friendly technologies of manure processing [11], but some of them lead to a significant loss of nutrients [1, 12], others are not sufficiently effective in terms of decontamination [13, 14].

Composition of pig manure is quite diverse and can satisfy not only the needs of crop plants in the main supply elements (nitrogen, phosphorus, potassium), but also in the macro and micronutrients such as iron, magnesium, calcium, zinc [10]. However, at the same time, manure and poultry droppings are objects that potentially hazardous to health of human, animal and plant because they can contain a huge number of microorganisms (including pathogenic), larvae and eggs of helminths, weed seeds. Contamination by pathogenic microorganisms which are in manure or poultry droppings includes the following areas for distribution: direct transfer of pathogens to humans, for example, the staff of the livestock enterprise; the spread of diseases affecting farm animals; water pollution from further migration of environmental chains; contamination of foodstuffs [15, 16].

Therefore, the use of manure and droppings as fertilizers should be accompanied by a number of measures to ensure their complete disinfection and desinvasion [17, 18].

If livestock intensification continues (and that is what we have seen), there is the need to develop technologies and control strategies of environmental problems associated with this issue.

The method of UHF processing of native manure and droppings can become such method that can solve simultaneously some aforesaid problems such as effective disinfection, preservation of value of waste as fertilizer, reducing processing times.

## 2. Materials and methods

The industrial installation «Volna-100K» (Ltd. «EkoMashServis», Russia) was used for UHF processing of native pig manure and poultry droppings. The investigated processing modes are presented in Table 1. They differed only by the exposure time. These modes were selected in advance, based on the technical possibilities of installation, compliance with energy efficiency conditions, as well as the need to bring the initial raw materials to the required level of humidity (30-50%).

**Table 1** – Options of investigated processing modes of pig manure and poultry droppings

| The processing option | Magnetron power, kW | Magnetron frequency, MHz | Processing exposure, sec |
|---|---|---|---|
| Initial pig manure (0) | 0 | 0 | 0 |
| UHF-processed pig manure (1) | 60 | 915 | 60 |
| UHF-processed pig manure (2) | 60 | 915 | 90 |
| UHF-processed pig manure (3) | 60 | 915 | 120 |
| Initial poultry droppings (0) | 0 | 0 | 0 |
| UHF-processed poultry droppings (1) | 60 | 915 | 60 |
| UHF-processed poultry droppings (2) | 60 | 915 | 90 |
| UHF-processed poultry droppings (3) | 60 | 915 | 120 |

UHF-processed pig manure poultry droppings organoleptically represent a flowing bulk and particulate material of brown color with inclusions (particles of food), low odor of the initial raw material, moisture content of 43.6% and 33.0%, respectively.

Research of microbiological and parasitic safety of UHF-processed organic fertilizers was carried out in the Test center of Kemerovo Interregional Veterinary Laboratory (Kemerovo).

After the UHF processing the manure and droppings were added into the soil in predetermined amounts with further hand application.

Field experience included the following options of application into the soil before planting:

1. Control without application;
2. UHF-processed poultry droppings, application rate of 5.0 t / ha;
3. UHF-processed poultry droppings, application rate of 10.0 t / ha;
4. UHF-processed poultry droppings, application rate of 15.0 t / ha;
5. UHF-processed pig manure, application rate of 5.0 t / ha;
6. UHF-processed pig manure, application rate of 10.0 t / ha;
7. UHF-processed pig manure, application rate of 15.0 t / ha;

Location of plots was consistent. Repeatedness was triple. The plot area was 1 $m^2$. The object of the study was spring wheat of variety Novosibirskaya 31. The sowing of wheat seeds was conducted on the 10th of May in 2016. The seeding rate was of 6 million germinating seeds per hectare.

Soil was presented by podzolized humus. Humus content was 8.6%. Phosphorus content was increased (approximately 150 mg / kg). Potassium content was high (170 mg / kg), nitrate nitrogen content was average (approximately 6 mg / kg) and $pH_{sol}$ was 5, 9.

Two months after the harvest (15.09.2016) the mycological study of spermosphere of received wheat grain was conducted. For this purpose the sowing of swabs from the surface of the seeds on a nutrient medium Saburo was carried out. The cultivation was carried out at a temperature of 30 ° C for 48 hours (preliminary registration of the results), followed by rearing and definition of micromycete. Systematic fungus belonging up to the genus was determined.

## 3. Results and their discussion

Danger level of studied native pig manure in the microbiological and parasitological respect is relatively low and is represented in four taxa, indicating a relatively high sanitary level that is maintained on livestock enterprise. The qualitative composition of conditionally pathogenic and pathogenic microflora, as well as larvae and eggs of helminths in the native sample of pig manure is not too diverse and represented, respectively, by two types of microorganisms and two kinds of worms. Among the identified microorganisms one of the species (*Citrobacter diversus*) refers to the coliform bacteria, representatives of the intestinal microflora of humans and animals. It is characterized as conditionally pathogenic species capable of causing a wide range of diseases both in humans [17] and farm animals and poultry [19]. There is evidence that bacteria of this type are much more susceptible to disinfection than intestinal protozoa and viruses [20]. The second type (*Staphylococcus hyicus*) is the causative agent of zoonotic diseases, some strains of which can cause exudative parakeratosis of pigs [21].

The data obtained indicate a strong influence of the electromagnetic field of the UHF on the microflora and juvenile stages of parasitic flora of the studied animal waste. Some selected modes can lead to a reliable disinfection of these types of raw materials and complete inactivation of mentioned above species. It is shown that bacterial microflora found in the initial pig manure was more resistant to UHF processing than helminths (Table. 2) which are killed by all studied modes of processing, including minimum exposure, whereas microbial component is stored in its entirety at the option of 60 sec exposure. However, microorganisms found in the native materials die with increasing time of exposure of UHF energy on the studied raw material up to 90 and 120 sec. Thus, disinfection of pig manure on microbiological and parasitological parameters can be achieved at satisfactory extent even at the level of average processing time as 90 sec.

The total number of microorganisms, larvae and helminth eggs identified in investigations of native poultry manure was even less than in manure research and amounted to three taxonomic units. One type of bacterium *Proteus* and one kind of staph *Staphylococcus xylosus* refer to conditionally pathogenic microflora. Also one species of parasitic roundworms of kind *Ascaridia galli* (Table. 3) was found. It is indicated in some papers that some strains of S. xylosus are absolutely harmless and even used in the food industry as starting cultures [22]. Other scientists argue that this species belongs to conditionally pathogenic microorganisms and has multidrug-resistant to various antibiotics [23].

There is evidence that some strains of this species can serve as sources of opportunistic infections of animals [24, 25]. Since it is unknown exactly what kind of staphylococcus was isolated in this case, we will adhere to the maximum harm hypothesis and assume that pathogenic strain was isolated.

**Table 2** – Results of microbiological and parasitological studies of pig manure

| Type of tested substrate | Microbiological / parasitological index | | | |
|---|---|---|---|---|
| | Coliform bacteria of species *Citrobacter diversus* | Pathogenic bacteria of species *Staphylococcus hyicus* | Larvae and eggs of nematodes of the genus *Strongyloides* | Larvae and eggs of nematodes of the suborder *Strongylata* |
| Initial pig manure (control) | + | + | + | + |
| UHF-processed pig manure (1) | + | + | – | – |
| UHF-processed pig manure (2) | – | – | – | – |
| UHF-processed pig manure (3) | – | – | – | – |

Studies on UHF processing of poultry manure showed similar tendencies observed for the experiment with pig manure. Bacterial microflora was more resistant to this type of radiation than helminths and withstood minute exposure at specified power and frequency parameters. However, conditionally pathogenic microorganisms *Proteus mirabilis* found in the initial raw material were less stable than staph and died already at 60-second duration. Two modes of decontamination of droppings were recognized as effective in which the entire pathogenic and conditionally pathogenic bacterial microflora, as well as larvae and eggs of helminths were completely destroyed. These modes were with exposition of 90 and 120 sec.

**Table 3** - Results of microbiological and parasitological studies of poultry manure

| Type of tested substrate | Microbiological / parasitological index | | |
|---|---|---|---|
| | Conditionally pathogenic bacteria of species *Proteus mirabilis* | Conditionally pathogenic bacteria of species *Staphylococcus xylosus* | Larvae and eggs of kind *Ascaridia galli* |
| Initial poultry droppings (control) | + | + | + |
| UHF-processed poultry droppings (1) | – | + | – |
| UHF-processed poultry droppings (2) | – | – | – |
| UHF-processed poultry droppings (3) | – | – | – |

The obtained data of field trials suggest that the formation of the qualitative composition of microflora of grain of spring wheat is affected by the type and dose of UHF processed animal waste.

Our studies indicate that the predominance of different micromycetes depends on the amount of applied organic fertilizer. Thus, the application of manure and droppings at a dose of 10 t/ha leads to the predominant development of the fungi of the *Rhizopus*. At the same time the growth of remaining fungi is inhibited.

Increase in the dose of applied UHF-processed poultry droppings up to 15 ton per hectare gives rise to the phytopathogen *Fusarium* among others.

In all the studied variants of application, including the control, the presence of such micromycetes as *Helminthosporium* and *Alternaria* recorded was recorded.

It should be noted such particularly of studied grain as the complete absence of the so-called «storage» molds – representatives of *Aspergillus* and *Penicillium*, which probably explains the small grain storage period, during which the phytopathogenic microflora has not had time to be replaced by the storage molds.

## 4. Conclusion

Two studied modes of UHF processing with medium and long duration of action (90 and 120 sec respectively) lead to the same results in terms of efficiency of disinfection of pig manure and poultry droppings by UHF energy. But it is more rational to use the version with a shorter duration of the processing since in this case the energy saving is achieved and it is possible to process larger quantities of raw materials per time unit.

The qualitative composition of microflora of the surface of the wheat grain is determined by edaphic conditions of the place of growth and also depends on the type and dose of applied UHF-processed animal waste. The safest part of the microflora of grain was marked with the application of the UHF-processed pig manure and poultry droppings in doses of 10 t/ha.

## References

[1] Recycling of livestock manure in a whole-farm perspective / S.O. Petersen, S.G. Sommer, F. Béline [et al.] // Livestock Science. 2007. 112. P. 180-191.

[2] A Cost-Benefit Analysis (Based on Recycling Economy) of Effective Utilization of Livestock Manure / Y. Hui, L. Cuixia // BioTechnology An Indian Journal. Vol. 10. 2014. P. 8031-8036.

[3] *Gumifikacija organicheskih veshhestv navoza pri kompostirovanii ih s nekondicionnymi fosforitami* [Humification of organic substances of manure when composting them with substandard phosphorite] / U.Sh. Temirov, A.M. Rejmov, Sh.S. Namazov, N.H. Usanbaev // Universum: tehnicheskie nauki. 2016. 8. P. 43-47.

[4] Treatment of swine manure: case studies in European's N-surplus areas / M.C. García-González, B. Riaño, M. Teresa, E. Herrero, A.J. Ward, G. Provolo, G. Moscatelli, S. Piccinini, A. Bonmatí, M.P. Bernal, H. Wiśniewska, M. Proniewicz. Sci. Agric. 2016. Vol.73, n.5. P. 444-454.

[5] Advanced swine manure treatment and utilization in Brazil / A. Kunz, M. Miele, R.L.R. Steinmetz. Bioresource Technology 100 (2009) 5485–5489.

[6] Riaño B.; García-González M.C. On-farm treatment of swine manure based on solid-liquid separation and biological nitrification-denitrification of the liquid fraction // Journal of Environmental Management. 2014. Vol. 132: 87-93.

[7] Optimisation of the anaerobic digestion of agricultural resources / A.J. Ward, P.J. Hobbs, P.J. Holliman, D.L. Jones // Bioresource Technology. 2008. Vol. 99. P. 7928–7940

[8] Martens W., Böhm R. 2009. Overview of the ability of different treatment methods for liquid and solid manure to inactivate pathogens. Bioresource Technology 100: 5374-5378.

[9] *Sovremennye tehnologii podgotovki ptich'ego pometa k ispol'zovaniju: analit. spravka* [Modern technology of preparation of bird droppings for use: analytical report]. M: FGNU «Rosinformagroteh», 2011. 40 p.

[10] Effects of different liquid-manure treatments on yield and quality parameters of second-crop silage corn under reduced tillage conditions / T. Akar, M. Kaplan, N. Sagir, A. Gelebur // Romanian agricultural research. 2014. 31. P. 193-203.

[11] 2009. Challenges and innovations on biological treatment of livestock effluents / Bernet N., Béline F. // Bioresource Technology, 100(22): 5431–5436

[12] L. Qing Research Progress on Resource Utilization of Livestock and Poultry Manure, Agricultural Science & Technology, 2, 338-343 (2013).

[13] Rassolov S.N. *Biologicheskij sposob utilizacii svinogo navoza* [Biological method of disposal of pig manure] / S.N. Rassolov, O.A. Bagno, K.V. Bespomestnyh. Vestnik KrasGAU. 2015. 11. P. 220-225.

[14] R. Nkoa. Agricultural benefits and environmental risks of soil fertilization with anaerobic digestates: a review / Agronomy for Sustainable Development. - 2014, 34, P.473-492

[15] Böhm R. Hygienic safety in organic waste management / In: Lens P., Hamelers B., Hoitink H., Bidlingmaier W. (Eds.) // Resource recovery and reuse in organic solid waste management. IWA publishing, London, 2004. P. 459-482.

[16] Guan T.T.Y. Hog Manure Management, the Environment and Human Health / T.T.Y. Guan, R.A. Holley // Kluwer Academic / Plenum Publishers, New York, 2003. 168 p.

[17] *Federal'nyj zakon ot 30 marta 1999 g. N 52-FZ «O sanitarno-jepidemiologicheskom blagopoluchii naselenija»* [Federal act of 30 March 1999 N 52-FZ "On the sanitary-epidemiological welfare of the population] (s izmenenijami ot 3.06.2016)

[18] Treated wastewater effluent as a source of microbial pollution of surface water resources / S. Naidoo, A.O. Olaniran // Int. J. Environ. Res. Public Health 2014, 11(1), 249-270.

[19] Aznabaev G.K. *Biologicheskie svojstva bakterij roda Citrobacter, vydelennyh pri mono- i associirovannyh bakterial'nyh infekcijah: avtoref. dis. ... kand. med. nauk* [Biological properties of bacteria of the genus *Citrobacter* isolated at mono - and associated bacterial infections]. Orenburg, 2003. 23 p.

[20] Ashbolt, N.J. Microbial contamination of drinking water and disease outcomes in developing regions. Toxicology 2004, 198, 229–238.

[21] *Antimikrobnoe dejstvie pektinov na uslovno-patogennuju mikrofloru kishechnika cypljat-brojlerov* [Antimicrobial effect of pectins on the pathogenic microflora of the intestine of broiler chickens] / Portjanko A.V., Zadorozhnaja M.V., Suncova O.A. [i dr.] // Uchenye zapiski KGAVM im. N.Je. Baumana. 2015. 222. P. 180-186.

[22] Vojtenko A.V. *Jeksperimental'noe zarazhenie porosjat toksin-producirujushhimi shtammami Staphylococcus hyicus* [Experimental infection of pigs toxin-producing strains of *Staphylococcus hyicus*] / A.V. Vojtenko, V.N. Skvorcov, A.A. Balbuckaja // Veterinarnaja patologija. 2006. 3. P. 77-79.

[23] Talon R. Bacterial starters involved in the quality of fermented meat products / R. Talon, S. Leroy-S´etrin, S. Fadda // Research advances in quality of meat and meat products. 2002. P. 175-191.

[24] Al-Mathkhury H.J.F. Colonization of *Staphylococcus xylosus* in the kidneys and bladder of mice / Um-Salama Science Journal. 2008. Vol. 5. P. 70-73.

[25] Inoculation of *Staphylococcus xylosus* in SJL/J mice to determine pathogenicity / Thornton V.B., Davis J.A., St. Clair M.B., Cole M.N. // Contemp. Top. Lab. Anim. Sci. 2003. Vol. 42. P. 49-52.

# Fibrous Concrete with Reduced Permeability to Protect the Home Against the Fumes of Expanded Polystyrene

**R S Fediuk [a], A K Smoliakov, R A Timokhin, N Yu Stoyushko, N A Gladkova**

Far Eastern Federal University,
ul.Sukhanova 8, Vladivostok, 690950

*E-mail:* [a] roman44@yandex.ru

**Abstract**. Designed the composition of the fiber-reinforced concrete on composite binder with high gas-, water- and steam impermeability. With 1.6% of reinforcing steel anchoring a fiber can be obtained in terms of the maximum physical and mechanical properties ($R_{compr}$ = 100.9 MPa). It was found that the combined effect of mechanical and chemical activation (the presence of limestone particles) increases the pozzolanic activity of acidic environment. It has a catalytic effect on the reaction activity of the surface of ash and sand during machining in vario-planetary mill. Furthermore, the introduction of limestone increases the alkalinity of the concrete, which leads to the formation of greater hydration products of cement per unit of time. Theoretical and experimental results can be recommended for enhanced implementation of the construction in various regions of the World, taking into account the availability of raw materials.

## 1. Introduction

In modern construction one of the most commonly used construction materials are concrete and reinforced concrete, and from heat-insulating materials - the expanded polystyrene. At the same time, a well-known fact that the foamed polymers materials have ecologically and fire hazard, and, in the absence of adequate protection, also have short time of exploitation. Undoubtedly, that is not allowed contact of polystyrene to the interior and from the exterior climatic conditions. The problem can be solved by the internal arrangement of insulation between two structural concrete (reinforced concrete) layers. At the same time, concrete - a porous material, and the evaporation of styrene from polystyrene occurs continuously, and when the temperature rises, significantly accelerated. Adequately, only the injection of special additives, bridging the pores in concrete and sealing the structure is able to neutralize the environmental and fire hazard of polystyrene, as well as to increase its durability by preventing the penetration of atmospheric moisture (which is very important for the monsoon climate of Asian-Pacific area). Thereby, the engineering of concrete, having high gas, water and steam impermeability, able to protect from moisture as expanded polystyrene, and styrene vapors from the room is an urgent task [1,2].

IOP Conf. Series: Earth and Environmental Science **66** (2017) 012026 doi:10.1088/1755-1315/66/1/012026

## 2. Materials and methods

Analysis of the literature suggests the possibility of producing concrete with high density. Note, however, that the implementation of the potential of concrete is only possible by create the optimum structure of the material, the formation of which is determined by such parameters as:

- types and quality of raw materials;
- technology preparation of concrete mixtures, etc.

The main factors determining the efficiency of designing of concrete and its construction units are technical features:

- obtaining the desired properties (strength, workability, and etc.) At the lowest possible consumption of costly and energy-intensive components;
- reducing the complexity of operations related to the selection of optimum compositions;
- verification of quality control samples and testing at the factory.

The properties of fine-grained concrete are characterized by the same factors as the conventional heavy concrete, but the lack of coarse aggregate leads to increased water demand in concrete, and in equally strength concrete cement consumption increased by 20...40%. To reduce the consumption of cement is advisable to use high-quality fillers, plasticizers, mixtures to produce a good seal [2,3,4].

In the manufacture of fine-grained concrete it is advisable to use natural and enriched sand, crushed and granulated slag and fly ash, waste industry, which also includes screenings of crushed granite. The properties of fine-grained concrete are highly dependent on the properties of the binder component in the concrete and aggregate used, grain structure, surface quality and water demand which determine the basic physical and mechanical properties of concrete. When receives a fine concrete aggregate used smaller size and higher surface area compared to conventional heavy concrete, the presence of these factors leads to an increase in cement consumption, which negatively affects the physical and mechanical characteristics of fine-grained concrete.

To achieve this goal have been developed composite binders obtained by co-grinding in the vario-planetary mill cement, fly ash limestone, and superplasticizer. It is advisable to use local raw materials, which are characterized by high activity silica-containing phases of man-made sand (in particular, elimination of crushed granite), obtained from crushing rocks.

Fine-grained structure, in addition to high homogeneity is also characterized by reduction of specific stress in the contact area and an increase of the integrated strength of adhesion between cement stone and aggregate.

Structure-forming role of filler is most developed with increasing surface of interaction, these conditions are realized in the fine-grained concrete with screenings of crushing of granite rubble of Wrangel field (Russian Federation) based on composite binders that due to the high surface intensify and accelerate the process of structure formation of concrete strength and seal structure.

As a component of the composite binder used the fly ashes of Russian largest thermal power plants: Vladivostok TPP and Artem TPP, Primorye TPP and Partizansk TPP An important factor is the ability to separate the dry ash that is currently implemented in these power plants. The use of man-made products in the manufacture of building materials contributes to the solution of the following main objectives: saving energy and raw resources; waste utilization; improvement the environmental situation in the regions. It should be noted that revenues in the seaside ash dumps are one of the largest in the Far Eastern Federal District of Russian Federation. In particular, the annual flow of ash in the ash dumps of the Primorsky Territory from 2.5 to 3.0 mln. tons per year, Khabarovsk Territory - to 1.0 mln. tons. Fly ash thermal power plants it is an effective material for the production of fine and active mineral additives. These residues do not require special preparation when injected into the concrete. Was applied mineral powder from limestone Dinnogorsk deposit (Russian Federation) production of "Spasskcement".

To study the effect of composition on the binding properties of fine-grained concrete samples of concrete slump were produced in the range of 10-12 cm in the developed binder, which were further investigated by standard methods for various technological parameters.

IOP Conf. Series: Earth and Environmental Science **66** (2017) 012026 doi:10.1088/1755-1315/66/1/012026

The properties of composite materials largely depend on the structure of disperse systems of which they are obtained. The structural strength of the dispersed system, its stability and the behaviour in the time, the rate of destruction and restoration of the structure are interconnected. Conglomerate building material - concrete refers to a class of composites. As a concrete matrix substance, successively ambitious outline levels are the cement-sand stone, cement stone (micro concrete), cementing substance, neoplasms cementing substance, solid phase neoplasms, essence single structural element of neoplasms, which corresponds to a large-scale order from the macro- to the nanoscale structure. Ordered structure of the composites due to the proportionality of scale levels of the structure - the corresponding properties of the composite at each scale level. The achievement of high strength concrete contributes to a combination of factors: increasing density of systems by optimizing the composition of the grain; reducing the number of long cement matrix by reducing the water-cement ratio; filling the pores between the cement particles and to improve the rheology due to the effect of lubrication; formation of secondary products of hydration in the pozzolanic reaction with $Ca(OH)_2$ for the introduction of concrete additives microfill effect.

To select the optimal method of grinding tests were carried out in a ball, vibration and vario planetary mill. The grinding in vario-planetary mill is more energy-efficient compared to the ball and vibration mills. In addition due to the combined action of the drum and centrifugal drum and abrasive efforts, it is possible to achieve a highly dispersed powders.

## 3. Results and discussion

A composition of ash and slag waste is determined by the quantitative ratio of contained in the material minerals that are dependent on initial mineralogical composition of fuel used.

Process optimization of structure by hydration composite binder components creates a dense structure of the matrix, it is necessary to solve the problem - the creation of a composite for the production of high-strength concrete. This can be realized by co-grinding of Portland cement, mineral admixtures and polyfunctional reducing the water-cement ratio of the concrete mix through use the hyper plasticizer.

To reduce the water demand of concrete mix produced powder hyper plasticizer selection of the six most common in the Far East market for construction materials. cement paste broke was measured using Hagermann cone. For grout used Spassky cement CEM I 42,5N. Water-cement ratio - 0.3. Dosage plasticizer - 0.3%. Time of measurement face breaking cone was recorded after the end of mixing the cement paste.

Achieving high values of cone face breaking noted on raw binder mixture using hyper plasticizer PANTARHIT PC160 Plv (FM) (Table 1).

**Table 1** – Cone face breaking of the cement paste with different superplasticizers.

| Time measurements, min | Melflux 1641 F, Germany | Melflux 5581 F, Germany | PANTARHIT PC160 Plv (FM), Russian Federation | FOX$^{TM}$-8H (Pwd), Russian Federation | PC-1030, China | JK-04 PPM, China |
|---|---|---|---|---|---|---|
| | Cone face breaking, mm | | | | | |
| 0 | 290 | 350 | 370 | 250 | 240 | 130 |
| 5 | 380 | 390 | 400 | 260 | 280 | 120 |
| 30 | 390 | 350 | 390 | 240 | 190 | 98 |

For further research 7 composite binders were developed. In each superplasticizer PANTARHIT PC160 Plv (FM) was added in an amount of 0.3%, the ratio of binder: sand - 1: 3. To determine the optimal number of components in the "cement-lime-ash" system carried them grind to a specific surface of 600 $m^2/kg$ at various ratios (Table 2).

**Table 2 -** The compositions and properties of composite binders

| No | Content cement, wt. % | Fly ash, wt. % | | Limestone, wt. % | Compressive strength, MPa | | |
|---|---|---|---|---|---|---|---|
| | | Vladivostok TPP | Artem TPP | | 3 d | 7 d. | 28 d. |
| 1 | 100 (without grinding) | – | – | – | 17 | 32.5 | 47.5 |
| 2 | 30 | – | 50 | 20 | 30.2 | 40.1 | 50.4 |
| 3 | 35 | 45 | – | 20 | 34.2 | 43.1 | 53.2 |
| 4 | 40 | – | 45 | 15 | 36.6 | 48.2 | 56.6 |
| 5 | 45 | 45 | – | 10 | 39.2 | 50.1 | 59.2 |
| 6 | 50 | – | 40 | 10 | 45.1 | 54.9 | 65.8 |
| 7 | 55 | 40 | – | 5 | 47.2 | 54.1 | 70.2 |
| 8 | 100 | – | – | – | 60.3 | 81 | 103.2 |

Note: The control formulation No 1 (without final grinding); number 2-8 compositions milled to $S_{sp}$ = 600 $m^2$/kg

The resulting composite powder had a specific surface area equal to 600 $m^2$/kg (determinations were performed on surface meter PSH-11), 0.15-500 micron particle size, the average particle diameter is shifted to 0.65-11.2 mm, density 930 kg/$m^3$ (Figure 1).

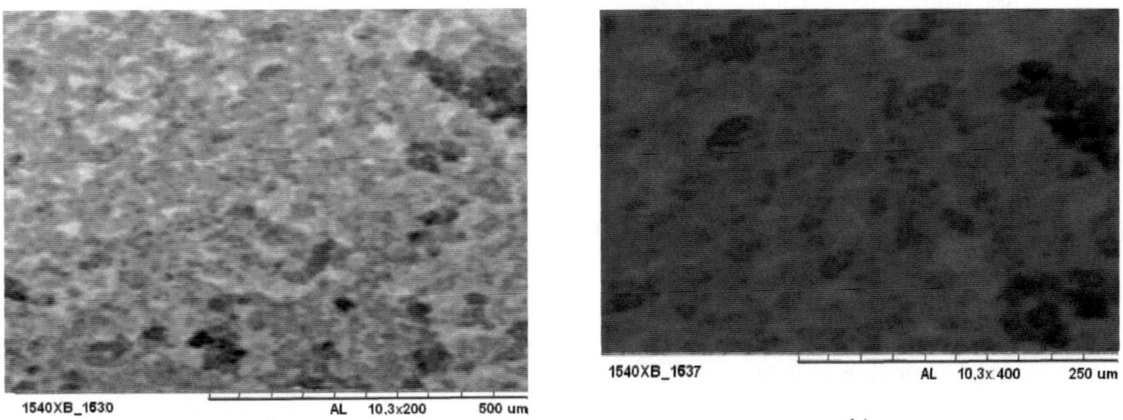


a) b)

**Figure 1** – General view of the composite binder (magnification 200 and 400 times)

Co-grinding of cement with ash and limestone can increase the activity of binding to 70.2 MPa. The increase in strength is due to the improvement of co-grinding of cement stone structure.

This is due to the presence the mill ground active mineral components that contribute to an earlier binding portlandite, intensify the process of hydration of clinker minerals.

At the same time, the larger particles act as nucleation sites and act as microfiller reducing shrinkage deformation, improve performance of the composite. A characteristic feature of the structure of cement stone on designed binders ground with additives is substantially smaller number of microcracks.

The structure of the cement paste on the composite binder denser than conventional Portland cement, it is a very dense packing of the grains in the total mass of neoplasms. Co-grinding of components leads not only to increase the ultimate compressive strength, but also increases the rate of curing of the samples with additives.

In order to determine the optimum particle size were produced co-grinding of cement, hyper plastisizer, ash and limestone (in a ratio of composition number 7) to different specific surface: 500, 550, 600, 700, 800, 900 $m^2$/kg.

The results of determination of compressive strength of cement specimens are shown in Table 3.

**Table 3** - The results of determination of compressive strength (MPa) cement stone samples

| Age of sample, | The specific surface area of the composite binder, $m^2$ / kg | | | | | |
|---|---|---|---|---|---|---|
| d. | 500 | 550 | 600 | 700 | 800 | 900 |
| 3 | 46.1 | 47.4 | 47.2 | 46.0 | 45.6 | 45.5 |
| 7 | 50.3 | 54.2 | 54.1 | 49.1 | 48.6 | 48.4 |
| 28 | 68.1 | 77.3 | 70.2 | 65.8 | 55.0 | 65.0 |

The optimal activity of the binder is 550-600 $m^2$ / kg. The Increase activity in excess of these values has a negative effect on the structure. Using a binder with increased activity significantly speeds up the process of "setting" - "setting the end" of the mixture ends after 35-40 min., while the developing temperature 95-97°C. Fast setting prevents the formation of raw uniformly distributed spherical particles in the macrostructure of cement stone. Electron microscopy revealed the presence of heterogeneity macrostructure and irregularly shaped cells.

Calculation of the composition produced in accordance with the recommendations on the selection of compounds and heavy fine-grained concrete.

Baseline data for the design of the composition:

- the desired strength - 100 MPa;
- the mobility of the concrete mix slump - 10-12 cm;
- activity composite binder - 77.3 MPa;
- screening of crushed granite with water demand 7% and the true density of 2.6 kg / l and a bulk density of 1.37 kg / l;
- sand true density of 2.63 kg / l.

$$W/C = \frac{A_1 R_{CB}}{\mathrm{R_c} - A_1 \cdot 0.5 R_{CB}}, \qquad (1)$$

coefficient $A_1$ is accepted by the reference literature [5].

$$W/C = \frac{0.43 \cdot 77.3}{100 - 0.43 \cdot 0.5 \cdot 77.3} = 0.40$$

According to the schedule in reference [5] water flow turned 220 $l/m^3$.

$$CB = \frac{220}{0.4} = 550 \kappa g/m^3 \qquad (2)$$

Emptiness dropping out of granite rubble:

$$E_{gr} = 1 - \frac{1.37}{2.6} = 0.47 \qquad (3)$$

According to the handbook [5] extendable coefficient α = 1.5.

The amount of filler:

$$F = \frac{1000}{0.47 \cdot 1.5/1.37 + 1/2.6} = 1623 \kappa g/m^3 \qquad (4)$$

Thus, an exemplary structure of a concrete mixture: composite binder - 550 $kg/m^3$, a filler - 1623 $kg/m^3$, water - 220 $l/m^3$. The optimal composition will be selected by experimental verification.

Research of physic-mechanical properties of fine-grained concrete showed that the use of the composite binder prepared by co-grinding of cement, fly ash and limestone, and hyper plasticizer allows to improve the characteristics of concrete, as compared to similar compositions made with other binding materials. This fact is explained more dense structure of cement stone developed composite binder lesser porosity due to the smaller amount of water in the concrete (Table 4).

**Table 4** - Physical and mechanical properties of fine-grained concrete depending on the composition of the binder

| Composition | Material consumption per 1 $m^3$ | | | | | | | Slump | The compressive strength R, MPa | Prism strength R, MPa | Elastic modulus |
|---|---|---|---|---|---|---|---|---|---|---|---|
| | Binder, кg | | | | | | | | | | |
| | Cement | Fly ash | Limestone | Hyper plasticizer | Screenings crushing of granite, kg | Sand, kg | Water, *l* | | | | |
| 1* | 550 | - | - | 1.2 | 1000 | 623 | 220 | 10-12 | 107.5 | 86.3 | 61.2 |
| 2 | 288 | 235 | 27 | | | | 240 | | 83.7 | 59.5 | 43.8 |
| 3 | 275 | 246 | 29 | | | | 241 | | 84.2 | 60.3 | 44.5 |
| 4 | 257 | 257 | 36 | | | | 242 | | 76.3 | 55.2 | 40.9 |
| 5 | 244 | 268 | 38 | | | | 243 | | 75.2 | 55.0 | 40.8 |
| 6 | 230 | 278 | 42 | | | | 244 | | 75.0 | 54.9 | 40.8 |
| 7** | 550 | - | - | | | | 215 | | 63.1 | 42.3 | 36.2 |

* binder of low water with a specific surface 550 $m^2$ / kg.
**binder based on Portland cement produced by JSC "Spasskcement".

The best physical and mechanical properties of the compositions showed No1 and No2: (cement - 48-51%, acidic ash - 43-45%, limestone - 4-9%).

It should be noted that with increasing amounts of ash and reducing the amount of cement to ensure equal mobility formulations (slump = 10-12), it is necessary to increase the amount of mixture introduced into the concrete mixing water [6,7,8].

Despite a number of advantages compared, for example, conventional heavy concrete, fine concrete has high shrinkage and reduced by 20-25% modulus [9,10].

Great opportunities to improve the performance opens disperse fine concrete reinforcing polymer and metal fibers.

In order to obtain high-density fiber-reinforced concrete was studied the effect of the introduction of the concrete matrix of reinforcing fibers. As a basis for the concrete matrix was adopted composition No 3 from Table. 4.

To determine the optimal percentage of reinforcement were fine steel fiber concrete formed concrete samples of the same composition with different steel fiber content (Table 5).

**Table 5** - The dependence of the strength of the fine steel fiber concrete reinforcement ratio

| Composition | Material consumption per 1 $m^3$ | | | | Reinforcement, % | $R_{compr}$, MPa |
|---|---|---|---|---|---|---|
| | Binder | Water | Filler | Fiber | | |
| 3-1 | 550 | 240 | 1623 | - | 0 | 94.2 |
| 3-2 | 550 | 240 | 1623 | 23.97 | 1 | 96.1 |
| 3-3 | 550 | 240 | 1623 | 28.76 | 1.2 | 97.3 |
| 3-4 | 550 | 240 | 1623 | 33.56 | 1.4 | 99.8 |
| 3-5 | 550 | 240 | 1623 | 38.35 | 1.6 | 100.9 |
| 3-6 | 550 | 240 | 1623 | 43.15 | 1.8 | 99.5 |
| 3-7 | 550 | 240 | 1623 | 47.94 | 2 | 99.6 |

For further optimization of structure and prevent crack initiation, increase strength and durability of the structure was used basalt fiber by domestic production (Table 6). It was found that by reinforcement 1.4-1.6% by volume is possible to obtain the maximum physical and mechanical properties. A further increase in the percentage of reinforcement is not advisable as it causes a decrease in strength and performance steel fiber concrete.

For further optimization of structure and prevent crack initiation, increase strength and durability of the structure was used basalt fiber domestic production (Table 6).

**Table 6** - The dependence of the strength of the fine basalt fiber concrete reinforcement ratio

| Composition | Material consumption per 1 m$^3$ | | | | Reinforcement, % | $R_{compr}$, MPa |
|---|---|---|---|---|---|---|
| | Binder | Water | Filler | Fiber | | |
| 2-1 | 550 | 240 | 1623 | - | 0 | 94.2 |
| 2-2 | 550 | 240 | 1623 | 1 | 0.0004 | 95.6 |
| 2-3 | 550 | 240 | 1623 | 1.2 | 0.0005 | 95.9 |
| 2-4 | 550 | 240 | 1623 | 1.4 | 0.0006 | 96.3 |
| 2-5 | 550 | 240 | 1623 | 1.6 | 0.0007 | 97.1 |
| 2-6 | 550 | 240 | 1623 | 1.8 | 0.0008 | 98.2 |
| 2-7 | 550 | 240 | 1623 | 2.0 | 0.0009 | 98.1 |

Comparing the experimental data on the Table 2 and Table 3 for further investigations take part steel fiber concrete No 5 (1.6% reinforcement).

The positive dynamics of growth of the strength of the composite binder under the joint influence of the fine constituents of ash, limestone crushing screenings and hyper plasticizers with the maximum increase in the activity of the binder is 62%.

Active mineral components of the composite binder contribute to the binding of $Ca(OH)_2$ produced during cement hydration additional amount hydrosilicate neoplasms. At the same time optimizing the process of structure formation is achieved by polydispersity composite components. Highly spherical ash particles act as nucleation sites and act as a nano- and micro- filler. In conjunction with the larger particles of the mineral component is more dense filling intergranular spaces cement concrete structure with a reduction in the number of pores and microcracks [11].

This is confirmed by micrographs of cement paste on the composite binder, the resulting joint grinding of clinker and industrial wastes of the Far Eastern region. Cement stone structure is a very dense packing of fine grains to the total weight of the crystalline neoplasms (Figure 2).

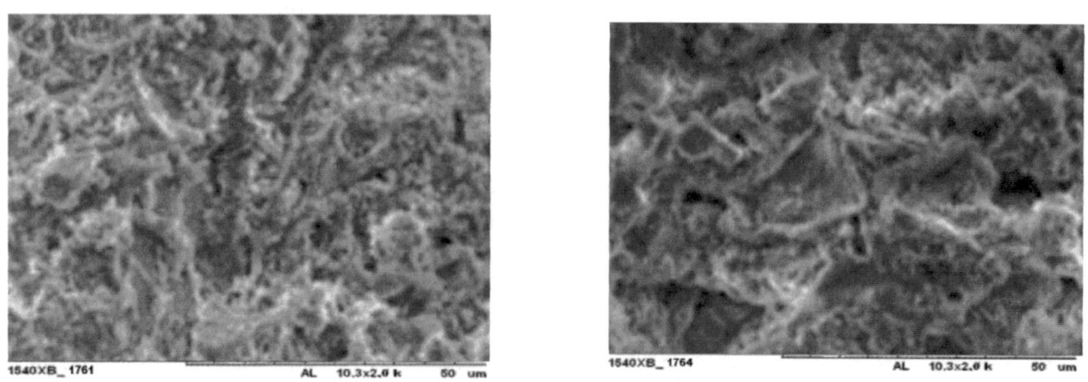


a) b)

**Figure 2 -** The microstructure the cement stone: on CEM I 42,5 N *(a)* on the composite binder (composition No 7 in Table. 2) *(b)*

The additional amount of hydrated crystalline phases is contributing to filling the voids at the micro level in the crystalline matrix of calcium hydrosilicates on the boundary of the contact area, increasing the degree of adhesion binder with filler.

## 4. Conclusions

Designed the binder composite obtained by co-grinding of cement (55%) of acid composition of ash (40%) and limestone (5%) to a specific surface of 550 kg / m2, the activity of 77.3 MPa.

It was found that the combined effect of mechanical activation increases pozzolanic activity of acidic sol, has a catalytic effect on the reaction activity of the surface of the ash and sand in the process of machining in the vario-planetary mill.

It was revealed that the additive ash and limestone crushing of waste composite binder for all dosages and reduce water permeability of concrete. Thus, it revealed a clear link between the properties of concrete and the characteristics of the cement stone structure - the increase in the number of neoplasms at the complex hydrosilicate reducing gel and capillary porosity, particularly at the molecular and submicroscopic levels that determine the growth and increasing strength of concrete impermeability.

Designed the fiber-reinforced concrete structure on the composite binder. At 1.6% of reinforcing steel anchoring a fiber can be obtained in terms of the maximum physical and mechanical properties ($R_{compr}$ = 100.9 MPa).

It was found that the developed concrete provides an effective diffusion coefficient, it allows you to protect the premises from the toxic effects of expanded polystyrene. Enough material low water absorption and low water vapor permeability values - 0.021 mg / (m • h • Pa) explains the features of the pore structure of space the cement stone.

Reduced permeability (porosity) of fiber-reinforced concrete improves the corrosion resistance, including for concrete using incorporates acidic ashes.

## References

[1] Fediuk R. S., Khramov, D. A. Research on porosity of the cement stone of composite binders. International Research Journal 2016 Volume 1 (43); pp. 77-79. DOI: 10.18454/IRJ.2016.43.039.

[2] Singh B., Jain K. Appraisal of steel fibers as minimum shear reinforcement in concrete beams. ACI Structural Journal 2014 Volume 111, Is. 5; pp. 1191-1202.

[3] Bhuvaneswari T., Elavarasi D. Effects of supplementary cementitious materials in slurry infiltrated fibrous concrete. International Journal of Applied Engineering Research 2014 Volume 9, Is. 16; pp 3335-3350.

[4] Khalifa E. S. Macro-mechanical strut and tie model for analysis of fibrous high-strength concrete corbels. Ain Shams Engineering Journal 2012 Volume 3, Is. 4; pp. 359-365.

[5] Bazhenov Yu M *Tehnologoya betonov* [Concrete technology]. Moscow: ASV, 2007. 501 p.

[6] Balaji, S; Thirugnanam, G.S. Flexural strengthening of reinforced concrete beams using precast SIFCON laminates. Journal of Structural Engineering (India) 2013 Volume 40, Is. 3, pp. 262-267.

[7] AlTaan, S A; Mohammed, A. A.; Al-Jaffal, A. A. Breakout capacity of headed anchors in steel fibre normal and high strength concrete. Asian Journal of Applied Sciences 2012 Volume 5, Is. 7, pp. 485-496.

[8] Ramli, M. B.; Alonge, O. R. Characterization of metakaolin and study on early age mechanical strength of hybrid cementitious composites. Construction and Building Materials 2016 Volume 121; pp. 599-611.

[9] Khan, S. U.; Ayub, T. Modelling of the pre and post-cracking response of the PVA fibre reinforced concrete subjected to direct tension. Construction and Building Materials 2016 Volume 120; pp. 540.-557.

[10] Fediuk, R.S.; Khramov, D. A. Physical equipment spectroscopic study of coal ash. Modern Construction and Architecture 2016 Volume 1; pp. 57-60. Doi: 10.18454/mca.2016.01.11.

[11] Antonius. Strength and energy absorption of high-strength steel fiber-concrete confined by circular hoops. International Journal of Technology 2015 Volume 6, Is. 2; pp. 217-226.

[12] Li J., Lu S., Cai Y., Xue H., Cai J. Impact of coal ranks on dynamic gas flow: An experimental investigation. Fuel 2017 Volume 194, pp. 17-26

[13] Glagolev E., Suleimanova L., Lesovik V. High reaction activity of nano-size phase of silica composite binder. International Journal of Environmental and Science Education 2016 Volume 11, Issue 18, pp. 12383-12389

# Aspects of Production Ecologization of Machine-Building Enterprises as Part of the System Approach

**T V Lazutina[a], Yu A Tempel[b], O A Tempel[c], N K Lazutin[d]**

Tyumen Industrial University,
38 Volodarsky St., Tyumen, Russia, 625000

*E-mail:* [a]Lazutinatv@yandex.ru, [b]Tempeljulia@mail.ru, [c]Tempel_o@mail.ru, [d]kpw72@yandex.ru

**Abstract**. The paper considers the role of the machine-building industry in Russia and the impact of its activities on the ecological situation in the country. As part of the problem we identified areas of environmental pollution from different production industries, including foundry, power, metal working and welding. In addition, the paper presents a strategy for production ecologization, based on a system approach viewed as a set of measures aimed at reducing the danger of technological processes for the environment and people.

## 1. Introduction

The machine-building industry is one of the strategic sectors of the Russian Federation; technical equipment of all branches of the economy and the country's defense largely depend on the level of its development. Machine building includes twelve major sub-sectors, the main ones being heavy engineering, power engineering, chemical engineering, transport engineering, petroleum engineering, as well as machine-tool building and instrument machine-tool building.

The main role of the industry in question is to ensure that other sectors are provided with the necessary machines, aggregates, various complexes and other equipment.

Today, machine building defines the level of development of the domestic industry. In 2014, the share of engineering products released was 22% of the total volume of the manufacturing industry in Russia. In addition, it has a social importance as well, since more than a third of all industry employees are involved in enterprises of the machine-building complex.

However, development of machine building, improvement in the competitiveness of products and provision of faster growth of production require a high consumption of raw materials, energy, natural resources, having an adverse impact on the ecology of the Earth. More than twenty thousand Russian machine-building companies with a progressive development of processing play a significant role in environmental pollution. Moreover, in some areas of the industry with the most hazardous production emissions exceed all specified requirements and standards.

Despite a variety of eco-innovations, advanced equipment for waste disposal and wastewater treatment, management systems focused on ecological safety and environmental protection in accordance with international standards of ISO series 140001 "Environmental Management", the ecological situation in the world remains complicated. The machine building industry largely affects this fact.

In view of the above, the purpose of the article is to analyze potential environmental pollution types from various productions of machine building and apply a system approach as part of the ecologization of machine-building enterprises.

## 2. Results and discussion

Manufacturing of engineering products produces hazardous solid waste polluting soil and negatively affecting the flora.

Figure 1 shows the main polluting elements contained in emissions from the production of machine-building enterprises. One of the most hazardous substances entering the atmosphere is hexavalent chromium. This carcinogen causes various forms of cancer if it enters the human body. Among the most dangerous and harmful substances that accumulate in the ground are lead, mercury and cadmium, adversely affecting the human body and flora. Water pollution with heavy metal ions also has an adverse effect on the human condition and fauna.

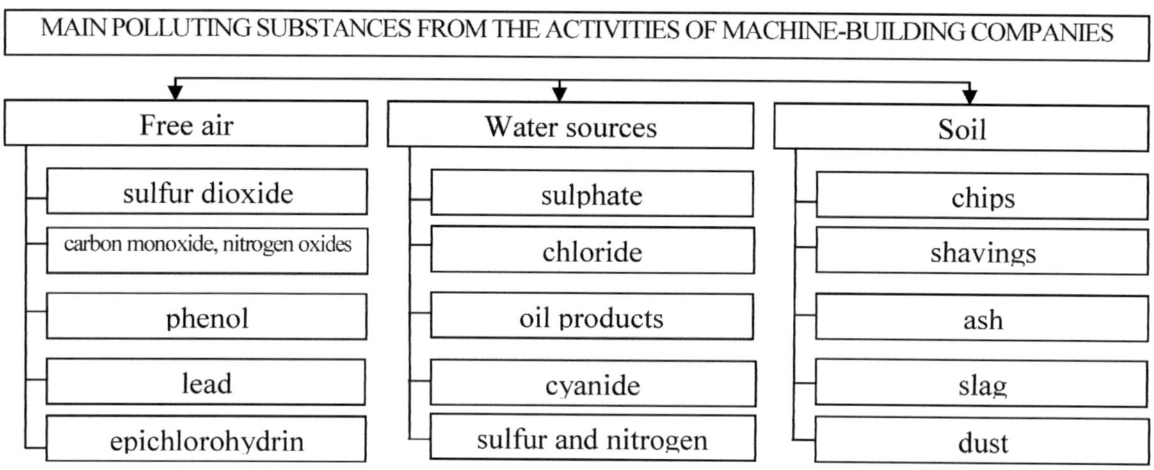


**Figure 1** – Main polluting substances from the activities of machine-building companies

By analyzing the information on the subject [1,2,3,4] it is possible to determine that the situation with the influence of various environmental pollutants is defined by the fact that modern factories of the machine-building industry use processes requiring production of harmful substances exceeding the permissible level of established standards. Depending on the type of production, various pollutants are identified, which are presented in Table 1.

Hazardous and noxious substances produced by different types of production pollute the environment and have a negative impact on human health. Hazardous waste solids occurring in each kind of production have carcinogenic, toxic, mutagenic and allergic properties which ultimately contribute to the development of various human diseases, among which the most dangerous are cancer and mutations. Besides, exceeding the established standard of pollutants in the air, soil and water leads to the extinction of animal and plant species, as well as disturbs the ecosystem as a whole.

According to statistics, emissions of harmful substances into the atmosphere from the Russian machine-building complex make up 32% of the entire industrial pollution from stationary sources [1], and only 30-50% of machine-building enterprises are equipped with wastewater treatment facilities.

Since anthropogenic factors are the main in a complicated environmental situation, it is necessary to reduce the risk of technological processes for the environment and humans with the help of a set of measures which, in general, is called production ecologization.

Ecologization activities are mainly aimed at limiting or reducing the environmental capacity of technological processes through the use of a highly efficient strategy of low-waste or cleaner production.

**Table 1** – Potential pollution from various productions of machine building

| № | Type of production | Technological process | Hazardous and noxious substances | Object of pollution |
|---|---|---|---|---|
| 1 | Foundry engineering | Technological processes of cast iron and steel details production | Carbon monoxide, dust, sulfur and nitrogen oxides, ammonia, phenol, formaldehyde and cyanide. Solid waste - spent sand | Soil, water bodies |
| 2 | Intrafactory power production | Technological fuel combustion processes | Hazardous gases and solids from the combustion products | Atmosphere |
| 3 | Metalworking | Technological processes associated with chip removal | Metal shavings, chips, dust | Air, soil, atmosphere |
| 4 | Electroplating production | Technological processes: nickel plating, cyaniding, chrome plating | Mercury, lead, cadmium, bismuth, nickel, zinc and others | Wastewater and rivers |
| № | Type of production | Technological process | Hazardous and noxious substances | Object of pollution |
| 5 | Paint and varnish production | Technological processes associated with coloring and varnishing | Lead, epichlorohydrin, tricresyl phosphate, dichloroethane and others | Soil, air |
| 6 | Welding production | Welding processes | Noxious welding fumes, manganese, copper, silicon, vapors of iron, zinc and nitrogen oxides, hexavalent chromium | Atmosphere |

To date, the scientific community [1,2,4,5,6] identified two ways of production ecologization. The first way - introduction of low-waste or non-waste technologies, the second way- application of a conditionally clean technology including the main production with special treatment facilities for waste disposal.

The above problems can only be solved if the production is viewed as a system which includes the following elements: personnel, machinery and equipment, technique and technology, product, management and control, each of which is aimed at minimizing the negative impact of industrial emissions on the environment of the country as a whole. The main areas of production ecologization of enterprises of the machine-building industry are presented in Figure 2 as an Ishikawa diagram.

One of the most important areas in the field of an eco-oriented machine building enterprise is personnel as one of the key elements of the production system. In this area it is necessary to attract new personnel for the modernization and ecologization of production, highly qualified and competent in the field of activity. To implement this directly, at first it is necessary to perform selection and evaluation of candidates by determining the selection criteria, and then organize adapting activities to the eco-innovation strategy of the enterprise. Key selection criteria or qualitative characteristics of applicants are divided into three groups: individual, professional and psychological.

Individual characteristics, most effectively contributing to the development of the eco-oriented enterprise concept, include: creativity – the ability to creatively tackle professional problems, think and act non-traditionally; flexibility – a qualitative characteristics of an applicant, which is the

IOP Conf. Series: Earth and Environmental Science 66 (2017) 012027 doi:10.1088/1755-1315/66/1/012027

readiness of an employee to accept innovations and to learn quickly; adaptability - the ability of a person to adapt quickly to changes in the external environment and to life conditions.

Professional qualities of candidates are their qualification, that is, the degree and type of their professional readiness for a certain employment; competence - a body of knowledge, skills, grasps and the ability to apply them in their work; erudition - in-depth knowledge and awareness in one or more areas of professional activity.

Of all the psychological characteristics of an eco-oriented person the most important in this sphere are:

- ecological loyalty - commitment to the principles of ecological production, preservation of environmental quality;
- ecological responsibility - the ability to practically apply environmental trends, views and ideas by investing tangible assets into procedures ensuring the protection of the environment [2].
- ecological discipline - the ability to comply with the principles and norms of behavior that are focused on achieving cleaner production preventing the ingress of polluting emissions into the environment.

For hired personnel and existing staff of enterprises it is expedient to develop training programs aimed at improving their skills, competence and knowledge in the area of production ecologization. In addition, it is required to carry out various activities in the field of environmental safety training and conduct training and retraining on the modernization and ecologization of machine-building production.

The most essential programs are programs of motivation in the considered area that promote the involvement of all personnel in the work on environmental safety and protection, the creation of conditions for effective and efficient operations of all staff.

When implementing measures in the field of the aforementioned production ecologization at the "personnel" level, apart from the listed positive factors there can arise risks associated with the development and implementation of inefficient personnel motivation systems which lead to losses in the financial part of the company's capital. In addition, substantial material expenditures are required for training and retraining, as well as selection of employees from the external reserve and their adaptation.

The next level of an eco-oriented machine building enterprise is "Management", as an element of the production system in the area of the eco-oriented strategy. It involves introducing systems for monitoring and control of the ecological state of the area, as well as hazard identification, assessment and management of professional risks. In addition to this, the best procedure in this case is to manage and exchange experience between enterprises in the field of ecology and ecological innovations and, as a result, implementation of best international and domestic practices in the field of engineering, technology and management of environmental safety. Development and implementation of the systems require current material and labor resources which are accompanied by costs to assess the efficiency and effectiveness of systems implementation, as well as the cost for their further maintenance.

The area "techniques and technology" is by far one of the top priority and is actually implemented, as humanity has accumulated extensive experience in the field of innovative technologies and practices for production ecologization [1,3,4,5,6,7,8,9,10,11]. The main ones include: introduction of eco-innovative technologies, including innovative technologies for monitoring and control of the environment, biotechnology, wastewater treatment and re-use in production; introduction of modern low-waste technologies; improvement of waste water filtration systems for enterprises and application of treatment technologies that reduce environmental pollution; processing of hazardous substances and waste disposal; use of energy-saving technologies that use natural resources sustainably.

These technologies and methods require material costs for environmental activities and improvement of intellectual potential of employees in the sphere of eco-innovations development.

To ensure eco-oriented concepts in the field of machine-building enterprises it is important to consider such an area as "machinery and equipment", which consists in upgrading the main assets of the enterprise, introducing innovative equipment for waste recycling, using energy-saving facilities. These activities are also financially expensive, since they require investment in the purchase and installation of new equipment.

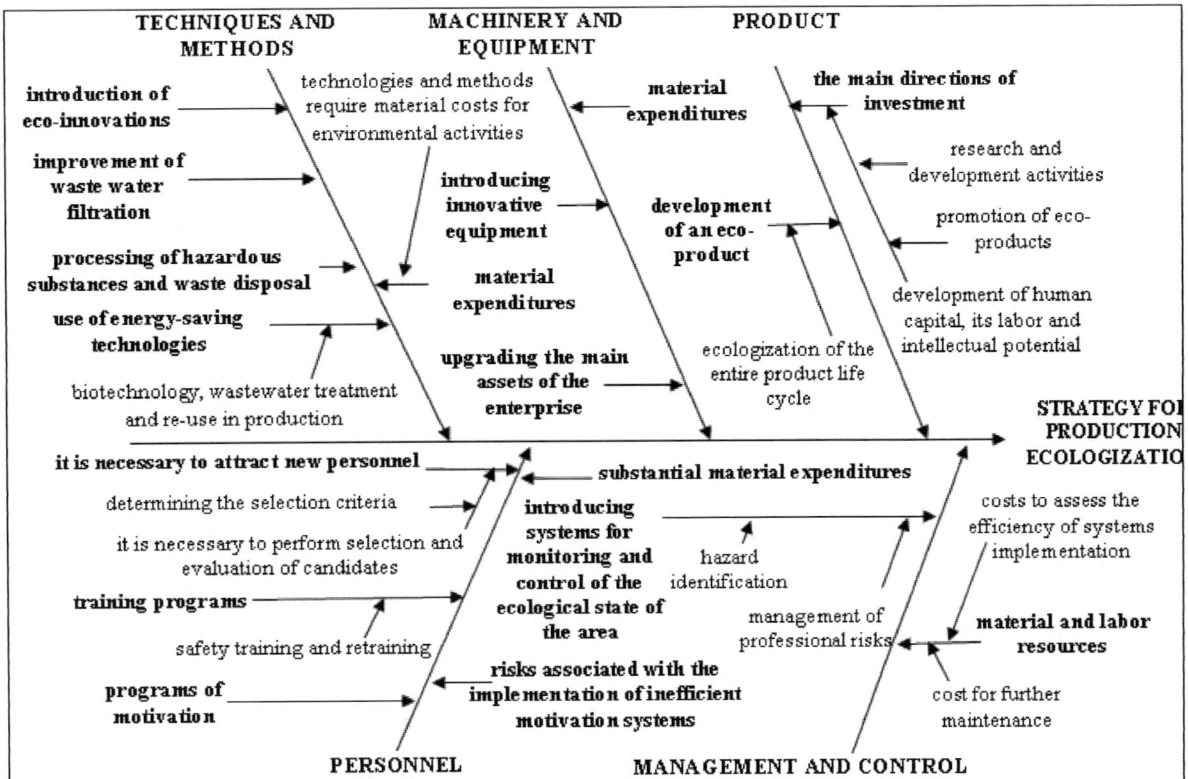


**Figure 2** – Main areas of production ecologization of machine-building enterprises

The area "product" includes ecologization of the entire product life cycle, from market research, research and development activities to its production, promotion and utilization. The main areas of investments for the realization of this area are as follows:

- development of human capital, its labor and intellectual potential;
- research and development activities in the field of technically feasible innovative solutions focused on the ecological principles of the enterprise concept;
- promotion of eco-products: study of supply and demand, analysis of the internal and external environment of the enterprise, advertising and marketing policy.

The end product is an object that satisfies the requirements of the Russian legislation and the demands of consumers, and in addition it does not violate the ecosystem and does not pollute the environment.

### 3. Conclusions

Thus, to solve the problems related to the influence of the production of machine-building enterprises on the ecological situation of the country, more attention should be paid to production ecologization.

Besides, it is not enough to use only non-waste or low-waste technologies, it is also necessary to consider the company as a system, a set of interrelated elements: personnel, equipment, management, technique, product. It is essential to introduce the eco-oriented production strategy which main goal is to ensure the safety and protection of the environment from exposure to harmful and hazardous emissions and waste.

The main aspects of this strategy, as described above in detail, are: improvement of the personnel competence and knowledge on the environmental safety; creation of conditions and methods of motivation and involvement of all employees in the efforts to ensure the protection of the environment; introduction of eco-innovations and modern technologies for processing of hazardous substances and waste through the use of innovative equipment; implementation of systems for monitoring and control of environmental condition of the area.

The reviewed complex of measures will identify hazards, assess and manage risks in the area of environmental safety and protection, and significantly reduce the impact of production on the country's ecology as a whole.

**References**

[1] Rudneva L.N., Pchelintseva I.G., Guryeva M.A. The indicative system of assessing the level of ecologization in the context of the region's sustainable development. International Journal of Economics and Financial Issues. 2016 Vol.6 p.227-232.

[2] Zakharova, E.N., Kerashev, A.A., Prokhorova, V.V., Gorelova, G.V., Mokrushin, A.A. Ecological innovations as a tool to provide the region's sustainable development // Mediterranean Journal of Social Sciences. 2015 Vol. 6 p.295-302.

[3] Hajej Z., Rezg N., Gharbi A. Ecological optimization for forecasting production and maintenance problem based on carbon tax. International Journal of Advanced Manufacturing Technology . 2017 Vol. 88 p.1595-1606.

[4] Moiseyeva V.A., Nazarov V.P., Zhuravlev, V.Y., Zhuykov, D.A., Kubrikov, M.V., Klokotov, Y.N Theoretical and experimental fundamentals of designing promising technological equipment to improve efficiency and environmental safety of highly viscous oil recovery from deep oil reservoirs. Izvestiya - Atmospheric and Ocean Physics. 2016 Vol. 52 p.753-759.

[5] Qiuli Y., Ke X.,Jun Z. Application of environmentally friendly materials in landscape infrastructure (Conference Paper). International Conference on Material Science and Engineering, ICMSE 2016; Guangzhou; China. 2017 Vol. 52 p.753-759.

[6] Zakirov D.G., Mukhamedshin M.A., Zakirov G.D., Gulyaev, V.E., Yangibaev, I.I. Problems of Yaregskoye field for oil and titanium extraction, possibilities of using secondary energy resources). Gornyi Zhurnal. 2013 p.59-61.

[7] Shotaevna Khetagurova V., Iakovleva Yurievna, I., Anatolyevna Bryukhanova, G., Aleksandrovna Denikaeva, Viktorovna Tolbuzina T. Ecological education as a basis for an ecological culture of the russian society. Journal of Environmental Management and Tourism. 2015 Vol. 6 p.78-90.

[8] Semenyuk E.P. Role of informatics in the ecologization of society. Scientific and Technical Information Processing. 2012 Vol. 39 p.1-12.

[9] Liu Y., Shi L. Critical review on industrial diversity indicators. Shengtai Xuebao/ Acta Ecologica Sinica. 2016 Vol. 36 p.7302-7309.

[10] Kasianenko, T.V., Nagornyi, Y.I. Grounding the approaches to estimation of ecologically oriented innovative development. Actual Problems of Economics. 2015 Vol. 167 p.313-320.

[11] Kalimullin A.M. Ecological problems of middle Volga in the second half of the XXth century: Historical and ecological analysis of leading industrial regions development. Review of European Studies. 2015 Vol. 7 p.86-93.

# Three-Loop Automatic of Control System the Landfill of Household Solid Waste

**T G Sereda**[1,2,a], **S N Kostarev**[1,2,b]

[1]National Research Perm Polytechnic University Affiliate,
pr. komsomol'skij 29, Perm, Russia, 614990
[2] Perm State Agricultural Academy named after academician D.N. Prianishnikov Affiliate, ul. Petropavlovskaja 23, Perm, Russia, 614990

*E-mail:* [a]stg@mail.ru, [b]iums@dom.raid.ru

**Abstract**. The analysis of models of governance ground municipal solid waste (MSW). Considered a distributed circuit (spatio-temporal) ground control model. Developed a dynamic model of multicontour control landfill. Adjustable parameters are defined (the ratio of $CH_4$ $CO_2$ emission/fluxes, concentrations of heavy metals ions) and control (purging array, irrigation, adding reagents). Based on laboratory studies carried out with the analysis of equity flows and procedures developed by the transferring matrix that takes into account the relationship control loops. A system of differential equations in the frequency and time domains. Given the numerical approaches solving systems of differential equations in finite differential form.

## 1. Introduction

Landfill is a complex biochemical system in which the flow of interrelated physical, mechanical, chemical, and biological processes happen [1]. The establishment of a system for managing ground MSW as a bioreactor with several adjustable values, is a challenge. Distributed management model single contour landfill MSW was reviewed in [2] (Figure 1).

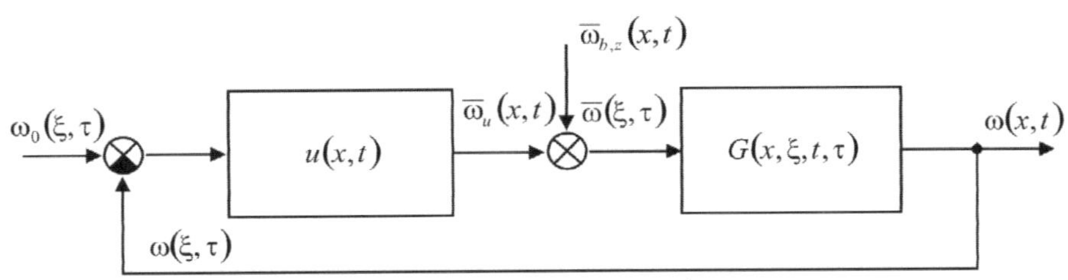


Figure 1 – Block diagram of a generic distributed control landfill MSW: $\omega_0(\xi,\tau)$ – humidity condition in the array specified wastes; $\omega(x,t)$ – deviation from the specified output distributed; $\overline{\omega}_u(x,t)$ – distributed control; $\overline{\omega}_{b,z}(x,t)$ – external perturbations; $\overline{\omega}(\xi,\tau)$ – standardizes function

The aim of this study was to investigate the influence of physico-chemical parameters of MSW landfill on the resulting emission products (filtrate and biogas) and development of a multi-circuit control system in multidimensional control systems, unlike the one-dimensional, outrage one of the loops can occur due to changes in other circuits.

## 2. Main factors that describe the dynamics of an object

Depending on external factors (age, physical-chemical processes) the object's State is characterized by steady or transient modes, which has changed management requirements object [2]. Available managed parameters monitoring processes in the array of MSW, suggested to use: purge air waste array hydration (leachate recirculation) and the addition of reagents to the array. Thus, the adjustable parameters will be: 1 – the presence of oxygen in the body of the landfill (B); 2 – humidity MSW array $\omega$; 3 – strong response (pH).

## 3. Block diagram models of multi-circuit control systems testing ground

For building control system ground MSW were used the system of linear differential equations or corresponding images by Laplace in the frequency domain of the complex variables [3]. Figure 2 shows a block diagram of a multi-dimensional system of regulation in landfill processes, including the systems with perturbations of the $k$, $l$-offices ($l = 3$) and $m$-observations (exits, $m = 3$) dependence (1):

$$y(s) = G(s)u(s) + G_d(s)d(s), \tag{1}$$

where $d(s)$, $u(s)$, $y(s)$ are vectors of indignation, management and exit, respectively

$$d(s) = \begin{bmatrix} d_1(s) \\ d_2(s) \\ \cdots \\ d_k(s) \end{bmatrix}; \quad u(s) = \begin{bmatrix} u_1(s) \\ u_2(s) \\ u_3(s) \end{bmatrix}; \quad y(s) = \begin{bmatrix} y_1(s) \\ y_2(s) \\ y_3(s) \end{bmatrix}; \tag{2}$$

$G(s)$ – transfer function matrix management system ($g_{lm}$ – matrix element that associates a vector: surveillance, control)

$$G(s) = \begin{bmatrix} g_{11}(s) & g_{12}(s) & g_{13}(s) \\ g_{21}(s) & g_{22}(s) & g_{23}(s) \\ g_{31}(s) & g_{32}(s) & g_{33}(s) \end{bmatrix}; \tag{3}$$

$G_d(s)$ – transfer function matrix system disturbance ($g_{lk}$ – element matrix linking: vector surveillance, outrage)

$$G_d(s) = \begin{bmatrix} g_{11}^d(s) & g_{11}^d(s) & \cdots & g_{1k}^d(s) \\ g_{21}^d(s) & g_{22}^d(s) & \cdots & g_{2k}^d(s) \\ g_{31}^d(s) & g_{32}^d(s) & \cdots & g_{3k}^d(s) \end{bmatrix}. \tag{4}$$

Formation of control action of $u(s)$ will be determined by multidimensional controller $G_c(s)$ taking into account the mismatch errors

$$\varepsilon(s) = y^*(s) - y(s),$$

where $\varepsilon(s) = [\varepsilon_1(s), \varepsilon_2(s), \varepsilon_3(s)]^T$.

Multidimensional controller $G_c(s)$ is represented by a diagonal matrix types (elements of $g_{ll}(s)$ corresponding to an one-plane metric (one-dimensional) regulators set certain control Act)

$$G_C(s) = \begin{bmatrix} g_{11}(s) & & 0 \\ & \cdots & \\ 0 & & g_{ll}(s) \end{bmatrix}. \tag{5}$$

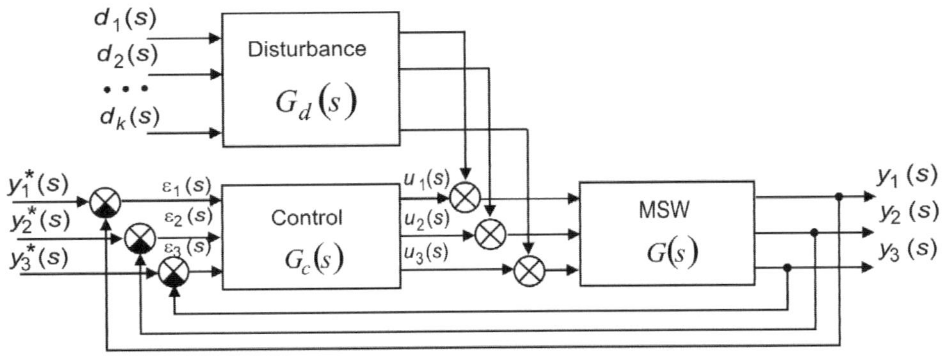


**Figure 2** – A block diagram of the system of management of MSW landfill

## 4. Physical modelling

### *4.1. The laboratory plant biodegradation of waste*

Laboratory installation was performed in the form of reactor simulating on the body of MSW landfill waste management of certain morphological structure [4,5]. Reactor allowed to regulate water-air regime of array wastes (Figure 3).

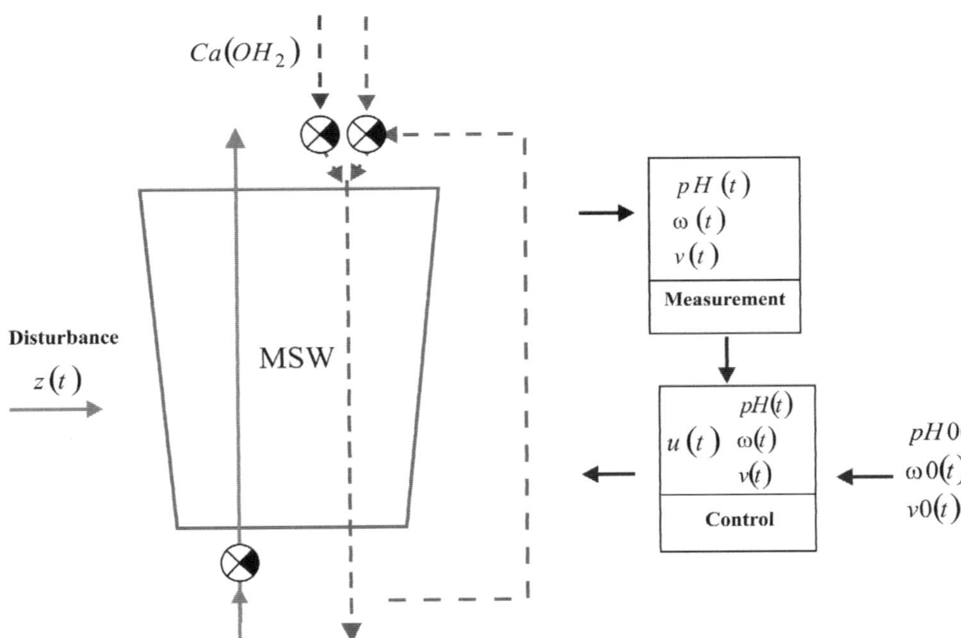


**Figure 3** – Schematic control laboratory installation: —— purge array of MSW; - - - leachate recirculation; $Ca(OH)_2$ – filing of limy milk

Challenge of studies was to assess the degree of biodegradation of waste within the reactor. In order to accelerate the biodegradation of waste there was carried out at the expense of the reactor control irrigation stabilized the seep liquid canister with the addition of lime milk and purge air. The main parameters measured were physical and chemical values: pH, concentration of ions of heavy metals, ammonia nitrogen in leachate, volume of leachate and landfill waste moisture content [6,7].

### *4.2. Study to determine the optimum parameters of biodegradable waste*

Ratio control actions (u) and monitoring parameters of emission flows shown in Table 1.

**Table 1** – Control actions and monitoring emission streams

| № | Control actions, *u* | Observable options, *y* |
|---|---|---|
| 1 | Purge array | $CH_4$ /$CO_2$ Ratio |
| 2 | Irrigation | pH |
| 3 | Adding Irrigation | Reagents concentration of heavy ions |

Research results have shown that the optimal physico-chemical parameters of the array are:
1 – The presence of oxygen and filling the pores of the landfill body air (B) (Air/water) – 40/60 %;
2 – humidity MSW array ( ω ) – 60%; 3 – strong response (pH) – 8-9,5.

### *4.3. Study the interrelations of the multidimensional control tasks*

In the synthesis of multidimensional control systems, a great problem difficulties linked to static and dynamic cross-linkages array between different inputs and outputs - *R* system. Research links control actions used to input the step signals that represent inertial link of the first order [8].

Experiment on research of mutual influence, irrigation and limy milk processing (Figure 4) with stepped change of control actions showed that the humidity was most important (space) factor that affects the amount of air pores and Ph; purge provided effects on pH (Figure 5) [9, 10].

Thus the transfer matrix (3) were not considered unimportant connection:

$$G(s)=\begin{bmatrix} g_{11}(s) & 0 & 0 \\ g_{21}(s) & g_{22}(s) & 0 \\ g_{31}(s) & g_{32}(s) & g_{33}(s) \end{bmatrix}. \qquad (6)$$

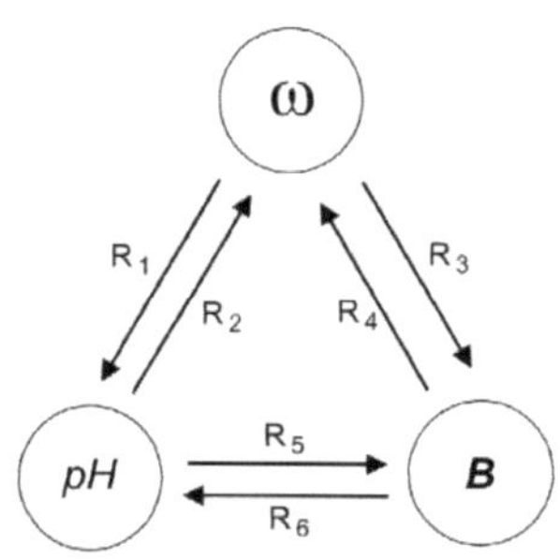


**Figure 4** – Scheme of the experiment on research of mutual influence, irrigation and processing of limy milk, R-cross-connection control actions

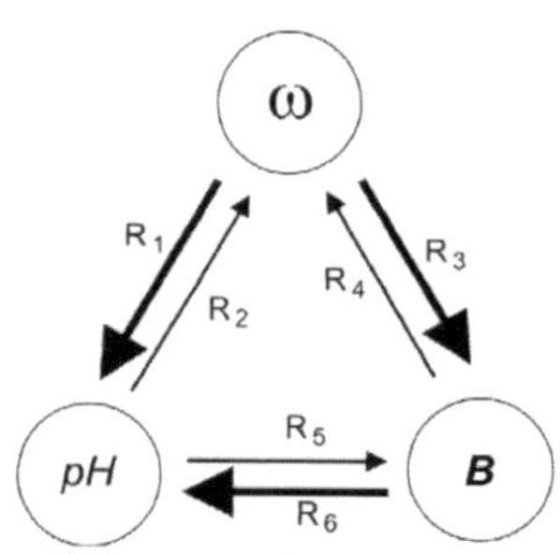


**Figure 5** – Distribution of dominant factors: moisture content, availability of oxygen (air), pH

## 5. Mathematical model of ground MSW

Equivalent model (model of the complex variable *s* (1)) of a linear system is a model in the temporary area

$$\begin{cases} \dfrac{dx}{dt} = Ax + Bu + Ld, \\ y = Cx, \\ x(t_0) = x_0. \end{cases} \quad (7)$$

where $x$ – $n$-dimensional State vector; $d$ – $k$-dimensional vector perturbations; $u$ – $m$-dimensional vector offices ($m = 3$); $y$ – $l$-dimensional vector of observations ($l = 3$).

The matrix $A$ is State, $B$ – offices, $C$ – observations and $L$ – perturbations related dimensions:

$$A = \begin{bmatrix} a_{11} & a_{12} & \dots & a_{1n} \\ a_{21} & a_{22} & \dots & a_{2n} \\ \dots & \dots & \dots & \dots \\ a_{n1} & a_{n2} & \dots & a_{nn} \end{bmatrix}; \quad B = \begin{bmatrix} b_{11} & b_{11} & \cdots & b_{1n} \\ b_{21} & b_{22} & \cdots & b_{2n} \\ b_{31} & b_{3n} & \cdots & b_{3n} \end{bmatrix}; \quad (8)$$

$$C = \begin{bmatrix} c_{11} & c_{12} & \cdots & c_{1n} \\ c_{21} & c_{22} & \cdots & c_{2n} \\ c_{31} & c_{32} & \cdots & c_{3n} \end{bmatrix}; \quad \Gamma = \begin{bmatrix} \gamma_{11} & \gamma_{12} & \cdots & \gamma_{1n} \\ \gamma_{21} & \gamma_{22} & \cdots & \gamma_{2n} \\ \cdots & \cdots & \cdots & \cdots \\ \gamma_{k1} & \gamma_{k2} & \cdots & \gamma_{kn} \end{bmatrix}. \quad (9)$$

An algorithm for solving problems (1–9) is reduced to the determination of the transfer function of the object (6), the transformation equations of frequency to the staging area. For computer simulation process multidimensional process control system requires conversion of differential equations to discrete finite differential form.

## 6. Conclusions

The work solved is an ongoing challenge, who in the development of the automated system of state management, landfill and the main results were obtained:

1. Relationship between reference and Identified internal factors affecting the state of the landfill that allowed the development of a transfer matrix, taking into account the cross-connection control loops of the landfill.
2. A system of linear differential equations in the frequency and time fields that implement the multicontour control system of the landfill.
3. Given the numerical approaches solving systems of differential equations infinite differential form.

## References

[1] Kostarev S.N., Sereda T.G. Automated process control of sanitary municipal solid waste landfill. World Applied Sciences Journal. 2013 Vol. 22, Issue SPL.ISSUE2. p. 64–69.

[2] Kostarev S.N., Sereda T.G., Elancheva E.N. Application Models of Continuum Mechanics Control Landfills. Applied Mechanics and Materials. 2014 Vol. 682. p. 251–255.

[3] Fakharudin A.S., Sulaiman M.N., Mustapha N. Modelling of biogas production from banana stem waste with neural networks learning strategies to optimse the production. Journal of Theoretical and Applied Information Technology. Volume 95, Issue 2, 2017, p. 285-291.

[4] Kostarev S.N., Sereda T.G., Elancheva E.N. Study safety environmental protection landfills using Models anaerobic digesters. Applied Mechanics and Materials. 2014 Vol. 682. p. 339-345.

[5] Garcia-Ivars J., Wang-Xu X., Iborra-Clar M.-I. Application of post-consumer recycled high-impact polystyrene in the preparation of phase-inversion membranes for low-pressure membrane processes. Separation and Purification Technology. 2017 Vol. 175. p. 340-351.

[6] Arev I.A., Marchuk A.S., Bogolyubov N.V. Monitoring of the quality of natural water by a 'Nayada' automatic station. Soviet Journal of Water Chemistry and Technology (English Translation of Khimiya i Tekhnologiya). 1983. Vol. 5, Issue 2. p. 107-110.

[7] Kunz N.C., Fischer M., Ingold, K., Hering J.G. Drivers for and against municipal wastewater recycling: A review. Water Science and Technology. 2016 Vol. 73, Issue 2. p. 251-259.

[8] Mboowa D., Quereshi S., Bhattacharjee C., Tonny K., Dutta S. Qualitative determination of energy potential and methane generation from municipal solidwaste (MSW) in Dhanbad (India). Energy. 2017 Vol. 123. p. 386-391.

[9] Awe O.W., Zhao Y., Nzihou A., Minh D.P., Lyczko N. A review of biogas utilisation, purification and upgrading technologies. Waste and Biomass Valorization. 2017 Vol. 8, Issue 2, p. 267-283.

[10] Supriyo P.T., Hanum F., Aman A., Bakhtiar T. Optimization in MSW management system: Site selection of the waste-to-energy facilities. Far East Journal of Mathematical Sciences. 2017 Vol. 101, Issue 3, p. 473-485.

# Advanced Technique and the Results of a Research of a Heat-Mass-Exchange Processes in Clothes Packages in the Subnormal Climate

**M V Rodicheva [a], A V Abramov[a], N V Kanatnikov[b], P A Kanatnikova[b]**

Federal State Budgetary Education Institution of Higher Education «Orel State University named after I.S.Turgenev», Orel, Naugorskoe shosse 29 Russia 302020

E-mail: [a] Ant-lin88@mail.ru, [b] NVKanatnikov1989@gmail.com

**Abstract.** Quality of clothes in the conditions of subnormal temperatures can be provided by using a scientifically based approach for completing of a set of materials. In the article, the method of a research of heat-mass-exchange in the conditions of a non-stationary heat-mass-exchange is stated; the results of a research of influence of materials on the efficiency of heat-protective clothes are considered.

## 1. Introduction

In the temperate climate zone, cold and transient periods of year last from four to nine months a year. The protection problem of the person in these conditions still remains urgent. All ranges of the lowered temperatures are divided on subnormal temperatures (+10°C ÷ -5°C), moderated (-10°C ÷ -25°C) and severe (lower than -25°C) cold. Theoretical bases of protection of the person in the conditions of moderate and severe cold are studied rather widely. Physiological features of heat exchange of the person are investigated by Burton A.C., Edholm O.G., Fanger P.O. et al [3, 5]. Properties of materials for clothing in cold are evaluated by Buzov B.A. Nikitin A.V., Zhikharev A.P. et al [4, 10]. Approaches to completing the sets of clothes are created by Afanasieva R.F., Koshcheev W.S. et al. [2, 6].

Subnormal temperatures are observed during transient periods of the year and are followed by the abrupt change of weather therefore the heat-mass-exchange in a set of materials proceeds in the non-stationary mode. However, approaches of stationary heat exchange are used when completing clothes in these conditions that leads to increasing the number of cold injuries. About 18% of cases of people's death of cooling occur in this range of temperatures while intense cold leads to a lethal outcome only in 4% of cases [6].

The multiple research of a non-stationary heat-mass-exchange in a set of materials will allow to increase the quality of heat-protective garment due to scientifically based approach to completing of clothes. Nowadays the research on processes of moisture transfer in the structure of a package of materials in the conditions close to subnormal is conducted by Gibson P., Rivin D., Weder M. [8, 9]. It should be noted that they neglect the impact of heat on moisture flows that demands carrying out the additional calculations at design of clothes. M. Rock, P. Yeo, and other researchers use these results

IOP Conf. Series: Earth and Environmental Science **66** (2017) 012029 doi:10.1088/1755-1315/66/1/012029

for completing the sets of materials of heat-protective clothes in the conditions close to subnormal temperatures [7].

## 2. Method of a research of a non-stationary heat-mass-exchange in the garment

The authors developed an experimental method, which allows to investigate processes of a heat-mass-exchange in the set of materials at subnormal temperatures [1]. The hypothesis that the person thermal status in these conditions is provided by the quasistationary mode of heat exchange is the basis for a method. The constant body temperature at change of weather is maintained due to regulation of the heat production level in an organism. The cylindrical device filled with water and equipped with an internal source of heat is used for modeling of heat exchange of the person in the quasistationary mode. On the surface of device the set of materials (pos. 1, fig. 1, a) in the form of the system of the enclosed concentric elements is located. The automated system maintains the target temperature of the cylinder due to adjustable, giving off heat and measures heat flux density on its surface.

Modern sets of garment exclude the possibility of penetration of moisture and a cold air under clothes. Therefore, the processes of a heat-mass-exchange at output of sweat from under the clothes to the environment are considered in the researches. For this purpose the moistening node is provided in the device (pos. 2, figure 1, a), giving moisture between thermal model of the person's body and a set of materials.

The temperature sensors are fixed on the each layer of a set (fig. 1, b), elementary samples are prepared for assessment of moisture increment (fig. 1, c). Before the start the experiment, the samples are fixed by the thread in order to provide a required contact of a set on site of cuts. The airflow hub is placed over an upper face end of thermal model (pos. 3, fig. 1, a); primary transformers of temperature, speed motion (pos. 4, fig. 1, a) and humidity of air (pos. 5, fig. 1, a) are placed in the neck of the hub.

All types of the works, which are carried out in the open air, can be divided into several categories of severity. Work of II "a" category means that the mass of a transferable load does not exceed several kilograms. Sweating of the person is insignificant. Works of II "b" category is for cargo transfer weighing up to 10 kg. Sweat is absorbed by the clothes and has an impact on the intensity of a heat-mass-exchange in a set of materials. At accomplishment of hard physical work (III category) in clothes with low vapor permeability, the essential increment of moisture weight is fixed.

For modeling of these features of heat exchange, the experiment is conducted during five different phases. On the first phase the performed works belong to the II "a" category, heat exchange proceeds in the stationary mode, moistening is not made. Second phase: intensity of work increases, the beginning of sweating is modelled by the inclusion of the moistening node. Supply of moisture is carried out until saturation of a set of materials for a research of the heat-mass-exchange proceeding in case of limit moistening. The fourth phase: abrupt decrease in weight of works that leads to the sweating termination. In the fifth phase, the set completely dries, and thermal properties of materials recover to the initial level.

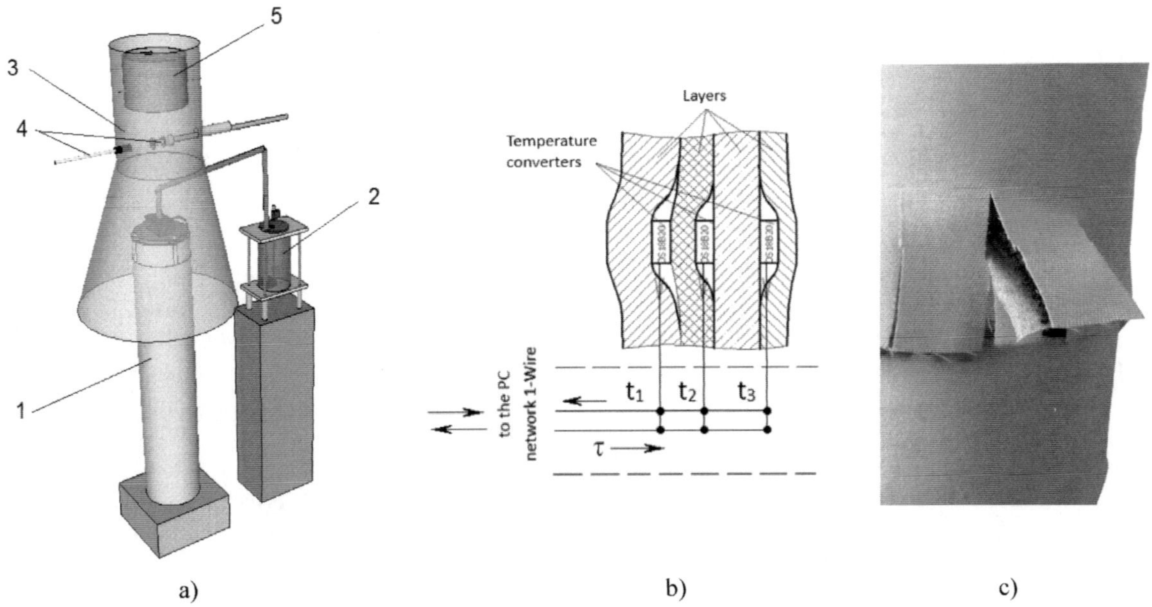


**Figure 1** – An illustration of a method of pilot study of complex processes of a heat-mass-exchange in a set of materials

## 3. Structure and properties of heat-protective garment materials in the conditions of subnormal temperatures

The processes of a heat-mass-exchange in the set of material for heat-protective garment in the conditions of subnormal temperatures are investigated by means of this method. In accordance with the regulatory requirements, the set of clothes consists of three parts - underwear, warming and upper layers. The characteristics of the most frequently used textile materials are presented in table 1.

Five options of set varying in option of one layer (table 2) were made by combination of the materials.

**Table 1** – Materials for clothes at subnormal temperatures

| No | Set layer | Material type | Fibre composition | Surface density, g/m$^2$ | Thickness, mm | Interweaving |
|---|---|---|---|---|---|---|
| 1 | Underwear | Knitted fabric | cotton, 100% | 240 | 0.6 | Jersey structure |
| 2 | | Knitted fabric | cotton, 100% | 220 | 0.5 | Jersey structure |
| 3 | Heating | Knitted fabric | 20% PA 80% Wool | 360 | 2.4 | Jersey patterning |
| 4 | | Knitted fabric | 20% PA 80% Wool | 310 | 1.4 | Jersey double patterning |
| 5 | | Knitted fabric | 20% PA 80% Wool | 280 | 1.2 | Jersey double patterning |
| 6 | Upper | Fabric | 60% cotton 40% PES | 430 | 0.45 | Twill, ½ |
| 7 | | Fabric | 60% cotton 40% PES | 462 | 0.35 | Twill, 1/3 |

IOP Conf. Series: Earth and Environmental Science **66** (2017) 012029 doi:10.1088/1755-1315/66/1/012029

**Table 2** – Options of completing of the set of materials

| Option of set completing | Layers options (in accordance with table 1) | | |
|---|---|---|---|
| | Underwear | Heating | Upper |
| Trad No 1 | Underwear knitted fabric (sample No1) | Warming knitted fabric (sample No3) | Cotton fabric (sample No6) |
| Trad No 2 | Underwear knitted fabric (sample No2) | Warming knitted fabric (sample No3) | Cotton fabric (sample No6) |
| Trad No 3 | Underwear knitted fabric (sample No1) | Warming knitted fabric (sample No4) | Cotton fabric (sample No6) |
| Trad No 4 | Underwear knitted fabric (sample No1) | Warming knitted fabric (sample No5) | Cotton fabric (sample No6) |
| Trad No 5 | Underwear knitted fabric (sample No1) | Warming knitted fabric (sample No3) | Cotton fabric (sample No7) |

**4. Results and discussion**

Results of researches are presented in figures 2-4 in the form of curves of dynamics of a heat flow, dynamics and kinetics of moisture. It allows to compare the nature of heat-mass-exchange processes in sets of materials of different completing and to evaluate its influence on a person thermal status.

Figure 2 illustrates the results received for sets with a different underwear layer. Thickness and area density are a little less in Trad No.2. Apparently, during the work of II "a" category of severity (the 1st phase of an experiment) distinction in values of heat flux density do not exceed 10% (fig. 2, a). At intensive sweating (increasing category of severity to III) in a set Trad No.1 accumulated moisture increases by 0.5 g by the end of the 2nd phase of an experiment (fig. 2, b); heat flux density increases on 80 W/m2 in comparison with Trad No. 2.

At the beginning of the fourth phase of an experiment, when level of physical activity decreases, Trad No. 2 dries quicker than a set Trad No. 1; however, by the end of a phase these distinctions become almost unnoticeable.

The choice of material of an underwear layer does not influence a person thermal status of the person during accomplishment the work with the low category of severity. Textile fabric of smaller thickness is more preferable when physical activity changes, for the additional weight of moisture of a set of materials and heat loss by the person in wet clothes will be less.

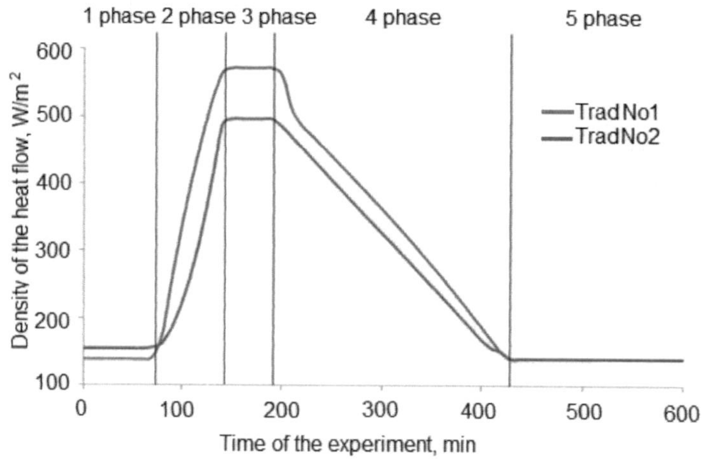


a)

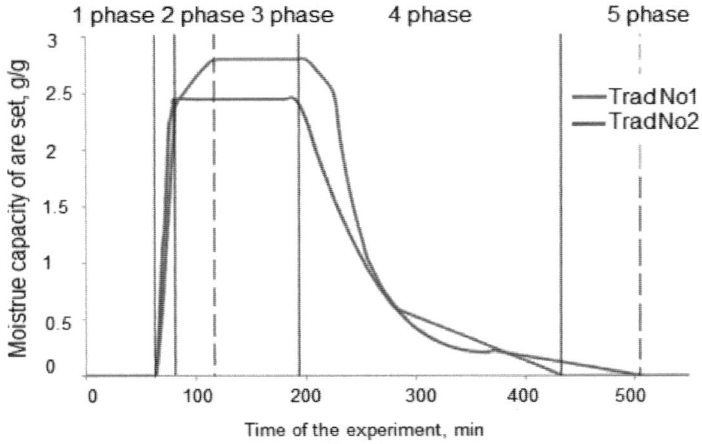


b)

**Figure 2** – Results of a research of heat-mass-exchange processes in the set of the materials varying in option of underwear layer

Results of a research of a heat-mass-exchange in set of materials with the reduction of area density and thickness of the warming layer are presented in the figure 3. Values of heat flow density in dry Trad No. 3 and Trad No. 4 are 178 W/m$^2$ and 192 W/m$^2$ respectively (figure 3, a). Thus, reduction of the thickness of the warming layer by 40 and 50% in the conditions of subnormal temperatures leads to increase of heat flow density on 20 and 30%.

In the conditions of intensive sweating, the value of incremented moisture weight in the sets with thinner heat-insulation decreases (by 15 and 17% for Trad No. 3 and Trad No. 4 by the end of the 2nd phase). At the same time, heat flux density in the third phase of an experiment in these sets is higher by 7 and 10% respectively. These distinctions are explained by the high percentage of woolen fibers in a Trad No. 1. It is known that wool better keeps warm in a wet state.

When the level of physical activity decreases, sets Trad No. 3 and Trad No. 4 dry for 9% and 24% faster than the set Trad No. 1. Thus, the duration of a drying-out period is proportional to the change of moisture capacity of the warming layer.

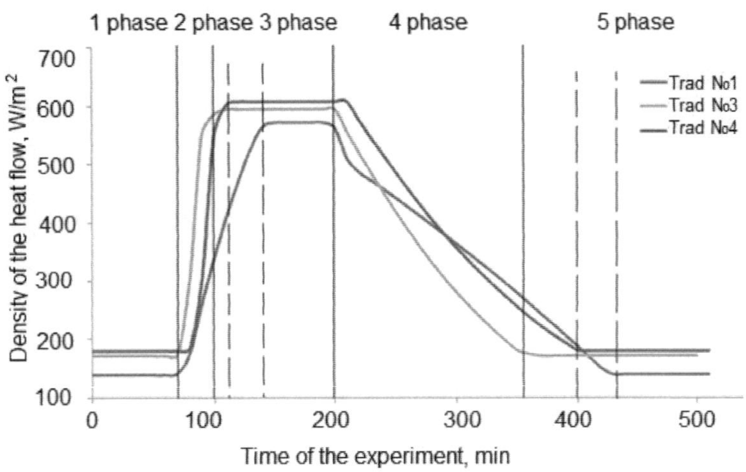


a)

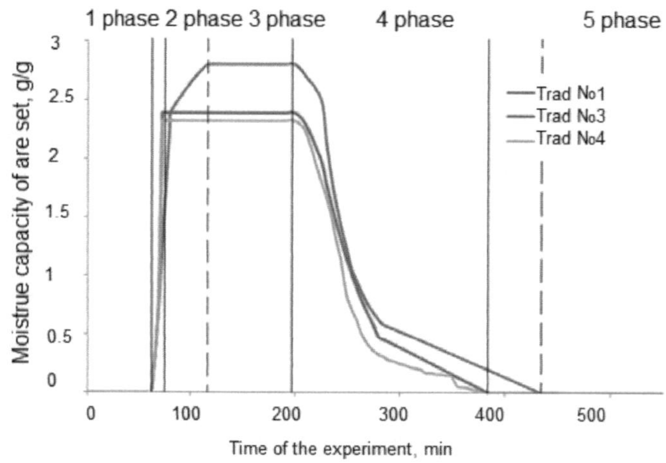


b)

**Figure 3** – Results of a research of heat-mass-exchange processes in the set of the materials varying in warming layer

Thus, at constant performance of work of low and high category of severity, it is more preferable to use materials with the bigger thickness of the warming layer if needed to preserve the heat.

When working with constantly changing category of severity where rapid recovery of initial heat-protective properties of clothes is necessary, sets with warming layer of smaller thickness are more preferable. Intensity of removal of moisture from a set with such cloths is higher by 20-25%, and distinction in the value of heat losses in comparison with a set of big thickness does not exceed 30% and 10% in a dry and wet status respectively.

The results of researches received for the sets Trad No. 1 and Trad No. 5 differing with an area density of upper layer fabric are presented in the figure 4. Apparently, change of mentioned parameter in the studied range of values does not affect the intensity of a heat-mass-exchange process in a set of materials.

Developed method will allow to receive rather detailed characteristic of a heat-mass-exchange processes in the set of materials. Obtained results will allow to increase the quality and efficiency of a heat-shielding of completing clothes in the conditions of subnormal temperatures.

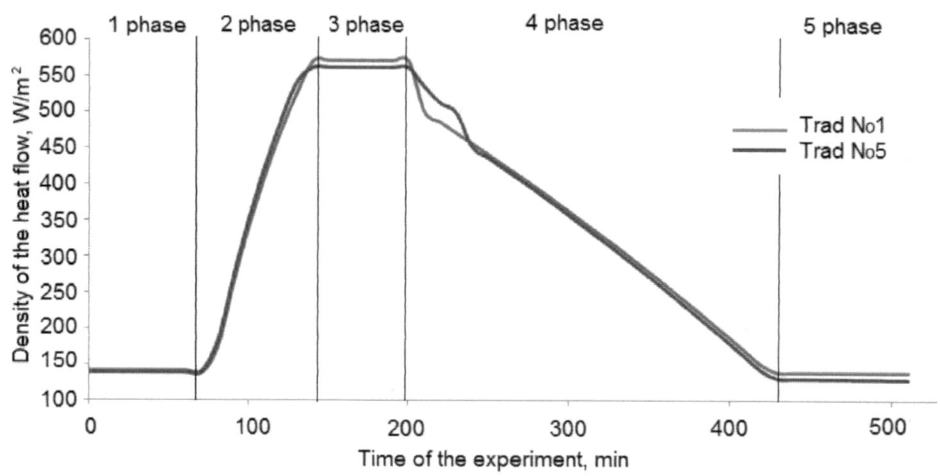


a)

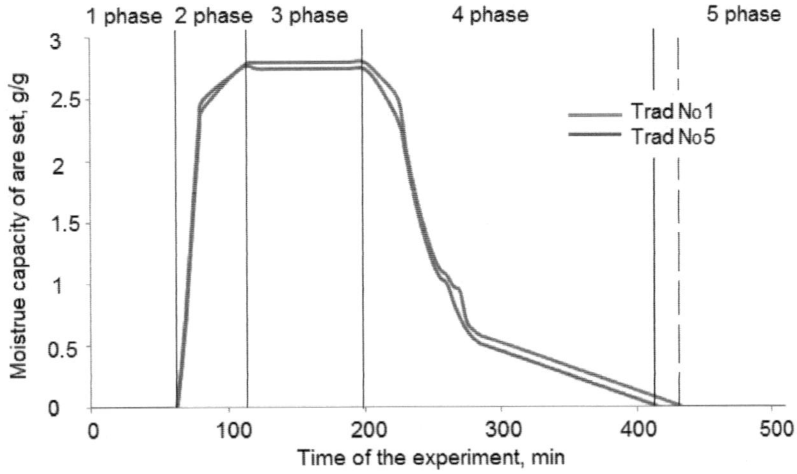


b)

**Figure 4** – Results of a research of heat-mass-exchange processes in the sets of the materials varying in the choice of upper layer

## 5. Conclusions

As a result of a system research of heat exchanging processes in the set of materials at the accomplishment of the work of different category of severity it was established that:

1. Using the more dense and thick cloths of the underwear and warming layers allows to reduce person heat losses in the conditions of subnormal temperatures by 10 - 30% during performing the work of IIa category.

2. At short-term increase of severity of the performed works, a set of materials with thin underwear and warming layers is more preferable. Intensity of moisture removal in these sets is higher by 10-25% in comparison with thicker layers, and the $q_{set}$ value is 15 - 30% lower; that allows to quickly recovering the initial level of heat-protective properties of clothes, promoting the improvement of a person thermal status. At the same time, at accomplishment of hard physical work for the long time, it is preferable to use sets with thicker and dense warming layer. Such set of clothes intensively absorbs moisture from the surface of a person's body and preserves the high level of heat-protective properties.

3. In the conditions of subnormal temperatures, it is also more preferable to produce a high layer of clothes of more dense and thick cotton fabrics. Influence of thickness and density of high layer fabric in the considered range of values does not influence the intensity of heat-mass-exchange processes in a set, at the same time, increment of these values increases a wind tightness of clothes.

## References

[1] Abramov A V, Rodicheva M V, Uvarov A V. Biotehnicheskie osnovy proektirovaniya odezhdy dlya zashchity ot neblagopriyatnykh temperature [Biotechnical bases of designing the clothes for protection against adverse temperatures], Orel, State University-ESPC., 2014, 137 p.

[2] Afanas'eva R.V. Gigienicheskie osnovy proektirovanija odezhdy dlja zashhity ot holoda Moskva [Hygienic bases of designing the clothes for protection against cold]: Legkaja promyshlennost', 1977, 136 P.

[3] Burton A.C., Edholm O.G. Man in cold environment: A study in physiology, Arctic Aeromedical Laboratory, 1956, 356 P.

[4] Buzov B.A., Nikitin A.V, Issledovanija materialov dlja zashhitnoj odezhdy v uslovijah ponizhennyh temperatur [Research of materials for protective clothes in the conditions of the lowered temperatures] M.: Legprombytizdat, 1985, 221 P.

[5] Fanger P.O. Thermal Comfort, New York McGrow-Hill, 1973, 244P.

[6] Koshcheev Fiziologija i gigiena individual'noj zashhity cheloveka ot holoda [Physiology and hygiene of individual protection of the person against cold] M: Medicia, 1981, 188 P.

[7] EP 0777978 A2 Composite textile fabric having moisture management characteristics, Priority date 1995-12-08, Publication date 1997-06-11, Inventor P. Yeh, 12 P.

[8] Gibson P., Rivin D., Berezin A., Nadezhdenskii A. Measurement of water vapor diffusion through polymer films and fabric/membrane laminates using a diode laser spectroscope. Polymer-Plastic Texchnology and Engineering. Vol. 38, No2. pp 221-239, 1999

[9] Weder M. Neutron Radiography measurement of moisture distribution in multilayer clothing systems. Tehtile Research J. Vol 74(8), pp 695-700, 2004

[10] Zhiharev A.P. Svojstva materialov [Properties of materials] M.: MGUDT, 2003, 163 P.

IOP Conf. Series: Earth and Environmental Science **66** (2017) 012030 doi:10.1088/1755-1315/66/1/012030

# Phytomining Perspectives in Rehabilitation of Mining and Industrial Areas of South Ural

**S S Timofeeva[1, a], D V Ulrikh[2], S S Timofeev[1]**

[1]Irkutsk National Research Technical University,
83 Lermontova St., Irkutsk, Russia, 664071
[2]South Ural State University (National Research University),
76 Lenin Prospekt, Chelyabinsk, Russia, 454080

E-mail: [a] timofeeva@istu.edu

**Abstract**. The ability of midland hawthorn (Crataegus laevigata (Poir)), common barberry (Berberis vulgaris), red elder (Sambucus racemosa), cinnamon rose (Rosa cinnamomea L.), couch grass (Elytrigia repens), crested wheat grass (Agropyron cristatum), meadow fescue (Festuca pratensis) and meadow grass (Poa pratensis) for phytoextraction of heavy metals from technogenic soil is proved in the article. The possibility of effective phytoextraction with the use of hawthorn and elder is shown. Maximum accumulation of zinc takes place in the surface mass of couch grass and meadow fescue. In regard to the conditions of South Ural, planting of elder and hawthorn with seeding of couch grass and meadow fescue is recommended for phytomining purposes.

## 1. Introduction

High concentration of manufacturing plants with extremely ineffective technological processes generates extremely tense ecological environment in Chelyabinsk oblast.

Chelyabinsk, Magnitogorsk and Karabash are in the black list of the most polluted Russian cities. According to space observations data, the main territories polluted by heavy metals are situated around Chelyabinsk and Magnitogorsk. 52% of the Oblast territory is polluted by the mill waste of iron and non-ferrous works, mining and coal industries. According to statistics data, pollutants emission into the atmosphere in a year per each citizen of Russia amounts to 372 kg of repugnant substances, in Chelyabinsk oblast this ratio varies from 200 to 2000 kg, and in Karabash it reaches 25 kg in a day. Concentration of heavy metals in soils of the mentioned territory varies in a wide range from 10 to 1500 and more mg/kg [1].

Restoration of technogenic landscapes is one of the pressing issues of modern age, especially in mining and industrial areas. The choice of treatment methods for such territories is determined by its climatic conditions, by the type, level and form of pollution, by features of the landscape and by the technologies used for obtaining the end product of a mining and processing complex.

In recent years, phytoremediation methods which provide phytoextraction (or phytomining) – metal removal by their accumulation in plant bodies – are met with wide recognition. This method is more beneficial compared to other methods of contaminated territories treatment: it is low-cost, and there is an opportunity to obtain additional products. Nowadays phytomining is defined as a usage of green plants for commercial extraction of metals of value from soil with their high concentration [2].

Aboveground parts of plants, from which the produced metals are extracted by the method of combustion, are used for phytomining.

Phytomining technology is currently under intensive development, and there are already some reports about successful testing [3-9].

Phytomining technology development is preceded by multiple research works on studying the regularities of metals accumulation in plant bodies, growing in contaminated soils. Research on developing ecological technologies (biological remediation of anthropogenic landscapes formed as a result of mining industry activities) and on their further practical use (especially in agriculture), have been conducted for almost half a century.

Phytomining is considered as a phytoextraction technology which allows extracting economically valuable metals from a substrate (in particular, from the dumps of mining industry). At that, the key role in phytoextraction and phytomining belongs to the plants – hyper-accumulators of metals. Such plants must absorb metals from the substrate, hold them in root tissues, and then transport them to the aboveground part.

The ability to form a big amount of aboveground biomass during one growing season, the simplicity of harvesting, the ability to grow and develop in unfavorable conditions are also important features of the plants – hyper accumulators of metals, primarily determined by the substrate's specificity [10].

The considered disadvantages of heavy metals phytoextraction are: comparatively shallow substrate cleaning depth depending on the rooting depth of the used plants, and the slow rates of extracting heavy metals from the substrate. But if we consider phytoextraction as a method of obtaining heavy metals from industrial wastes, then the effectiveness and speed of phytoextraction are determined from economic point of view [11].

Different types of plants can be used as phytoextractors. Specifically, the authors have elaborated and implemented technologies for phytoaccumulation of heavy metals from waste water [12] of shrubs and grass, which are sustainable to industrial pollution and which are capable to accumulate heavy metals of high concentrations inside their tissues and organs.

In present article, the ability of local plant community representatives to accumulate heavy metals while growing on technogenic soils in the region of Karabash city is studied.

## 2. Objects and methods of research

Object of research in present article is the soil from territory of Karabash Mining and Metallutgical Combine. The properties of technogenic soils selected at the Karabash Mining and Metallurgical Combine, and also the amount of heavy metals contained in them, are presented in Table 1.

Midland hawthorn (Crataegus laevigata (Poir)), common barberry (Berberis vulgaris), red elder (Sambucus racemosa), cinnamon rose (Rosa cinnamomea L.), couch grass (Elytrigia repens), crested wheat grass (Agropyron cristatum), meadow fescue (Festuca pratensis) and meadow grass (Poa pratensis) had been tested as accumulators of soil metals.

Research had been conducted for three years: from 2012 till 2014. Botanic areas with single-crop sowing of sample herbs and shrub planting had been created in the field environment. Surface of each area amounted 1m2. The top soil was hoed to a depth from 10 to 15 cm. Studies were conducted in three replications.

In the vegetation process, metals accumulation in the dried biomass of the plants had been assessed every 15 days. Content of metals in plant ash had been determined by the method of ISP plasma spectroscopy on the basis of standard practice.

**Table 1** – Physical and chemical properties, agrochemical properties and heavy metals content in the soils, selected at the territory of Karabash Mining and Metallurgical Combine

| Parameter | Soil from the Combine territory | Soil from tailing dump | Soil from the Rizhy ruchey (Red Creek) territory |
|---|---|---|---|
| pH of aqueous suspension | 4.6 | 2.7 | 3.5 |
| pH of saline suspension | 4.1 | 2.5 | 3.2 |
| $P_2O_5$, mg/100 g | 10.2 | - | - |
| $NO_3^-$ ,mg/100 g | 1.9 | 0.7 | 1.1 |
| Dense deposit of aqueous extract (before the washing process), % | 0.34 | 0.56 | 1.98 |
| Dense deposit of aqueous extract (after the washing process), % | 0.27 | 0.42 | 1.31 |
| Organic matter (according to Turin), % | 18.2 | 0.87 | 9.8 |
| Gross content of metals | | | |
| $Fe_2O_3$, mg/kg | 15063.7 | 49983.2 | 5157.0 |
| Cu, mg/kg | 3075.0 | 6687.8 | 583.0 |
| Ni, mg/kg | 456.3 | 525.7 | 30.2 |
| Zn, mg/kg | 2639.5 | 6112.2 | 234.0 |
| Pb, mg/kg | 241.3 | 379.6 | 215.0 |
| Cd,mg/kg | 1.23 | 3.5 | 2.83 |
| Content of mobile fractions | | | |
| $Fe_2O_3$, mg/kg | 215.2 | 2195.5 | 172.6 |
| Cu, mg/kg | 35.7 | 110.7 | 44.8 |
| Ni, mg/kg | 3.6 | 10.3 | 2.8 |
| Zn, mg/kg | 8.1 | 120.5 | 2.03 |
| Pb, mg/kg | 2.26 | 11.2 | 2.36 |
| Cd, mg/kg | 1.2 | 1.3 | 0.95 |

In Table 2 common physical and water-physical properties of soils are given.

**Table 2.** Common physical and water-physical properties of soils

| Sample | Density, g/m$^3$ | Particle density, g/m$^3$ | Total porosity, % | Capillary moisture capacity, % |
|---|---|---|---|---|
| Soil from the Combine territory | 1.27 | 2.05 | 57.36 | 21.74 |
| Soil from tailing dump | 1.02 | 3.15 | 67.39 | 27.32 |
| Soil from the catchment territory of the Red Creek | 1.19 | 2.53 | 59.84 | 24.96 |

### 3. Results and discussion

Average research results on soil metals accumulation in plants on 105 days' exposure are given in Table 3.

As it follows from the given data, shrubs show the highest accumulation ability. Accumulation coefficient of zinc and copper in them is practically 10-20 times higher than in the reference area. Accumulation coefficient of metals in the roots is 20-45% higher than in the leaves.

**Table 3.** Accumulation of heavy metals in green plants, mg/kg of dry weight

| Plant | Content, mg/kg of dry substance | | | | | |
|---|---|---|---|---|---|---|
| | $Fe_2O_3$ | Cu | Ni | Zn | Pb | Cd |
| Shrubs | | | | | | |
| Midland hawthorn Crataegus laevigata (Poir) (leaves) | 52.4 | 84.2 | 10.53 | 64.23 | 2.35 | 0.13 |
| Midland hawthorn Crataegus laevigata (Poir) (roots) | 102.36 | 100.24 | 11.25 | 112.5 | 7.36 | 3.07 |
| Common barberry Berberis vulgaris (leaves) | 85.12 | 93.54 | 14.15 | 68.23 | 12.85 | 2.12 |
| Common barberry Berberis vulgaris (roots) | 142.3 | 120.6 | 15.2 | 145.34 | 16.34 | 3.87 |
| Red elder Sambucus racemose (leaves) | 64.12 | 88.41 | 13.18 | 87.12 | 10.64 | 3.14 |
| Red elder Sambucus racemose (roots) | 150.3 | 126.4 | 14.36 | 135.3 | 15.43 | 6.35 |
| Cinnamon rose Rosa cinnamomea L. (leaves) | 47.15 | 80.54 | 9.57 | 74.1 | 7.96 | 0.19 |
| Cinnamon rose Rosa cinnamomea L. (roots) | 98.7 | 112.2 | 13.2 | 125.0 | 12.21 | 2.36 |
| Reference areas | | | | | | |
| Midland hawthorn leaves /roots | 4.6/7.8 | 3.2/6.4 | 09/2.8 | 0.7/1.5 | -/- | -/- |
| Common barberry leaves /roots | 7.3/12.2 | 6.4/10.4 | 1.2/4.1 | 0.9/2.7 | -/- | -/- |
| Red elder leaves /roots | 6.9/14.1 | 5.6/12.2 | 1.4/5.2 | 1.3/2.4 | -/- | /-/ |
| Cinnamon rose leaves /roots | 3.3/6.4 | 3.4/9.8 | 0.8/4.5 | 1.2/1.8 | -/- | -/- |
| Permanent grasses | | | | | | |
| Couch grass Elytrigia repens (aboveground part) | 215.6 | 7.3 | 9.65 | 50.23 | 2.01 | 2.1 |
| Crested wheat grass Agropyron cristatum (aboveground part) | 179.6 | 27.3 | 8.69 | 49.6 | 1.23 | 2.78 |
| Meadow fescue Festuca pratensis (aboveground part) | 220.2 | 26.4 | 10.2 | 56.2 | 2.03 | 3.16 |
| Meadow grass Poa pratensis (aboveground part) | 140.12 | 20.2 | 7.2 | 50.3 | 1.36 | 1.63 |
| Reference areas | | | | | | |
| Couch grass | 10.3 | 2.1 | 0.9 | 2.4 | -/- | -/- |
| Crested wheat grass | 8.9 | 5.8 | 0.7 | 3.8 | -/- | -/- |
| Meadow fescue | 12.6 | 5.1 | 1.2 | 5.7 | -/- | -/- |
| Meadow grass | 18.2 | 3.6 | 0.4 | 4.6 | -/- | -/- |

Herbaceous plants possess far lesser accumulation ability; accumulation coefficient does not exceed.

Plant biomass, after planting on contaminated areas, can be used as biological ore material after burning.

As it is stated in preliminary research, the combustion of plant biomass can be conducted at a temperature 300 degrees. The obtained ash can be processed with the use of pyrometallurgical method as refuse burnouts, in accordance with the reference book ETS 3-2015, Copper Industry.

## 4. Conclusion

The ability of midland hawthorn (Crataegus laevigata (Poir)), common barberry (Berberis vulgaris), red elder (Sambucus racemosa), cinnamon rose (Rosa cinnamomea L.), couch grass (Elytrigia repens), crested wheat grass (Agropyron cristatum), meadow fescue (Festuca pratensis) and meadow grass (Poa pratensis) for phytoextraction of heavy metals from technogenic soil had been tested.

The possibility of effective phytoextraction with the use of hawthorn and elder is shown. Maximum accumulation of zinc takes place in the surface mass of couch grass and meadow fescue.

In regard to the conditions of South Ural, planting of elder and hawthorn with seeding of couch grass and meadow fescue can be recommended for phytomining purposes.

The work was supported by Act 211 Government of the Russian Federation, contract No02.A03.21.0011.

## References

[1] Timofeeva S.S., Ulrikh D.V. The modern condition of tailing dump in Karabash city, and its impact on technogenesis of the surrounding territory. Ecology and Industry of Russia, 1, 2015, pp. 56-59

[2] Patra H.K., Mohanty M. (2013). Phytomining: an innovative phytoremediation management technology // The Ecoscan: special issue, vol.11, pp. 15-20

[3] Mohanty M. (2016) Post-Harvest Management of Phytoremediation Technology. J Environ Anal Toxicol 6: 398. doi: 10.4172/2161-0525.1000398

[4] Piccinin R.C.R., Ebbs S.D., Reichman S.M., Kolev S.D., Woodrow I.E., Baker A.J.M. (2007) A screen of some native Australian flora and exotic agricultural species for their potential application in cyanide-induced phytoextraction of gold. Minerals Engineering 20, pp. 1327-1330.

[5] Anderson C., Moreno F., Meech J. (2005) A field demonstration of gold phytoextraction technology. Minerals Engineering 18, pp. 385-392.

[6] Sheoran V., Sheoran A.S., Poonia H., Phytomining: A review. Minerals Engineering, 2004, vol.22, pp.1007-1019 Available from: http://dx.doi.org/10.1016/j.mineng.2009.04.001

[7] Robinson B.H., Banuelos B., Conesa H.M., Evangelou M.W.H., Schulin R. (2009) The Phytomanagement of Trace Elements in Soil. Critical Reviews in Plant Science 28, pp. 240-266.

[8] Fernando E., Quimado M. & Doronila A. (2014). Rinorea niccolifera (Violaceae), A new, nickel-hyperaccumulating species from Luzon Island, Philippines, *PhytoKeys,* *37* 1-13. DOI: 10.3897/phytokeys.37.7136

[9] Cecchi L., Gabbrielli R., Arnetoli M., Gonnelli C., Hasko A. & Selvi F. (2010). Evolutionary lineages of nickel hyperaccumulation and systematics in European Alysseae (Brassicaceae): evidence from nrDNA sequence data, Annals of Botany, 106(5) 751-767. DOI: 10.1093/aob/mcq162

[10] Levkin N.D., Mukhina N.E. Phytoremediation of soils contaminated y heavy metals. Izvestiya of TSU. Earth Sciences. 2014 no.1, pp.9-14

[11] Timofeeva S.S., Ulrikh D.V. Phytoremediation of contaminated and technologically loaded soils at the territory of copper-ore plants of South Ural. Mining informational and analytical bulletin MIAB №3, 2016 pp. 341-349

[12] Timofeeva S.S., Timofeev S.S., Ulrich D.V Environmental phytotechnologies in Eastern Siberia and South Ural. 14th International Multidisciplinary Scientific GeoConference & EXPO: Albena, Bulgaria, 17 - 26 June, 2014. pp. 259-266

IOP Conf. Series: Earth and Environmental Science **66** (2017) 012031 doi:10.1088/1755-1315/66/1/012031

# Technique of an Assessment of Industrial Safety of Paint and Varnish Productions

**E V Omelchenko[1], E A Trushkova[1], S L Pushenko[1], S V Sitnik[2], V A Firsov [1]**

[1]Architectural and construction academy of the Don State technical university
st. Socialisticheskaya, 162, Rostov-on-Don, Russian Federation, 344000
[2]Rostov State transport university
sq.Rostov Rifle Regiment of the National Militia, 2, Rostov-on-Don, Russian Federation, 344038

E-mail: ayurvedaom@mail.ru

**Abstract**. In article the technique of an assessment and the analysis of industrial safety of paint and varnish productions is described, the main measures of prevention, the main reasons for traumatism and action for decrease in risks of influence of dangerous and harmful production factors on workers are given.

## 1. Introduction

Industrial traumatism, now, despite introduction of technologies, more modern and safe for the person, represent a considerable problem in all industries. The organization of work in the sphere of ensuring industrial safety of a production activity consists in a choice and formation of such structure of management at the enterprise which in the best way would correspond to performance of the main task - to creation of safe and healthy working conditions for the working personnel.

## 2. Technique

We developed a technique of an assessment of fire safety of paint and varnish productions on the example of the paint and varnish enterprise JSC "Empils" of Sochi of Krasnodar area, the analysis of a security status and labor protection of this paint and varnish production was made.

The technique includes the following stages: an assessment of number of the personnel of the enterprise for a group sign, the analysis of the reasons of operational injuries, an assessment of influence of harmful and dangerous production factors, definition of degree of professional risk, a quantitative assessment of conditions of activity on factors of harm and injury risk, an assessment of security of means of individual protection (MIP), development of measures of prevention for ensuring industrial safety.

According to investigation phases by us it was established that the number of the personnel in 2014-2016 years. I increased on 48 people in connection with increase in production. The analysis of personnel structure showed that the greatest number of the personnel – workers, men, at the age of 30-40 years. At the enterprise workers with an experience of 10-20 years and prevail over 20 years. Presence of the skilled personnel promotes increase not only prestige of the enterprise, but also affects the level of safety of production.

IOP Conf. Series: Earth and Environmental Science **66** (2017) 012031 doi:10.1088/1755-1315/66/1/012031

The analysis of a condition of industrial safety and labor protection of the organization is carried out according to statistical and accounting reports for 2014 - 2016 years. Were the main reasons for traumatism in JSC "Empils": the organizational; the operational; design violations. The analysis of a condition of working conditions was carried out according to certification of workplaces on the basis of an assessment of influence of harmful and dangerous production factors. The major factors influencing workers of JSC Empils were: chemical factor; microclimatic conditions; noise; radiations; illumination; weight and intensity of work [1-10].

Distribution of workplaces of painters on classes of working conditions on JSC "Empils" following: optimum and admissible working conditions fall on 38% of workplaces; for 34% - harmful to the first degree; for 28% - harmful to the second degree. The greatest contribution on increase of a class of working conditions on workplaces is made by such production factor as weight of work. Working conditions with a class 3.1 and 3.2 are established for 62% of workplaces.

The analysis of results of certification of workplaces which is carried out on JSC "Empils" allowed to define concrete workplaces which are subject to influence of harmful production factors, these factors have the prevailing value and are shown more that gives the grounds further properly to control observance of requirements of labor protection, in those productions which have most a high risk [2]. A measure of the actual danger of each operation is the frequency of emergence of a dangerous situation of traumatizing workers.

The method of arrangement of priorities allows to define degree of professional risk [3]. At calculation of risk of influence of harmful and dangerous production factors, depending on a class of working conditions on workplaces of painters with high risk, the major factor introducing this risk is revealed, it is weight of work. On the offered strategy of management the risk with high is reduced to an average [2].

It is established that working at the paint and varnish enterprises organize the work according to employment policies and procedures. The persons which passed preliminary medical examination, induction and primary instructing in labor protection on a workplace and who passed examination in the profession are allowed to these works. Persons under 18, pregnant women and nursing mothers aren't allowed to works with the paints and varnishes containing toxic substances, solvents and lead connections. The quantitative assessment of conditions of activity on factors of harm and injury risk on life expectancy of the painter showed reduction of life expectancy for 5,5 years. The maximum (maximum permissible) mass of combustible liquids, safe for people, vapors of easily flammable liquids (EFL), kg, at explosion of explosive mix in JSC "Empils" is determined by a formula:

$$m_p^e = \frac{150\, V_f \rho_a (t_a + 273)}{Q_L^p z} \tag{1}$$

where $V_f$ - the free volume of the room, $m^3$; $\rho_a$ - air density indoors before explosion at a temperature of $t_a$, $kg/m^3$; $Q^P_L$ - the lowest warmth of combustion of substance, J/kg; z – the fuel share in explosion accepted for the painting camera – 0.3; for shop – 0.5.

Recognizing that the operator of production of JSC "Empils" works in electrowire footwear at a floor with an electrowire covering, the persons having the III qualification group on electrical safety, instructed about strict observance of measures of fire safety, acquainted with ways of a call of fire protection and able to apply practically available fire extinguishing means to elimination of the fires and fire are allowed to work on service of installation of manual [11].

All works connected with preparation of paint and varnish mixes, and also with dilution by their solvents are carried out strictly according to technological instructions in the special, well ventilated room. It is forbidden to apply paints and varnishes and solvents of unknown structure, to replace one solvents with another, not provided technical processes in compliance with Safety Rules "A safety rule of paint and varnish productions".

Apply materials which fire-dangerous characteristics - flash temperature, temperature limits of ignition, spontaneous ignition temperature, tendency to self-ignition, weight or volume area of ignition, toxic properties and precautionary measures at their application - are specified in standardly - technical documentation (Set of Rules 2.13130.2009) to painting works. The analytical passports attached to each consignment of paints and varnishes, powder polymeric paints, solvents, hardeners, semi-finished products for preparation of the washing, degreasing and polishing structures, contain data on percentage of extremely dangerous substances and flying part on separate components. Painting sites and platforms are equipped with effective local ventilation and have protections of an explosive zone. Coloring is made only in the rooms equipped stitched - exhaust ventilation according to the Set of Rules 7.13130.2009. All painting equipment (sprays, tanks) is grounded. As dangerous factors of the fire at the paint and varnish enterprises are: explosion - and fire danger of vapors of solvents and paints; compressed air; toxic components of paints and varnishes, lighting in preparatory and painting offices it is carried out in fire and explosion safe execution. Manual electropainting installations have so that possibility of mechanical damage of a cable for tension supply to a spray of paint supply and air hoses was excluded. The delivery tank with paint have out of the painting camera. Surfaces of the high-voltage equipment are kept clean. In the presence on its surface of dust and impurity installation isn't included. Cleaning is made by the electrotechnical personnel. Giving of a high voltage on the electropainting camera is followed by simultaneous automatic signals: the shining banners with inscriptions light up located at an entrance door, at an aperture and near the control panel: "The high voltage is included", "Not to enter". All works connected with repair of the painting equipment are made at the removed stress. The working zone is provided with fire-prevention posters and inscriptions. In case of the fire or its signs (emergence of a smoke, characteristic smell) it is immediately reported about fire or the fire in fire protection and start fire extinguishing. Fire extinguishing is made carbon dioxide fire extinguishers, the small centers of a flame liquidate use of sand or covered from nonflammable materials. At emergence of situations, according to the Set of Rules 1.13130.2009 which can lead to accident and accidents, stop work, disconnect supply of electricity, bring out of a dangerous zone of people, delete explosive and flammable materials and the equipment and report about the arisen situation to the master or the foreman [12-15].

High degree of protective efficiency and convenience at operation are provided by means of protection (overalls cotton; boots leather; mittens with slips; the headdress (takes); when coloring by an dip method an apron rubberized; rubber gloves; a gas mask, a jacket cotton on the warming laying; trousers cotton on the warming laying).

As the assessment of fire safety of the enterprise is part of system approach to decision-making, procedures and practical measures in the solution of problems of management of process of safety, a basis of a technique were the physical and mathematical modeling of the most technical system and its working processes including interactions of the main components of system, operators, the personnel with environment in regular and emergency situations. In the analysis of fire safety by the developed technique scenarios of emergence and development of accidents and accidents with application of the main defining equations and criteria of mechanics, physics with implementation of a complex of works on safety at all stages of life cycle of technical systems are formed and described:

- development of measures of protection from the fires, accidents and accidents;
- monitoring of danger of functioning in the field of industrial safety;
- extension of a resource of safe operation;
- modernization in process of increase of safety requirements;
- a safe conclusion from operation, storage and utilization of elements of the technological scheme.

Information on fire risk received during research of paint and varnish production is a basis for management of risk – development and optimization of organizational and technical actions for decrease in risk up to the set size (table 1). A problem of management of risk is development of plans of action for decrease and control of risk, an operating time of alternative options, and also an assessment of efficiency of these plans and development of recommendations for adoption of administrative decisions, up to refusal of the planned activity. The results are shown in Table 1.

**Table 1** – Main types of accidents, factors of danger and a consequence of accidents on dangerous components of JSC "Empils"

| Type of accident | Dangerous factors | Consequences of accidents |
|---|---|---|
| 1. Dangerous components of this object on which turn out, used, processed, are stored, transported, destroyed dangerous substances | | |
| – the destruction of gas pipelines, failure of the gas equipment which caused emission of natural gas with its subsequent fire (jet burning, explosion of air-gas mix);<br>– the explosions of gas in fire chambers and gas flues of coppers, devices which caused them local destructions or shutdowns;<br>– destruction of tanks, pipelines, devices with flammable and (or) combustible liquids, failure of the equipment about EFL and combustible liquids (CL), which caused emission of combustible substances, formation of a zone of flood with its subsequent ignition, explosion of air-gas mix | – dangerous factor of mechanical influence (fragments and splinters of the equipment, communications, excessive pressure);<br>– dangerous thermal factor (burns);<br>– dangerous toxic factor (emissions of dangerous substances, formation of toxic products of incomplete combustion) | – infliction of harm to health of the personnel;<br>– death of the personnel;<br>– material damage – infliction of harm and (or) destruction of property of natural and legal entities;<br>– ecological damage |
| 2. Dangerous components of this object on which the equipment working under excessive pressure more than 0.07 MPas is used | | |
| – destruction or damage (gap) of devices, units, the pipelines working under pressure;<br>– explosion of the vessels working under pressure | – dangerous factor of mechanical influence (fragments and splinters of the equipment, communications; excessive pressure);<br>– dangerous thermal factor (burns, frostbites) | – infliction of harm to health of the personnel;<br>– death of the personnel;<br>– material damage – infliction of harm and (or) destruction of property of natural and legal entities;<br>– ecological damage |
| 3. Dangerous components of this object on which permanently installed load-lifting mechanisms are used | | |
| – destruction or damage (deformation) of a metalwork of load-lifting cars or their elements;<br>– falling of load-lifting cars;<br>– falling of freight | – dangerous factor of mechanical influence (fragments of the equipment, designs, communications);<br>– dangerous electric factor (defeat by electric current) | – infliction of harm to health of the personnel;<br>– death of the personnel;<br>– material damage – infliction of harm and (or) destruction of property of natural and legal entities |

On the basis of the developed technique the analysis of engineering calculations is carried out and the complex of organizational, sanitary and household, treatment-and-prophylactic and technical actions for improvement of working conditions in the construction organization is offered [4,16,17].

The shown technique is a fundamental element of a basis of insurance protection, and assessment of fire safety is not a casual event, but legally reasonable condition here. The general logical sequence of steps (actions) to methodologies consists of performance of the following stages:

1. Justification of the purposes and tasks of the analysis of industrial safety.
2. Planning and organization of works.
3. Analysis of technological features of production object.
4. Identification of sources of risk and conditions under which they can make negative impact.
5. Determination of frequency (or probabilities) emergence of undesirable events.
6. Definition of characteristics of sources of influence of dangerous factors (totals, intensity and duration: emissions, drains, energy allocation) for all range of undesirable events.
7. Justification of models and calculation of existential transfer and distribution of initial factors of danger in environment.
8. Creation of fields of potential risk round each of the allocated danger sources.
9. Calculation of straight lines and indirect consequences (damages) of negative impact of sources of danger on various subjects (recipients) or groups of risk.
10. Estimation of fire risk. Calculation of indicators.
11. Comparison with criteria of the acceptability and assessment of the importance of a technique.

The assessment of sizes of the specified factors was carried out on the basis of the analysis of the physical phenomena proceeding at fire-dangerous situations, the fires, explosions (table 2). The results are shown in Table 2.

**Table 2** – Results of calculation of areas of coverage of the major striking factors at jet burning of the expiring compressed gas (a horizontal torch)

| Pressure in the device, MPa | Diameter of an opening, mm | Mass expense of the expiration of gas G, kg/s | Torch length at jet burning | Torch width d, m | Area of defeat of the person in a horizontal torch, area of impact of a horizontal torch on the next equipment (30° the sector limited to the radius equal to *LF*), m | The zone sizes with intensity of thermal radiation from a horizontal torch of 10 kW/sq.m | Area of possible impact of the fire flash (30° the sector limited to the radius equal *LF*), m |
|---|---|---|---|---|---|---|---|
| до 0,03 | 100 | 1.237 | 13.61 | 2.04 | 13.61 | 20.41 | 13.61 |
| | 50 | 0.309 | 7.81 | 1.17 | 7.81 | 11.72 | 7.81 |

Thus the following processes arising at realization of accidents or being their consequences were considered (depending on type of the equipment and combustible substances addressing on object):

– the expiration of liquid from an opening at destruction of the equipment;
– spreading of liquid and formation of a zone of flood;
– evaporation of liquid from the passage;
– emission of superheated liquid at destruction of the equipment;

– formation of a gas-steam-air cloud (gases and vapors it is heavier than air);
– combustion of gas-steam-air mix in open space;
– combustion of gas-steam-air mix in processing equipment or the room;
– thermal radiation at the fire of the device, transport capacity, the passage;
– fire indoors (building).

Decisions on the prevention of emergence of accidents on components of this object as a result of possible dangerous thermal influences of a hard frost and a strong heat provided passive and active ways of protection [7,18,19].

Passive ways include use of the equipment, designs of the buildings and constructions capable to provide the functions in the conditions of the possible temperature range of environment.

Active ways include use of systems of heating, the ventilation and conditioning providing a necessary microclimate in rooms of these buildings.

## 3. Result

The analysis of risk factors, conditions of emergence of accidents and their scenarios on other objects around placement of this object of object showed – accidents on nearby the located existing objects of JSC "Empils" don't make impact on the condition of this object exceeding similar indicators at possible accidents on components of the most this object. For this object the main consequence of accidents on nearby the located existing objects of JSC "Empils" which are followed by the fires, explosions will be need of a stop of functioning of shop of varnishes for the period before the end of suppression of the fires, carrying out rescue, emergency and recovery, repair work in the territory of the enterprise and completion of investigation of the reasons of accidents. The main action for ensuring industrial safety and the prevention of accidents on this object as a result of accidents on other objects around placement of this object will be observance of gaps between the buildings, constructions established by normative documents in the field of fire and industrial safety.

## 4. Conclusions

Application of a technique of an assessment of fire safety of paint and varnish productions on the basis of data of accounting and statistical reports, allows to create the systematized complex of the actions directed on improvement of quality and working conditions to reduce risk of influence of dangerous and harmful production factors, to reduce material, social and economic production losses.

## References

[1] Chernoplekov A.N, etc. The analysis of traumatism in construction and suggestions for improvement of working conditions.// Health and safety, 2005, No. 8. p.126-130.

[2] Staseva E.V., Pushenko S. L., Strakhova N. A.. Improvement and increase of efficiency of the organization of labor protection in construction on the basis of a control system of risks: Monograph. Rostov-on-Don, 2012. 240 p.

[3] Professional risk for health of workers (The management under the editorship of N. F. Izmerov and E. I. Denisov. M.: Trovant. 2003. 196 p.

[4] Omelchenko E.V., Trushkova E.A., Sidelnikov M.V., Pushenko S.L., Staseva E. V. Algorithm research exposure dust emissions enterprises of building production on the environment //IOP Conference Series: Earth and Environmental Science. 2017 Vol. 50, Article number 012018. p. 1-5

[5] Andreeva E E 2016 K ocenke riskov dlja zdorov'ja rabotnikov pri klassifikacii ob#ektov sanitarno-jepidemiologicheskogo nadzora i planirovanii proverok (na primere g. moskvy) Analiz riska zdorov'ju ed E.E. Andreeva 84-92 № 2 (14)

[6] Habibulina V M 2015 Zdravoohranitel'naja politika: pragmatichnost' na osnove otkrytyh dannyh medicinskoj statistiki Materialy Mezhdunarodnoj mezhdisciplinarnoj konferencii ed V.M. Habibulina 380-390
[7] Trushkova E.A. Harmful factors of production environment. Part I: tutorial. / E.A. Trushkova, E.V. Staseva, N.Y Volkova Rostov-on-Don: Rostov State Civil Engineering University, 2014. 103 p.
[8] Kannel W B 1991 Left ventricular hyperthrophy as a risk factor: the Framingham experience J. of hypertension 9. P. 3-9
[9] Pushenko S.L. Health and Safety. Part 2 Production Sanitation and health: Textbook. S.L. Pushenko, D.V. Deundyak, E.V. Omelchenko, A.V. Nihaeva , A.S. Pushenko, E.A. Trushkova, E.V. Staseva, E.V. Fedina, E.S. Fil - Rostov n / D: Height. state. building. University Press, 2014. 94 p.
[10] Denisov Je I 2013 Problema real'noj jeffektivnosti individual'noj zashhity i privnosimyj risk dlja zdorov'ja rabotnikov Medicina truda i promyshlennaja jekologija ed Je.I. Denisov, T.V. Morozova, E.E. Adeninskaja and N.N. Kur'erov 18-25 4.
[11] Trushkova E.A. Development of measures on preventing injures and occupational diseaseson JSC 'Rostvertol' / E.A. Trushkova, N.Y. Volkova // Scientific review. 2014. 10. S. 550-553.
[12] Zaharenkov V V 2013 Ocenka professional'nogo riska dlja zdorov'ja rabotnikov promyshlennyh predprijatij na osnove medicinskoj tehnologii Akademicheskij zhurnal Zapadnoj Sibiri 9 ed V.V. Zaharenkov, A.M. Oleshhenko, I.P. Danilov, D.V. Surzhikov, V.V. Kislicyna and T.G. Korsakova 8. 2 (45)
[13] Zaharenkov V V 2013 Primenenie novoj medicinskoj tehnologii dlja ocenki professional'nogo riska zdorov'ju rabotnikov promyshlennyh predprijatij Vestnik Kuzbasskogo nauchnogo centra ed V V Zaharenkov, A M Oleshhenko, I P Danilov, D V Surzhikov, V V Kislicyna, T G Korsakova and I Ju Motuz 28-30. 17
[14] Resnik L M 1993 Ionic basis of hypertension, insulin resistance, vascular disease and related disorders Am.J Hypertens. 6. P. 123-134
[15] Zeljaeva N V 2014 Ocenka zabolevaemosti rabotnikov promyshlennyh predprijatij i puti ee snizhenija (na primere predprijatij otrasli mashinostroenija g. Nizhnego Novgoroda) Fundamental'nye issledovanija ed N V Zeljaeva 698-705. 7-4
[16] Kislicyna O A 2015 Nezashhishhennost' v sfere truda rabotnikov v Rossii kak faktor riska plohogo zdorov'ja i povedenija s riskom dlja zdorov'ja: gendernyj aspekt Upravlenie zdravoohraneniem ed O.A. Kislicyna 66-77. 1 (43)
[17] Artem'eva A A 2014 Ocenka vlijanija faktorov proizvodstvennoj sredy na sostojanie uslovij truda i professional'nuju zabolevaemost' rabotnikov Federal'noe agentstvo zheleznodorozhnogo transporta, FGBOU VPO «Moskovskij gosudarstvennyj universitet putej soobshhenija (MIIT) ed A.A. Artem'eva 15-18.
[18] Demyanova V S 2006 The disperse filled cements on the basis of lithotripsy. Ecological aspects of production M.: The Young Guard ed V S Demyanova and G N Kazin
[19] Daviglus M.L., Liu K., Yan L.L. et al. Body mass index in middle age and health-related quality of life in order age: the Chicago heart association project in industry study. Arch. Intern. Med. 2003.Nov. 10; 163 (20):2448-55.

# Algorithm of Research of Influence of Professional and Ecological Risk Factors for Various Groups of the Population

**E V Omelchenko, E A Trushkova, G N Sokolova, A V Nikhayeva, A G Hvostikov**

Architectural and construction academy of the Don State technical university
st. Socialisticheskaya, 162, Rostov-on-Don, Russian Federation, 344000

E-mail: ayurvedaom@mail.ru

**Abstract**. The article is given algorithm of research of influence of professional and ecological risk factors for various groups of the population, considering professionally caused diseases as polietiology group of diseases, a natural environmental risk as probability of manifestation of many adverse natural phenomena by means of two components, and expediency of application in a security system of a method of stochastic management is proved.

## 1. Introduction

In the last decades the Russian hygienists, ecologists, pathologists study questions of influence of separate production factors at some enterprises of the Russian Federation, in total researches of a professional and environmental risk in the conditions of the Russian Federation aren't carried out.
Development of systems of safety demands the solution of a number of problems of big dimension, constant increase in adequacy of mathematical models, the accounting of uncertainty and accident when using initial information. Today initial information for planning, design and management of security systems needs completion.

## 2. Technique

Continuous change of public requirements, change of climatic conditions, natural disasters, opening of new fields, the work of equipment accompanied with unforeseen emergencies which statistical regularities can be not always defined and considered at calculation of the operating influences define existence of a factor of uncertainty.

Carrying out the stochastic analysis allows to study indirect communications - the mediated factors (in case of impossibility of definition of a continuous chain of direct link). The stochastic analysis can act as the instrument of deepening of the determined analysis of factors [1-10].

Stochastic modeling of interrelations and quantitative estimates of various parties of safe activity of the enterprise is based on generalization of regularities of a variation of values of various indicators – quantitative characteristics of factors and results.

Quantitative parameters of communication come to light on the basis of comparison of values of the studied indicators in total of objects of the enterprise and the periods.

Thus, the first prerequisite of stochastic modeling is opportunity to make set of supervision, i.e. opportunity repeatedly to measure parameters of the same phenomenon in various conditions [3]. We at creation of algorithm of research of influence of professional and ecological risk factors used model of the stochastic analysis which gathered by removal of empirical data, considering coincidence of numerical characteristics of communications and initial supervision. It is established that change of values of indicators happens within unambiguous definiteness of the qualitative party of the phenomena which characteristics are the modelled technological indicators (within a variation there shouldn't be a quantum leap in character of the reflected phenomenon).

Generally ecological danger to the population living in the territory with the industrial enterprises is defined by two factors: damage from the actual danger and risk (potential danger) in case of emergencies. Therefore the size of ecological danger in relative indicators it is possible to present functions [1] in the form:

$$G_d = f(Y_d, R_d), \tag{1}$$

where $G_d$ – risk of ecological danger in relative (dimensionless indicators); $Y_d$ – damage in dimensionless relative indicators; $R_d$ – risk in dimensionless relative indicators.

The choice of concrete weight coefficients of direct $P_{Y.dir}$ and indirect $P_{Y.indir}$ depends on extent of influence of environment on activity of the population living in this territory. In case of noticeable influence of environment on a condition of human life it is possible to accept model of equivalent factors where weight coefficients are equal and matter 0.5. In case of weak influence of environment on a condition of human life it is necessary to choose model of the allocated factors and to accept in particular: $P_{Y.dir} : P_{Y.indir} = 9 : 1$. [5, 12, 23].

Objects of ecology are: systems of organisms, population systems and ecological systems (ecosystems), and, each of these groups is analyzed as system of the live organisms interacting with each other and with the environment of their dwelling and components such unity (system) within which process of transformation of energy and organic substance is carried out. The ecosystem is the unity including all organisms on this site of the territory and interacting with the physical environment in such a way that the stream of energy creates accurately expressed trophic structure (power-supply circuit), a specific variety and circulation of substances. The concept of an ecosystem very wide, and not its size, but functional unity of the components existing on this site of the territory and shown in relationship, interdependence and relationships of cause and effect, determines key parameters of ecosystems. The main task of biological systems of any level is their survival at this conjuncture.

Understanding ecology as science about the relations of vegetable and animal organisms among themselves and with the environment surrounding them, considering also, as the person is a live organism, it is possible to allocate a number of the scientific directions. Within bioecology is an ecology of individuals and the types made by them (physiological ecology); ecology of populations and communities (biocenoses) – a demoecology; ecology of ecosystems (biogeocenoses) – a sinecology. From the point of view of biology as set of sciences about wildlife in this scheme should consider also the person (as a look and as society). However in safety issues and risk there was a division of concepts: the risk for health of the person caused by a state of environment – medicobiological (medico-ecological), and for environment components – an environmental risk and ecological safety. Sometimes instead of the term "medicobiological risk" use "an environmental risk for health of the population" that does the term clear in this developed look. Thus the conceptual framework of the legislation existing in the sphere of industrial safety treats risk for the certain person as individual, and for group of people – as collective and (or) social [20-25].

At the present stage the ecology from strictly biological science turned into a considerable complex of the knowledge including sections of geography, chemistry, geology, sociology and other sciences. Thus, technosphere safety questions in general and ecological – in particular, – demand application of a number of disciplines that confirms our judgments about interdisciplinary nature of safety.

Ecology subject – preservation of functional and structural integrity of that central object which is allocated in the course of researches. It can be both a microstructure, and macro - and even the megaphenomenon. Classification of bioecology by hierarchical levels of biotic systems begins with an molecular ecology, including at the gene level, and ecology of cages and comes to an end with an individuals, groups, types, – to global ecology of level of the planet and space communications.

Reckoning to a classical ecological cycle of sciences about conservation and environmental protection made ecological knowledge extensive set of disciplines. Development of ecological subject put forward concepts of ecopolicy, ecodevelopment, ecological safety. Communication of ecological sciences with economy generated emergence of a number of hybrid disciplines, including an economy of environmental management. In the international practice of existence of society such tendency partly corresponds to strategy of a sustainable development. [6,14,15].

According to a number of scientists fundamental environmental problems are in area of social culture and structure of society and are united by one general science – social ecology. Now a subject of social ecology is studying of interaction of society with global environment in all variety of its technogenic transformations. The purpose of this science consists in developing the theory of compatibility of society with environment of its existence. The social ecology is connected not only with interaction of the nature and society in general, but also with groups of people and with each person separately. As the directions of human activity are extremely diverse, the social ecology, besides the general-theoretical section of knowledge, is subdivided into a number of applied areas. First of all, the ecology of the person with its medicobiological, sanitary, economic, social and psychological and other aspects is allocated here. Below we will speak about social sources and the reasons of accidents, in particular, about social risk (risk for a certain group of people), about social and psychological causes of failures (mistakes) of the person operator. Important social problems of ecological education and education of culture of health and safety arose and are solved.

Such types of applied ecology as geoecology, agricultural, transport, industrial (engineering) are separately considered. The last can be defined as discipline which investigates technogenic impact on environment. The wide range of sources, types and levels of the dangers generated by technical objects in the form of emissions and drains of emergency or usual operational character is the one side of the problem. Natural objects recipients of a biotic and abiotic origin with which environment is sated now, are numerous and various. Aspects of influence of "stressor" on them (the influencing factors) technospheres and protection against the last in many respects are beyond the scope of known and set serious problems for the decision to environmental engineers, probably, not of one generation.

In this connection, we developed the algorithm of research of influence of professional and ecological risk factors for various groups of the population allowing to reveal the main criteria of diseases and to determine their professional origin due to allocation of the following investigation phases:

- establishment of existence of communication with concrete production factor (for example, dust – dust bronchitis);
- definition of existence of relationships of cause and effect with the production environment and a profession (for example, a dust aerosol - the miner);
- identification of excess of the average level of incidence at a certain professional group of persons (frequency rate in relation to full amount of this category).

In a basis of classification of occupational diseases we underlie the principle of etiology. Besides, allergic diseases and new growths are separately allocated.

As a result of application of algorithm - it is proved that professionally caused diseases - group of diseases - polietiology by the nature in which emergence production factors make an essential contribution [16,19].

For these diseases are characteristic: big prevalence; insufficient study of quantitative indices of the working conditions defining development of diseases; considerable social consequences - negative influence on demographic indicators (mortality, life expectancy, frequent and long diseases with temporary disability).

The developed automated complex by a hygienic assessment of working conditions included an assessment of parameters of a microclimate, levels of production noise, local and general vibration, air pollution of a working zone aerosols of mainly genetic fibrosis action and toxic substances, weight and intensity of work on ergonomic indicators. The assessment of risk of professional incidence of the working was made according to "The guide to an assessment of professional risk for health of workers. Organizational and methodical bases, the principles and criteria of an assessment", an environmental risk for health of the population - according to "The guide to a risk assessment for health of the population at influence of the chemicals polluting environment" [17, 26].

By us it was established that the natural environmental risk is connected with probability of manifestation of many adverse natural phenomena, such as earthquakes, a volcanism, mudflows, floods, a tsunami. The algorithm of research of influence of ecological risk factors considers features of a geological structure (property of rocks, existence or lack of breaks, etc.), a relief (strengthening of risk of pollution in hollows), landscapes (degree of their resistance to technogenic loadings), the neighbourhood of valuable and unique natural objects, territories of a specific mode of protection, population density, nature of perception the population of events. It is known that catastrophic consequences of accidents and the spontaneous natural phenomena sharply increase in result of emotional and psychological unavailability of the population to such events. As the frequency of realization of a potential environmental risk creates a real environmental risk, it has two components:

1. the bioecological - risk for wildlife;
2. the anthropoecological - risk for the person.

At research negative impact of technogenic factors on human health with formation of professionally caused pathology (is revealed by production of tanning products and textile fiber glass). For the XX century of change of life expectancy, a state of health and death rate in the Russian Federation, as well as in other countries, were defined by the universal process which received the name of epidemiological transition. The former diseases of acuity which had more often the exogenous nature and affecting representatives of all age groups, especially children are replaced ecologically caused, mainly endogenous etiology, first of all cardiovascular, allergic and oncological. Continuous growth of number of patients of a hay fever is noted around the world, and especially fast – in economically developed countries.

The analysis of influence of industrial pollyutant on incidence of a hay fever revealed a direct connection unsuccessful on a dust content and a gas contamination of the environment with growth of an allergic pathology. The comparative characteristic of number of the patients who addressed concerning respiratory allergic diseases in various allergic office of the Russian Federation confirms adverse influence of pollyutant of environment on properties of pollen of plants. As a result the number of patients with the respiratory allergosis including caused by influence of pollen anti-genes increases [10, 13, 18].

It is established that the most favorable is the regional approach to a problem of hay fever assuming complex acquaintance with local climatic conditions, allergenic properties of pollen and features of a clinical course of diseases. A taxonomical variety and quantitative balance in air of pollen and dispute depend on climatic conditions: character of a vegetable cover, pollen efficiency of plants, mode of meteorological factors. Therefore hay fever are estimated as group of diseases which distribution should be studied from the point of view of the integrated knowledge of geography, biogeometeorology and ecology [6-8].

In the analysis of the emotional sphere of the worker the analysis was made on an assessment of emotions, feelings, a self-assessment, uneasiness. Studying of emotional development has the theoretical importance and is defined by need of detection of the general regularities of ontogenesis of human mentality and creation of the complete concept of mental development. Still the problem of emotional development remains theoretically the least developed.

The behavior of the person is always followed by emotions. The abundance of emotions provokes the nervous psychological tension (NPT). According to T.A. Nemchin, NPN happens three types: NPN-1 (weak), NPN-2 (average expressed), NPN-3 (which is sharply expressed, excessive).

It is established that for weak NPN emotional processes aren't strongly expressed, mood usual, sensitivity isn't changed, a dream, the speech, mental qualities – in norm. The average expressed NPN the positive emotions promoting activization of adaptable mechanisms, mood thus raised characterize, sensitivity increase – moderate, a dream - strong, refreshing, there is a sense of responsibility, positively painted "rage", resistance to stress raises, speech activity amplifies, working capacity grows, mental qualities: memory, attention and others improve. At excessively expressed NPN the bright prevalence of negative emotions accompanied with strong deterioration of adaptation, mood - lowered, suppressed is observed, are observed sensation of fear, despair, continuous expectation of failures, sensitivity sharply increases, a dream – uneasy, resistance to stress – is lowered, the speech - silent, with pauses, slowed down, mental qualities – worsen, tension doesn't fall down even after disappearance of a difficult situation.

## 3. Result

Therefore when studying influence of professional risk factors objective registration of emotional states on resistance of skin and heart rate was carried out. Such simple methods, possessing objective character, help to carry out reliable diagnostics of shift of an emotional state during a production activity and at the correcting influences: pharmacological (medicinal) influence, physiotherapeutic means and by methods of psychocorrection.

## 4.Conclusions

Thus, application of algorithm of research of influence of professional and ecological risk factors allows to establish that it is possible to reduce incidence in various ways: primary prevention of an illness, early diagnostics and improvement of medical actions. Prevention of the professional and professionally caused diseases at the enterprises of the Russian Federation needs to be systematized and directed on decrease in risk of development of deviations in a state of health and diseases of workers, including actions of medical, economic and legal characters.

## References

[1] P 2.2.2006-05. Guide to a hygienic assessment of factors of a working environment and labor process. Criteria and classification of working conditions.

[2] P 2.2.1766-03. The guide to a risk assessment for health of workers. Organizational and methodical bases, principles and criteria.

[3] Professional risk for health of workers (The management under the editorship of N. F. Izmerov and E. I. Denisov. M.: Trovant. 2003. 196 p.

[4] Balabanov I.T. Analysis and planning of finance of an economic entity. M.: "Finance and statistics". 1998. 236 p.

[5] Omelchenko E.V., Trushkova E.A., Sidelnikov M.V., Pushenko S.L., Staseva E. V. Algorithm research exposure dust emissions enterprises of building production on the environment //IOP Conference Series: Earth and Environmental Science. 2017 Vol. 50, Article number 012018. p. 1-5

[6] Trushkova E.A. Development of measures on preventing injures and occupational diseaseson JSC 'Rostvertol' / E.A. Trushkova, N.Y. Volkova. Scientific review. 2014. 10. P. 550-553.

[7] Artem'eva A A 2014 Ocenka vlijanija faktorov proizvodstvennoj sredy na sostojanie uslovij truda i professional'nuju zabolevaemost' rabotnikov Federal'noe agentstvo zheleznodorozhnogo transporta, FGBOU VPO «Moskovskij gosudarstvennyj universitet putej soobshhenija (MIIT) ed A.A. Artem'eva 15-18

[8] Staseva E.V., Pushenko S. L., Strakhova N. A. Improvement and increase of efficiency of the organization of labor protection in construction on the basis of a control system of risks: Monograph. Rostov-on-Don, 2012. 240 p.

[9] Demyanova V S 2006 The disperse filled cements on the basis of lithotripsy. Ecological aspects of production M.: The Young Guard ed V S Demyanova and G N Kazin

[10] Pushenko S.L. Health and Safety. Part 2 Production Sanitation and health: Textbook. S.L. Pushenko, D.V. Deundyak, E.V. Omelchenko, A.V. Nihaeva , A.S. Pushenko, E.A. Trushkova,

E.V. Staseva, E.V. Fedina, E.S. Fil - Rostov n / D: Height. state. building. University Press, 2014. 94 p.

[11] Professional risk for health of workers (The management under the editorship of N. F. Izmerov and E. I. Denisov. M.: Trovant. 2003. 196 p.

[12] Zeljaeva N V 2014 Ocenka zabolevaemosti rabotnikov promyshlennyh predprijatij i puti ee snizhenija (na primere predprijatij otrasli mashinostroenija g. Nizhnego Novgoroda) Fundamental'nye issledovanija ed N V Zeljaeva 698-705. 7-4.

[13] Habibulina V M 2015 Zdravoohranitel'naja politika: pragmatichnost' na osnove otkrytyh dannyh medicinskoj statistiki Materialy Mezhdunarodnoj mezhdisciplinarnoj konferencii ed V.M. Habibulina 380-390.

[14] Trushkova E.A. Harmful factors of production environment. Part I: tutorial. / E.A. Trushkova, E.V. Staseva, N.Y Volkova Rostov-on-Don: Rostov State Civil Engineering University, 2014. 103 p.

[15] Kislicyna O A 2015 Nezashhishhennost' v sfere truda rabotnikov v Rossii kak faktor riska plohogo zdorov'ja i povedenija s riskom dlja zdorov'ja: gendernyj aspekt Upravlenie zdravoohraneniem ed O.A. Kislicyna 66-77 1 (43)

[16] Andreeva E E 2016 K ocenke riskov dlja zdorov'ja rabotnikov pri klassifikacii ob#ektov sanitarno-jepidemiologicheskogo nadzora i planirovanii proverok (na primere g. moskvy) Analiz riska zdorov'ju ed E.E. Andreeva 84-92. 2 (14)

[17] Daviglus M.L., Liu K., Yan L.L. et al. Body mass index in middle age and health-related quality of life in order age: the Chicago heart association project in industry study. Arch. Intern. Med. 2003.Nov. 10; 163 (20):2448-55.

[18] Kannel W B 1991 Left ventricular hyperthrophy as a risk factor: the Framingham experience J. of hypertension 9. P. 3-9

[19] Resnik L M 1993 Ionic basis of hypertension, insulin resistance, vascular disease and related disorders Am.J Hypertens. 6 P. 123-134

[20] Zaharenkov V V 2013 Ocenka professional'nogo riska dlja zdorov'ja rabotnikov promyshlennyh predprijatij na osnove medicinskoj tehnologii Akademicheskij zhurnal Zapadnoj Sibiri 9 ed V.V. Zaharenkov, A.M. Oleshhenko, I.P. Danilov, D.V. Surzhikov, V.V. Kislicyna and T.G. Korsakova 8. 2 (45)

[21] Denisov Je I 2013 Problema real'noj jeffektivnosti individual'noj zashhity i privnosimyj risk dlja zdorov'ja rabotnikov Medicina truda i promyshlennaja jekologija ed Je.I. Denisov, T.V. Morozova, E.E. Adeninskaja and N.N. Kur'erov 18-25. 4

[22] Haustov A P. Management of environmental management / Haustov A P, Redina M M. M.: The higher school, 2006.

[23] Trushkova E.A. Results of researches of acoustic safety equipment problem on JSC 'Mechanical Repair Factory of Krasnokamsk' / E.A. Trushkova, N.Y. Volkova // Scientific review. 2014. 11. P. 528-531.

[24] Oleynikov P.D. The main constructive and technological actions for decrease in noise level on workplaces / E.A. Trushkova, Oleynikov P.D. //Urgent problems of health and safety and ecology. The collection of scientific works of the second international scientific and practical conference with school of sciences for youth. Tver state technical universitet.2016. P. 66-68.

[25] Izrael Yu. A. Review of the status and pollution of environment in Russia in 2012 / Yu. A. Izrael, G. M. Chernogaeva. Moscow: Roshydromet, 2013. 176 p.

[26] Tatyanyuk T.K., Sineva E.P. Influence of working conditions on the health of workers at the enterprise experimental mechanical engineering // Regional problems of hygiene and public health strategy: Sat. scientific. tr. StaryOskol, 2004. P . 382-385.

# Soil Tillage as a Factor of Soil Conservation

**D V Sherer[a], N N Chumanova[b]**

FSBEI HPO Kemerovo State Agricultural Institute
Markovtseva Street, 5, Kemerovo, Russia, 650056

E-mail: [a]d.scherer@mail.ru, [b]Chumanova0203@mail.ru

**Abstract.** The work describes the question of the soil treatment system influence on agro-physical and microbiological properties of gray forest soils, and yield of barley in Western Siberia. Research works were carried out in 2013-2014 in Yaya region of the Kemerovo region. Tillage affects soil structure. The water stability in zero tillage conditions was poor (15.7%). Soil density corresponding to the optimum rate for barley is formed by the zonal processing system, while at the zero tillage soil remains solid. The best indicators of phosphataze, catalysis and amylase activity are formed with minimum processing system. In the experiment the highest yield of barley was obtained with minimum tillage - 12.1 c/ha.

## 1. Introduction

At present, all over the world for the purpose of energy and resource saving is very relevant to develop the so-called saving (preservative) agriculture, and the priority is given to the replacement of the traditional intensive technologies of grain and other crops cultivation on saving soil protection ones. The technologies of minimum and zero tillage for soil protection are among the specific and important ways to the economy of resource and energy saving in agriculture. [1,2,3,4].

In connection with the massive development of energy saving in the cultivation of crops in South-West Siberia, based on minimizing tillage, economical usage of fertilizers and plant protection products, there was a need for research to determine the reproduction of soil fertility with non-traditional technologies implementation.

The research task was to study the effect of the minimum (Mini-Till) and zero soil tillage (No-Till), and comparison of their agrophysical properties, soil biological activity and the formation of spring barley yields to the results of zonal tillage implementation.

## 2. Equipment and devices used in studies

The testing site: an experimental plot in the forest-steppe zone of the Kemerovo region in 2013-2014. The plot's soil - gray forest podzolized, medium heavy loam soil. The humus content in the soil is low - 4.3%, that is typical for the gray forest soils. The labile phosphorus and potassium concentration is average. While the nitrate nitrogen presence at the sowing time is very low - 2.4 mg / kg. The degree of base saturation is low - 70%. The reaction of the soil solution is medium acid - 4.6 pH.

The experiment repetition was conducted four times; plot allocation was stepped and systematical. The total area of the experiment site was 59 hectares. Forecrop was annual grasses. Optimal sowing time for barley in the natural climatic zone was May 21-22$^{nd}$, 2013-2014. The seeding rate was 4.5 million germinating seeds per hectare.

During the growing season the following surveys and observations were made:

- The yield components (number of germinating and surviving plant, weight of 1000 grains, number of grains per ear, grain weight from 1 $m^2$) were estimated by the procedure of the State Agricultural Crop Variety Testing of crops [5];
- The mass of 1000 grains - according to State All-Union standard 12042-80;
- Soil density by cutting ring method [6];
- An enzymatic activity in the field and laboratory conditions [7].

Meteorological conditions during the growing season are varying from year to year. The hydrothermal index was 1.34 in 2013 and 1.78 in 2014, that shows a sufficient moisture proportion for the crop sector.

Thus, this year, weather conditions were quite favorable according to the moisture availability and the effective heat sum (Figure 1 and 2). There were some unfavorable factors such as protracted rainfall showers that led to the soil overcrust and crust formation on the soil surface at the beginning of the growing season.

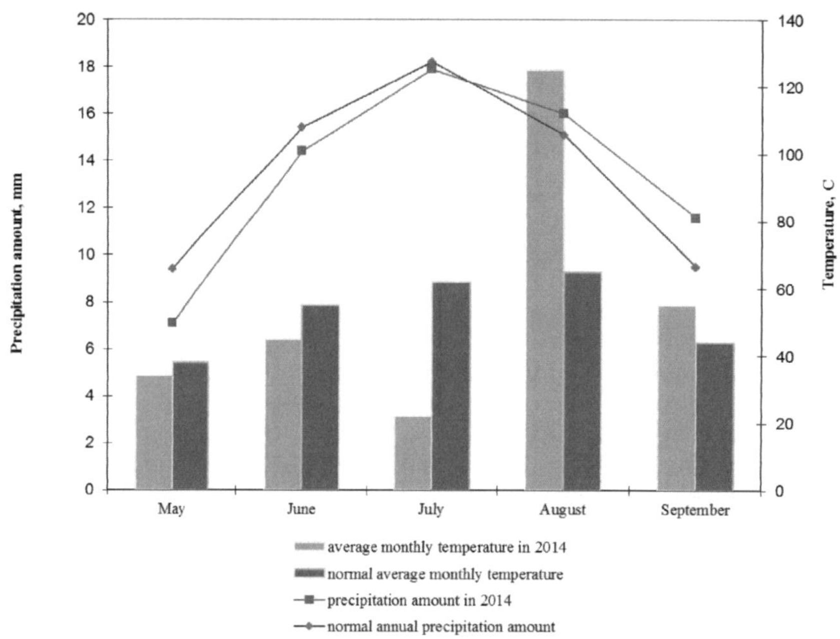


**Figure 1** – Climatograph of the growing season 2013

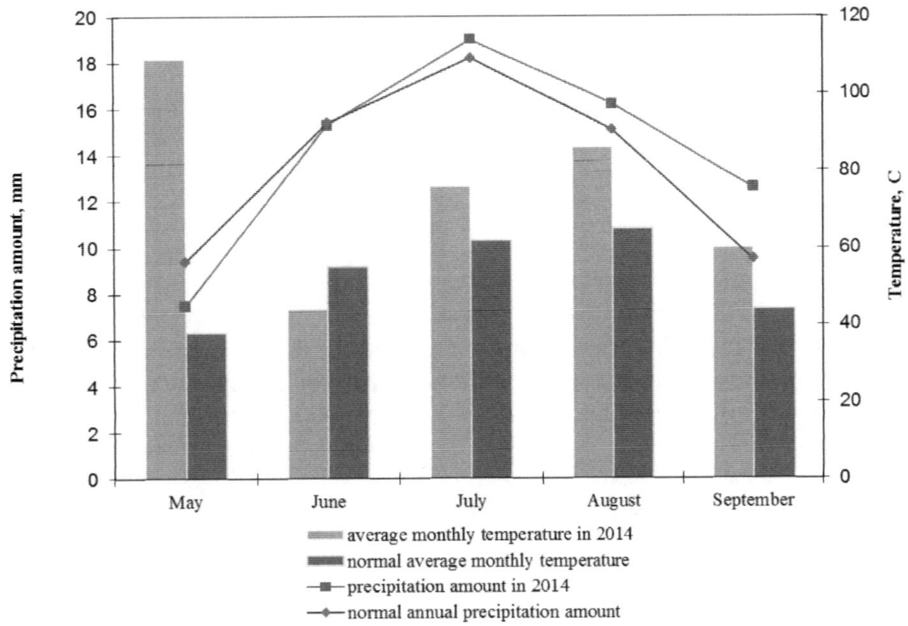


**Figure 2** – Climatograph of the growing season 2014

**Objects of the study.**

1. Zonal tillage (controlled experimental plot)

a) cultivation (autumn cultivation, rotary cultivator RUBIN) 13-15 sm;
b) dragging BZTS -1
c) preplanting cultivation SMARAGD 9/400 and fertilization (8-10 sm);
d) sowing SZP – 3.6;
e) post-emergence dragging

2. Minimal tillage

a) Air Drill John Deere 1820, sowing + fertilization+ seed rolling (depth 5-6 sm);
b) post-emergence dragging

3. Zero tillage

a) drill-machine Primera 602, sowing + fertilization (depth 5-6 sm);
b) post-emergence dragging

4. Spring barley "Odessa 100" (is middle, large-kerneled barley, 1000 grains weight – 50-53 g.; yield production 44.4 c / ha, resistant to lodging).

## 3. Results and discussion

Decision of use of the soil tillage or rejection of it depends on many factors, including soil and climatic conditions as well as technological level of production. The most costly element of the farming system is different tillage systems. To improve the competitiveness of agricultural sector it is necessary to introduce high technologies with low energy costs that can maintaine soil fertility parameters [8,9].

Soil structure resistant to mechanical stress and its ability to maintain the structure in soil moistening conditions (water resistance) define its conservation of favorable soil consistency in the context of repetitive tillage and moisturizing [10]. The experience results (table 1) show that water stability of aggregates is very low and does not reach 20%, which characterizes the soil state as unsatisfactory and liable to soil overcrust and crust formation on the soil surface, which indirectly affects the crop productivity.

IOP Conf. Series: Earth and Environmental Science **66** (2017) 012033 doi:10.1088/1755-1315/66/1/012033

**Table 1** – Influence of processing systems on the structural aggregate composition of 0-30 cm soil layer

| tillage | aggregate composition, % | | | |
|---|---|---|---|---|
| | blocky | macrostructure | microstructure | water resistance |
| Zonal | 14.7 | 83.66 | 1.98 | 18.86 |
| Minimal | 7.54 | 89.36 | 3.09 | 17.36 |
| Zero | 18.58 | 79.11 | 2.31 | 15.68 |

The received data on the content of more than 10 mm fractions differed significantly. The maximum content of blocky fraction observed in the soil with zero tillage (18.58%). The same trend in heterogeneity of soil aggregates, depending on the processing methods was observed in the work by G. Lili, E. Becker, 2017 [11]. The content of agronomically valuable aggregates in all experimental soil plots was optimal. The unsatisfactory water resistance index can not be stabilized in 2 years. To implement the methods of the minimum conservation tillage requires a special transition period. Tillage on the same depth requires use of variable-depth equipment (subsurface plow and chisel tools). To increase water resistance it is necessary to enrich the crop rotation with perennial grasses, green manure crops and to leave stubble and straw, etc. [12].

Soil density was determined before sowing and before harvesting in the layer of 0-60 cm.

The received data characterizing the soil condition by using various processing techniques are shown in Table 2.

**Table 2** – Soil density depending on soil tillage system, g/sm$^3$

| Soil layer, sm | Zonal | | Minimal | | Zero | |
|---|---|---|---|---|---|---|
| | beginning of the vegetation period | end of the vegetation period | beginning of the vegetation period | end of the vegetation period | beginning of the vegetation period | end of the vegetation period |
| 0-10 | 1.21 | 1.19 | 1.24 | 1.23 | 1.22 | 1.36 |
| 10-20 | 1.39 | 1.35 | 1.20 | 1.29 | 1.24 | 1.46 |
| 20-30 | 1.39 | 1.34 | 1.07 | 1.39 | 1.20 | 1.36 |
| 30-40 | 1.31 | 1.30 | 1.17 | 1.38 | 1.28 | 1.41 |
| 40-60 | 1.28 | 1.45 | 1.19 | 1.41 | 1.36 | 1.49 |

The data of the studies show that the soil is characterized as dense and very dense, both for the whole vegetative period and for the soil crossover. The autumn tillage of the soil did not help to reduce the soil density at the beginning of vegetation. Summer precipitation did not affect the decrease in density to the harvest period. Minimal tillage did not affect the density during the years of research. While zero tillage, despite a significant amount of summer precipitation, contributed to an increase in density at the end of the vegetation period by 1.36 in 0-10 cm of the upper layer. Thus, the gray forest soil optimal density for barley is provided by the zonal tillage system.

We believe that at the current stage of agricultural management, the technology of cultivating agricultural crops should be adaptive, contain a selection option for chosing a variant from a bank of technologies meeting requirements of being the most adaptable to natural, soil and production conditions. And it is not always necessary to refuse from deep soil treatment [13].

The process of mineralization of organic substances predominates over their renewal In the gray forest soil, as a result there is a strong depletion of soil humus. Soil treatment affects the microbiological processes taking place in it. In our studies soil biological activity is represented through the enzymatic soil activity.

IOP Conf. Series: Earth and Environmental Science **66** (2017) 012033 doi:10.1088/1755-1315/66/1/012033

Enzymatic activity, especially the activity of a large range of enzymes of different classes (oxidoreductase, hydrolases, etc.), gives valuable characteristics of soils and the ability to comprehend some specific soil features, which is successfully used in the diagnosis of soil fertility.

In soils there is a large amount of phosphorus in the form of organic compounds that are introduced with the dying remnants of plants, animals and microorganisms. The release of phosphoric acid from these compounds is carried out by a comparatively narrow group of microorganisms having specific phosphatase enzymes (Table 3).

**Table 3** – Phosphatase activity of soil plough-layer depending on the type of soil tillage

| Sample Collection Session | Type of soil tillage | Diameter of the decomposition zone, mm | Color density of the decomposition zone |
|---|---|---|---|
| May | Whole field | 25.0 | +++ |
| July | Zonal | 27.0 | ++++ |
| | Minimal | 27.0 | +++ |
| | Zero | 26.9 | +++ |
| September | Zonal | 26.3 | +++ |
| | Minimal | 26.3 | ++ |
| | Zero | 23.7 | + |

+ very weak
++ weak
+++ saturated
++++ very saturated

Analyzing the obtained data it can be said that the activity of phosphatase was dynamic during the vegetative period. The phosphatase activity of soils was maximal in active phases of plant growth at high soil temperature and sufficient humidity in July. The diameter of the decomposition zone was 26.9-27.0 mm. There was a significant decrease in phosphotase activity in the variant with zero tillage.

Catalase activity largely depends on agronomic techniques.

The catalase activity was determined by a gasometric method based on a change in the decomposition rate of hydrogen peroxide in its interaction with the soil (Table 4).

**Table 4** – Catalase soil activity depending on the type of soil tillage (ml $O_2$/g of soil in 3 min)

| Sample Collection Session | Type of soil tillage | ml $O_2$/g of soil in 3 mi |
|---|---|---|
| May | Whole field | 2.8 |
| July | Zonal | 2.9 |
| | Minimal | 3.6 |
| | Zero | 3.2 |
| September | Zonal | 3.7 |
| | Minimal | 3.7 |
| | Zero | 3.5 |

Data on the catalase soil activity indicate a uniform distribution of the enzyme in the arable layer of the soil processed with the zonal tillage. The minimal tillaged soil showed an increase in the activity of enzymes in the plow layer. The decrease in catalase activity by the end of vegetation period was not revealed. The enzyme amylase carries out the hydrolysis of starch, which is a part of the organic remains into the soil; therefore, the higher the amylase activity is (Table 5), the higher the rate of decomposition of organic matter and the higher the fertility of the soil are.

**Table 5** – Amylase activity depending on the type of soil tillage (diameter of the decomposition zone/1 g of soil / day)

| Sample Collection Session | Type of soil tillage | Diameter of the decomposition zone (1 g of soil /day), mm |
|---|---|---|
| May | Whole field | 18.3 |
| July | Zonal | 17.6 |
| | Minimal | 18.3 |
| | Zero | 16.5 |
| September | Zonal | 17.8 |
| | Minimal | 18.5 |
| | Zero | 17.8 |

Under the conditions of the experiment, the stable amylase activity was characterized by the minimal tillage.

Enzyme urease is involved in the regulation of nitrogen exchange in the soil. This enzyme catalyzes the hydrolysis of urea to ammonia and carbon dioxide, and its activity directly depends on the meteorological conditions of the year (Manoshkina NV, 2009).

The determination of urease activity was carried out under laboratory conditions (Table 6).

**Table 6** – Urease activity depending on the type of soil tillage

| Sample Collection Session | Type of soil tillage | Maximal pH | Time, hours |
|---|---|---|---|
| May | Whole field | 9.0 | 65 |
| July | Zonal | 9.5 | 68 |
| | Minimal | 10.0 | 91 |
| | Zero | 9.4 | 69 |
| September | Zonal | 9.5 | 72 |
| | Minimal | 10.0 | 78 |
| | Zero | 9.5 | 73 |

The maximum pH value 10.0 was noted in the variants with minimal soil tillage. Thus, the urease activity of the soil in the arable layer was positively affected by the lack of mechanical treatment, despite the fact that the plant remains in this case are concentrated on the soil surface.

The yield of barley is largely determined by a combination of factors: the genetic peculiarity of the variety, the elements of technology, and the indices of fertility. In our studies, the yield indicator in the experiment varied from 5.9 to 12.1 centners per hectare (Table 7). A variant with minimal soil tillage was highlighted. Such a low yield of barley is explained with the high density and low acidity index of the soil-absorbing complex of gray forest soil.

This fact has affected the number of risen plants, preserved to harvest, to form a spike. Despite the sufficient amount of precipitation in the tillering phase, productive bushiness is generally very low; the plants did not form a spike.

**Table 7** – Yield formula of barley

| Tillage | Plants, pcs./m$^2$ | Stems with a spike, pcs./m$^2$ | Tilling capacity | Grains per a spike, pcs. | Mass of 1000 grains, g | Biological productivity, c / ha |
|---|---|---|---|---|---|---|
| Zonal | 311 | 312 | 1.01 | 15.8 | 34.3 | 11.4 |
| Minimal | 241 | 240 | 1.0 | 15.8 | 36.0 | 12.1 |
| Zero | 157 | 152 | 0.97 | 13.7 | 31.6 | 5.9 |
| | | | | | | HCP $_{0.05}$=0.71 c / ha |

## 4. Conclusions

The maximum content of particles of the aggregate fraction - 18.58% was noted in the soil with zero tillage. The water resistance of aggregates is very low, does not reach 20%, which characterizes the state of the soil as unsatisfactory and capable of soil overcrust and crust formation on the soil surface.

The use of zero tillage promotes an increase in soil density by the end of vegetation period to 1.36 in the upper 0-10 cm of the soil layer. The optimum density of gray forest soils for barley was provided by the zonal processing system.

Minimization of tillage had a positive effect on the enzymatic activity of the soil in the experimental sites. The decrease in the thickness of the biologically active soil layer during zonal processing is associated with the redistribution of plant remains.

Soil with a minimum tillage showed high yield productivity of barley - 12.1 c / ha.

## References

[1] Crovetto, C. Stubble over the Soil / C. Crovetto. – Madison: American Society of Agronomy Inc., 1996. 248 p.

[2] Koller, K. Erfolgreicher Ackerbau ohne Pflug / K. Koller, M. Linke. Frankfurt am Main: DLG-Verl., 2001. 125 p.

[3] Pronin, D. Einfluss unterschiedlicher Bodenbearbeitungs – und Bestellverfahren auf die vertikale Differenzierung von Bodenkennwerten auf lehmigem Sand (Brandenburg) und auf Schwarzerde (Novosibirsk) sowie auf ausgewählte Pflanzenmerkmale / D. Pronin. Aachen: Shaker, 2003. P. 88–90.

[4] Ryabov, E.I. Resource-saving technologies of agricultural crops cultivation (Minimal soil-protecting treatment, fertilizers, pesticides, machines and tools) / E.I. Ryabov. Stavropol: Publishing house of StGAU "Agrus", 2003. 152 p.

[5] Methodology of the State Variety Testing of Agricultural Crops./ Ed. M.I. Fedina. M., 1989. 174 p.

[6] Vadjunina A.F., Korchagina Z.A. Methods for studying the physical properties of the soil. Moscow: Agropromizdat, 1986. 416p.

[7] Khaziev F.Kh. Methods of soil enzymology. M., 2005. 189 p.

[8] Abramov N.V., Rzaeva V.V., Fedotkin V.A., Fisunok N.V. Resource-saving technologies in agriculture // Plant growing and selection. Tyumen: Publishing House TGSHA, 2002. p. 128 - 132.

[9] Modelling soil tillage and mulhing effects on soil water dynamics in raised bed vegetable rotations / F. Alliaume, W. A. H. Rossing, P. Tittonell // European jowinal of agronomi. 2017. Vol. 82 p. 286-281.

[10] Rucknagel J., Rademacher A., Gotre P. Uniaxial compression behavior and soil physical guality of topsoil inder conventional and sonservation tillage // Geoderma. 2017. Vol. 286. P. 1-7.

[11] Lili G. Becker E., Guopend L. Effect of different tillage systems on aggregate structure and inner distribution of organic carbon // Geoderma. 2017, Vol. 288. P. 97-104.

[12] Tormena, Cassio A., Karlen, Douglas L.,Logston, sally. Corn stover harvest and tillage impacts on near – surface soil physical guality // soil Tillage Research. 2017. V. 166. P. 122-130.
[13] Kazakov G.I. Efficiency of soil treatment minimization in crop rotation in the forest-steppe zone of the trans-Volga region / G.I. Kazakov, A.A. Markovskiy // Izvestia of the Samara State Agricultural Academy. 2010. 4. p. 5-9.

IOP Conf. Series: Earth and Environmental Science **66** (2017) 012034 doi:10.1088/1755-1315/66/1/012034

# Usage of Farm Animal Waste for Biogas Production

**O V Sankina[a], A P Chernysh[b], A S Sankin**

Kemerovo State Agriculture Institute
Markovceva st. 5, Kemerovo, Russian Federation, 650053

E-mail: [a]olga_vk_06@mail.ru; [b]tmrm@mail.ru

**Abstract.** The article considers problems connecting with the development of cattle breeding in Russia, especially the utilization of animals and poultry waste products. Basing on the foreign scientists' experience, it has been proposed different solutions to this problem in terms of the Russian Federation, conducted the study, and presented the results of the undertaken experiments. Recommendations on the use of substances, that speed up fermentation processes at certain temperatures, has been developed.

## 1. Introduction

In Russian economy farming is the fastest growing sector. It is continuously improving animal and poultry breeding enterprises, and intensifying crop production by increasing sowing areas and yield productivity, so agricultural enterprises perform diversified highly mechanized commercial farm unit. However, the current farming technology is not applicable to the utilization of waste products, such as animal and bird wastes, livestock manure, vegetable remains and other organic materials, which are an integral part of the agricultural sector and increase respectively to the growth of the farming sector.

Traditionally livestock manure, as a result of animal and poultry breeding enterprises activity, is collected and stored in giant artificial pits – "dung pits". After being kept for 8-12 months it loses its dangerous qualities and gradually turns into a fertilizer that can be used to increase soil fertility. Although, many dung pits are outdated and even do not meet environmental standards. Most of the acting cattle-breeding complexes have been working for 25-30 years without upgrading their pollution control equipment; while such activity should be performed every 10-15 years due to its rapid wear. In addition, the ever-increasing number of livestock and poultry manure generates additional extra dung pits volume, which entails the removal of the "raw" manure on the fields and as a result lead to a number of problems.

Firstly, transportation of a huge waste quantity (solids content is 2-5%) requires additional resources. Secondly, raw litter and manure in soil are dangerous because of possible soil contamination with different infectious and parasitic diseases, ecotoxicants (heavy metals, pesticides, mycotoxins, etc.) and others. Soil largely seeds with microflora and weeds after application of organic manure, this situation can be environmentally and healthy dangerous too. Agricultural runoffs are one of the sources of pathogens that lead to over then 100 animal and human diseases. Such diseases include: aphthous fever, brucellosis, anthrax, leptospirosis, salmonellosis, encephalitis, erysipelas and swine fever, coccidiosis and many others. Animal waste is also a potential source of helminth eggs, molds and fungis. In addition it is continuously enriched with manure opportunistic pathogens, permanent inhabitants of animal digestible tract, such as Escherichia coli, Streptococcus, Pseudomonas Aureginosa and others. These microorganisms, passing through a lot of animal bodies, enhance its pathogenicity and cause colibacillosis, streptococcosis, pseudomonosis and other diseases. In the third, raw manure application into the soil leads to the local accumulation of nitrates, copper and zinc in the grain, grass, and water sources.

Other countries produce biogas from various wastes, e.g. China uses food wastes for biomethane production [1]. Not only food waste can be used for this purpose; other organic and inorganic plants' wastes can be transformed into biogas (as in Italy and Denmark) instead of being recycled as the law demands in many foreign countries [2, 3, 4]. Biogas production is economic due to the development of production system in foreign countries that minimize operating costs and cost of raw material transportation. Denmark biogas producing system is based on the use of fertilizers; Denmark researchers found that the benefits of a biomethane large-scale production and of its operating costs cover increasing transport expenses on the fertilizer and gas collection. To increase the yield, they use waste in biogas production [3].

One of the most rational solutions to this problem for Kuzbass and other regions in Russia could become a manure biotech processing in anaerobic conditions with a release of two by-products: alternate energy source (biogas) and environmentally friendly, plant readily available organic fertilizer (biosludge).

In the light of the above information, the development of biogas production technology in cold regions is very urgent and requires a detailed scientific approach.

## 2. Equipment and devices used in studies

The studies were held on the base of the Kemerovo State Agriculture Institute in the Scientific Research Laboratory "AgroBiotechnologies" with the use of 20 L Bio Gas Plants. The raw material for the study was cattle manure.

## 3. Results and discussion

Effect of different factors on the process of biogas release has been investigated several times by many researchers from different countries and covered in many scientific works, however, there is no consensus on this issue [5,6,7]. This is due to the fact that the studies were conducted in different climatic conditions and the raw materials were of varying quality and type.

A great number of strains of microorganisms produce biogas, but the most important of them are methane microorganisms because they emit methane that we are interested in. That is why the reproduction of such a complex and multi-step biological process as methanogenesis demands special attention to optimal environmental conditions for this group of microorganisms. But the other groups of bacteria participating in the process are also important; they gradually prepare a nutritious substrate for methane microorganisms in the process of anaerobic decomposition of organic matter and biogas release.

Thus, the process of biogas release is a complex multistep biological process, which involves a huge number of micro-organisms that differs by their nature, functional activities and even optimal living conditions, so it becomes complicated and practically implemented to create universal living conditions for all groups of bacteria.

Optimal temperature regulation is one of the most important factors in the fermentation process. Under natural conditions, the biogas formation takes place at temperatures from 0° C to 97° C. Most methane microorganisms are mesophiles, they grow at 30-40° C, while 25 °C is a lower limit for their functioning. But for some strains the optimal temperature zone is shifted down toward a lower limit (25 ° C) or up to 55-65° C.

It is considered that the higher the ambient temperature is, the more rapidly chemical reactions proceed. But this is only partly true for the processes of decomposition and metabolism since involved in the process microorganisms exist at different optimum temperatures. The temperature abuse can lead to the microorganisms' inhibition and even cause irreparable damage to them in extreme conditions.

IOP Conf. Series: Earth and Environmental Science **66** (2017) 012034 doi:10.1088/1755-1315/66/1/012034

Micro-organisms are very sensitive to temperature changes, in particular to its sudden drop, and slow down their metabolic and reproductive activity. The degree of the sensitivity, in turn, depends on the temperature boundaries of raw materials processing activity. The fermentation process can be performed at different temperature ranges:

- psychrophilic: ± 2 °C per hour;
- mesophilic: ± 1 °C per hour;
- thermophilic: ± 0.5 °C per hour.

This aspect performs a very big problem for the biogas production in the conditions of Siberian sharply continental climate, where winter temperature often shifts by 10-12° C in two hours. Therefore, from this perspective it is advisable to use psychrophilic mode of anaerobic decomposition.

One more practical aspect should be mentioned, the self-heating effect, that was registered by the scientists the Kemerovo State Agriculture Institute during the experiments. This effect appears with adding substrates having in its structure carbohydrates, such as manure with particles of plant litter and roughage. Self-heating is explained by the fact that certain groups of micro-organisms in the process of decomposition of carbohydrates produce heat. This leads to the temperature increase from mesophilic mode up to 43-48 °C. Intensive analytical support and control over the process of temperature changes can be carried out with short-term, minor drop in gas produced. But without human intervention in the process, these microorganisms can not adapt to changes in temperature and even stop biogas producing.

An intensive analytical support and the process temperature changes regulation can make temperatures change be a short-term phenomenon to minimize the decrease of gas production. Absence of the temperature management leads to the microorganisms' inadaptability to the temperature shifts and even to cessation of gas production

The situation with pH value is similar to the situation with temperature. Different microorganisms need different pH values to ensure optimum growth, e.g. the optimum pH value for hydrolysing and acid-forming bacteria should be acid about 4.5-6.3, the bacteria forming acetic acid and methane can live only at a neutral or slightly alkaline pH level 6.8-8. At the same time, the metabolic activity and the level of reproduction of methane bacteria is lower than that of acid-forming bacteria. Increase in the amount of organic substances can turn out into an excess of volatile acids that reduces the methane bacteria activity as soon as the pH drops below 6.5. When pH reaches 4.3 methanogenesis stops as a result of poisoning of microflora with toxic undissociated volatile fatty acids.

It is valid for all bacteria that the pH level exceeds the optimum rate, their vital activity slows down.

In the process of biogas production, the pH value within the system is established automatically under the influence of alkaline and acidic metabolic products that are formed during anaerobic decomposition, regardless of whether the process is single or multi-stage. But the degree of the balance sensibility to changes depends on the buffer capacity of the system itself.

The level of pH can be corrected by the amount and variety of added substrate. Rapidly oxidize substrates lead to a sharp pH drop; so they should be added gradually only in limited quantities.

Substrates differ in their ability to absorb the pH value. If the concentration of H + increases, the substrates can align it in a limited amount and bind free ions to themselves. Due to this, the pH level in general remains stable. Only depletion of the binding and leveling capacity (when too many organic acids have accumulated), the pH value drops down.

As a result, the inhibitory effect of hydrogen sulphide and propionic acid increases, that can lead to the reactor's overturning in a very short time. On the other hand, the pH can increase due to the decomposition of organic nitrogen compounds, that release ammonia from which ammonium is formed as a result of the reaction with water. As a result, the inhibitory effect of ammonia is increased. In any case, such a slowly measurable change in the content of $H^+$ leads to delays in the development of bacteria and thus to a violation of gas generation. Regarding the process control, it should be mentioned that due to the pH value inertness it can be used only relatively. However, pH measurements, again due to inertia, lag behind the actual situation. While it is a cheap way to control the system, the timely management of the process based only on measuring the pH level, is impossible. More effective is the measurement of buffer properties.

To maintain the stability of the anaerobic process, alkalis can be used in concentrations up to 6000 mg/l. It is considered desirable that the ratio of the fatty acids number to the amount, for example, CaCO3, should be at least 1: 6.

For growth and vital activity of bacteria, it is necessary to have organic and mineral nutrients in the raw materials. Each variety of anaerobic microorganisms differs by a specific need for macro-substances, microelements and vitamins. The concentration and availability of these components affect the rate of growth and activity of different micropopulations.

There are minimal and maximum concentrations, depending on the type of microorganisms, which are difficult to determine due to the diversity of microorganisms and their adaptive capacity. To obtain methane as much as possible from the used substrates, it is necessary to ensure optimal supply of microorganisms with nutrients. As a result, the amount methane that can be obtained from the used substrates is determined by the content of proteins, fats and carbohydrates. These factors equally affect the specific nutritive requirement.

Stability of the process requires a balanced ratio of nutrient macro-substances and trace elements. In this case, the activity of the microbial reaction is largely determined by the ratio of carbon and nitrogen. Microorganisms require both nitrogen and carbon to assimilate them into their cellular structure. Therefore, the C / N ratio in the substrates is important. If this ratio is too high (many C and little N), the available carbon can not be completely processed due to insufficient metabolism, so that amount of methane do not reach its maximum. In the opposite case, the excess of nitrogen lead to excess amount of ammonia (NH3), which even in small concentrations slows the growth of bacteria and can even result in death of the entire population of microorganisms.

Therefore, it is important to continuously maintain the C / N ratio in the range of from 10 to 30 to ensure the process stability. This can be achieved by the substrate supplying with a corresponding mass fraction of the missing substance to the reactor. Raw materials of different origin are significantly different from each other by the content of carbon and nitrogen and their ratio. An approximate ratio of carbon to nitrogen in different types of feedstock is given in Table 1.

**Table 1** – the C / N ratio in different types of raw materials for biogas production

| Biofermentable material | Nitrogen N(%) | the C / N ratio |
|---|---|---|
| A. livestock manure | | |
| cattle | 1.7 – 1.8 | 16.6 –25 |
| poultry | 3.7 – 6.3 | 7.3 – 9.65 |
| hourse | 2.3 | 25 |
| pig | 3.8 | 6.2 – 12.5 |
| sheep | 3.8 | 33 |
| B. plant dry wastes | | |
| corn ear | 1.2 | 56,6 |
| cereals straw | 1 | 49.9 |
| wheat straw | 0,5 | 100 – 150 |
| corn stover | 0.8 | 50 |
| oat straw | 1.1 | 50 |
| soya | 1.3 | 33 |
| lucern | 2.8 | 16.6 – 17 |
| sugar beet pulp | 0.3 – 0.4 | 140 – 150 |
| C. others | | |
| grass | 4 | 12 |
| sawdust | 0.1 | 200 – 500 |
| fallen leaves | 1 | 50 |

Along with carbon and nitrogen, phosphorus and sulfur are also important nutrients. Sulfur is an integral part of amino acids, and phosphorus compounds are necessary for the formation of energy sources of ATP (adenosine triphosphate) and NADP (nicotineamide adenine dinucleotide phosphate). To ensure sufficient supply of microorganisms with nutrients, the C: N: P: S ratio in the reactor should be 600: 15: 5: 3 relatively.

Along with nutrient macro-substances for microorganisms, a sufficient availability of individual trace elements is vital. Most agricultural biogas plants cover the need for micronutrients by the use of animal excrement. Methanogenic archaea require cobalt (Co), nickel (Ni), molybdenum (Mo), selenium (Se), and partly tungsten (W). Ni, Co and Mo perform an essential part of cofactors in metabolic reactions. Also important micronutrients are magnesium (Mg), iron (Fe) and manganese (Mn), they are necessary for the transport of electrons and the functioning of certain enzymes.

In the world practice of biogas production [8, 9, 10], to accelerate the fermentation process or to increase the gas yield and improve its quality, various substances are added to the substrate during fermentation with a purpose to gain the best results. Origin and composition of these substances are the most diverse, ranging from simple chemical compounds to complex high-molecular organic systems. Regardless of their origin, composition and fuction in the process of methanogenesis, all these compounds can be determined as a catalyst.

All existing catalysts for biogas plants and their technological schemes have been developed and are being implemented in Europe, where the methods of anaerobic fermentation of organic waste processing are widespread and widely used. Currently there are very few such biogas plants in Russia, since such methods of utilization of organic materials are not widely used due to harsh climatic conditions. One of the options for solving the problem of inappropriate for the sharply continental climate technology of biogas producing is the use of catalysts.

In this regard, a group of scientists, graduate students and students of the Kemerovo State Agricultural Institute, basing on the data on the application of various catalysts in Germany collected by the University of Weihenstephan (Trisdorf, Germany), conducted a series of experiments to produce biogas from livestock waste with the addition of various catalysts under conditions of rigorous climate. The data of these experiments are given below.

Substrate before its introduction into the reactor was brought to the necessary humidity, taking into account the water contained in the whey, and subjected to preliminary dispersion to a particle size of 0.5-0.7 mm.

At loading the reactors, milk whey was added there in the volumes: 1, 5, 10 and 15% of the weight of the loaded substrate. For the fermentation of cattle manure, a psychrophilic fermentation regime with a temperature of $18 \pm 1.5$ ° C was chosen. The test results are shown in Figure 1.

Milk whey due to its rich chemical composition, the presence of virtually all microelements and vitamins that are significant for the biogas release process has a significant impact on it. The results of the experiment showed that for the maximum production of biogas the most effective dose of whey application is 5%. At this concentration, the highest and constant biogas yields are observed, but the quality of the gas deteriorates significantly, the methane content in it is only 47%. Probably, this is due to the fact that this concentration of substances stimulates more hydrolytic microorganisms, which release carbon dioxide in large quantities in the course of their vital activity, that leads to an increase in the total biogas yield, as well as to its "dilution" and deterioration in its quality.

At the same time, small concentrations of milk whey influence directly on methanogenic archaea, partly inhibiting the remaining strains of microorganisms and increasing the proportion of methane, but with little effect on the overall yield of biogas during the first stages of fermentation, and subsequently depressing the whole process.

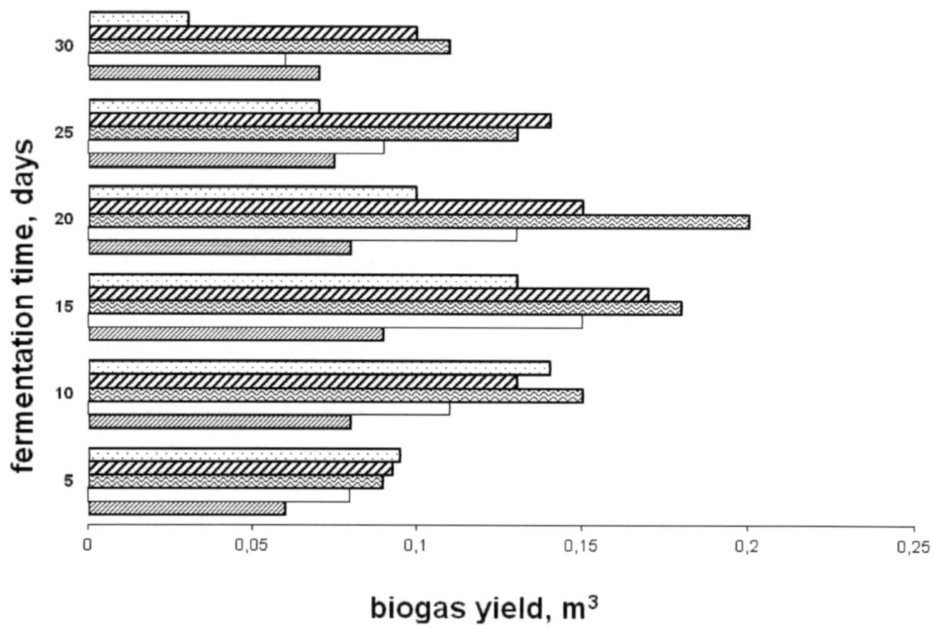


**Figure 1** – Change in biogas yield during anaerobic processing of cattle manure with the addition of various concentrations of milk whey

The most rational dose of milk whey is 10% of the weight of the raw material, since it showed an optimal combination of the gas yield and its quality. The methane content of the biogas sample from this reactor was 58%, and the volume released was only 0.2 $m^3$ less than when 5% of the whey was added. Probably, it is this dose of milk whey that combines the optimal ratio of nutrients, microelements and vitamins for all groups of bacteria involved in the process of biogas formation.

## 4. Conclusions

The work gives an observation of the process of biogas release, in which microorganisms are one of the most important factors, and one of the determining factors is the methanogens that directly release the final product. Metanogenesis, parameters and conditions of the technological environment of its production have been studied. Optimum temperatures of the fermentation process are examined: natural conditions and Siberian zone with sharply continental climate.

Stability of the anaerobic fermentation process is also ensured by using alkali in concentrations up to 6000 mg/l. For the growth and stability of the bacteria life, it is necessary to obtain organic and mineral nutrients in the raw material and a balanced ratio of nutrient macro-substances and microelements.

One of the solutions to the problem for the sharply continental climate is the development of technologies for producing biogas using catalysts. One of these catalysts is milk whey.

During the experiments whey was added in the reactors in volumes: 1, 5, 10 and 15% of the weight of the loaded substrate. For the fermentation of cattle manure, a psychrophilic fermentation regime with a temperature of $18 \pm 1.5$ ° C was chosen.

The results of the experiment showed that for the maximum production of biogas the most effective dose of whey application is 5%. At this concentration, the highest and constant biogas yields are observed.

## References

[1] De Clercq, Djavan; Wen, Zongguo; Fan, Fei *Performance evaluation of restaurant food waste and biowaste to biogas pilot projects in China and implications for national policy* [JOURNAL OF ENVIRONMENTAL MANAGEMENT] / De Clercq, Djavan; Wen, Zongguo; Fan, Fei. - Tome: 189 P.: 115-124 Published: MAR 15 2017

[2] Rotunno, Paolo; Lanzini, Andrea; Leone, Pierluigi *Energy and economic analysis of a water scrubbing based biogas upgrading process for biomethane injection into the gas grid or use as transportation fuel* [RENEWABLE ENERGY] / . Rotunno, Paolo; Lanzini, Andrea; Leone, Pierluigi. - Tome: 102 P.: 417-432 Part: B Published: MAR 2017

[3] Skovsgaard, Lise; Jacobsen, Henrik Klinge *Economies of scale in biogas production and the significance of flexible regulation* [ENERGY POLICY] / Skovsgaard, Lise; Jacobsen, Henrik Klinge. - Tome: 101 P.: 77-89 Published: FEB 2017

[4] Abadi, Nigussie; Gebrehiwot, Kindeya; Techane, Ataldti; и др. *Links between biogas technology adoption and health status of households in rural Tigray, Northern Ethiopia* [ENERGY POLICY] / Abadi, Nigussie; Gebrehiwot, Kindeya; Techane, Ataldti; и др. - Tome: 101 P.: 284-292 Published: FEB 2017

[5] [Huang, Zhe; Lu, Lu; Jiang, Daqian; и др. *Electrochemical hythane production for renewable energy storage and biogas upgrading* [APPLIED ENERGY] /_Huang, Zhe; Lu, Lu; Jiang, Daqian; and other. - Tome: 187 P.: 595-600 Published: FEB 1 2017

[6] Kim, Min Jee; Kim, Sang Hun *Minimization of diauxic growth lag-phase for high-efficiency biogas production* [JOURNAL OF ENVIRONMENTAL MANAGEMENT] / Kim, Min Jee; Kim, Sang Hun. - Tome: 187 P.: 456-463 Published: FEB 1 2017

[7] Budzianowski, Wojciech M.; Postawa, Karol *Renewable energy from biogas with reduced carbon dioxide footprint: Implications of applying different plant configurations and operating pressures* // 10th Conference on Sustainable Development of Energy, Water and Environment Systems (SDEWES) - Dubrovnik, CROATIA Published.: SEP 27-OCT 02, 2015 [RENEWABLE & SUSTAINABLE ENERGY REVIEWS] /_Budzianowski, Wojciech M.; Postawa, Karol. - Tome: 68 P.: 852-868 Part: 2 Published: FEB 2017

[8] Rodriguez, Cristina; Alaswad, A.; Benyounis, K. Y.; и др.*Pretreatment techniques used in biogas production from grass* // 10th Conference on Sustainable Development of Energy, Water and Environment Systems (SDEWES) - Dubrovnik, CROATIA Published: SEP 27-OCT 02, 2015 [RENEWABLE & SUSTAINABLE ENERGY REVIEWS] / Rodriguez, Cristina; Alaswad, A.; Benyounis, K. Y.; and other. - Tome: 68 P.: 1193-1204. Part: 2 Published: FEB 2017

[9] Levdansky, Valeri; Izak, Pavel *Membrane separation of gas mixtures under the influence of resonance radiation* [SEPARATION AND PURIFICATION TECHNOLOGY] / Levdansky, Valeri; Izak, Pavel. - Tome: 173. P.: 93-98 Published: FEB 1 2017

[10] Yan, Hu; Zhao, Chen; Zhang, Jiafu; и др.*Study on biomethane production and biodegradability of different leafy vegetables in anaerobic digestion* [AMB Express] / Yan, Hu; Zhao, Chen; Zhang, Jiafu; and other. - Tome: 7. Number article: 27. Published: JAN 25 2017

**Institute of Physics**
Dirac House, Temple Back
Bristol BS1 6BE UK

ISSN: 1755-1307
ISBN 978-1-5108-4072-0

9 781510 840720

CURRAN ASSOCIATES INC.
proceedings.com